Pattern Recognition and Neural Networks

Pattern Recognition and Neural Networks

B. D. RIPLEY
University of Oxford

CAMBRIDGE UNIVERSITY PRESS
Cambridge, New York, Melbourne, Madrid, Cape Town,
Singapore, São Paulo, Delhi, Mexico City

Cambridge University Press
The Edinburgh Building, Cambridge CB2 8RU, UK

Published in the United States of America by Cambridge University Press, New York

www.cambridge.org
Information on this title: www.cambridge.org/9780521717700

First published 1996
Seventh printing 2004
First paperback edition published 2007
Reprinted 2009

A catalogue record for this publication is available from the British Library

Library of Congress Cataloguing in Publication Data
Ripley, Brian D., 1952–
Pattern recognition and neural networks / B. D. Ripley.
    p.   cm.
Includes bibliographical references and index.
ISBN 0 521 46086 7
1. Neural networks (Computer science)   2. Pattern recognition
systems.   I. Title
QA76.87.R56   1996
006.4–dc20   95-25223 CIP

ISBN 978-0-521-46086-6 Hardback
ISBN 978-0-521-71770-0 Paperback

Additional resources for this publication at www.cambridge.org/9780521717700

Contents

Preface

Pattern recognition has a long and respectable history within engineering, especially for military applications, but the cost of the hardware both to acquire the data (signals and images) and to compute the answers made it for many years a rather specialist subject. Hardware advances have made the concerns of pattern recognition of much wider applicability. In essence it covers the following problem:

> 'Given some examples of complex signals and the correct decisions for them, make decisions automatically for a stream of future examples.'

There are many examples from everyday life:

Name the species of a flowering plant.
Grade bacon rashers from a visual image.
Classify an X-ray image of a tumour as cancerous or benign.
Decide to buy or sell a stock option.
Give or refuse credit to a shopper.

Many of these are currently performed by human experts, but it is increasingly becoming feasible to design automated systems to replace the expert and either perform better (as in credit scoring) or 'clone' the expert (as in aids to medical diagnosis).

Neural networks have arisen from analogies with models of the way that humans might approach pattern recognition tasks, although they have developed a long way from the biological roots. Great claims have been made for these procedures, and although few of these claims have withstood careful scrutiny, neural network methods have had great impact on pattern recognition practice. A theoretical understanding of how they work is still under construction, and is attempted here by viewing neural networks within a statistical framework, together with methods developed in the field of *machine learning*.

One of the aims of this book is to be a reference resource, so almost all the results used are proved (and the remainder are given references to complete proofs). The proofs are often original, short and I believe show insight into why the methods work. Another unusual feature of this book is that the methods are illustrated on examples, and those examples are either real ones or realistic abstractions. Unlike the proofs, the examples are not optional!

Those hardy perennials, the 'exclusive or' and 'two spirals' problems, do not appear in this book.

The formal pre-requisites to follow this book are rather few, especially if no attempt is made to follow the proofs. A background in linear algebra is needed, including eigendecompositions. (The singular value decomposition is used, but explained.) A knowledge of calculus and its use in finding extrema (such as local minima) is needed, as well as the simplest notions of asymptotics (Taylor series expansions and $O(n)$ notation). Graph theory is used in Chapter 8, but developed from scratch. Only a first course in probability and statistics is assumed, *but* considerable experience in manipulations will be needed to follow the derivations without writing out the intermediate steps. The glossary should help readers with non-technical backgrounds.

A graduate-course knowledge of statistical concepts will be needed to appreciate fully the theoretical developments and proofs. The sections on examples need a much less mathematical background; indeed a good overview of the state of the subject can be obtained by skimming the theoretical sections and concentrating on the examples. The theory and the insights it gives are important in understanding the relative merits of the methods, and it is often very much harder to show that an idea is unsound than to explain the idea.

Several chapters have been used in graduate courses to statisticians and to engineers, computer scientists and physicists. A core of material would be Sections 2.1–2.3, 2.6, 2.7, 3.1, 3.5, 3.6, 4.1, 4.2, 5.1–5.4, 6.1–6.4, 7.1–7.3 and 9.1–9.4, supplemented by material of particular interest to the audience. For example, statisticians should cover 2.4, 2.5, 3.3, 3.4, 5.5, 5.6 and are likely to be interested in Chapter 8, and a fuller view of neural networks in pattern recognition will be gained by adding 3.2, 4.3, 5.5–5.7, 7.6 and 8.4 to the core.

Acknowledgements

This book was originally planned as a joint work with Nils Lid Hjort (University of Oslo), and his influence will be immediately apparent to those who have seen Hjort (1986), a limited circulation report. My own interest in neural networks was kindled by the invitation from Ole Barndorff-Nielsen and David Cox to give a short course at SemStat in 1992, which resulted in Ripley (1993). The book was planned and parts were written during a six-month period of leave at the programme on 'Computer Vision' at the Isaac Newton Institute for the Mathematical Sciences in Cambridge (England); discussions with the participants helped shape my impressions of leading-edge pattern recognition problems. Discussions with Lionel Tarassenko, Wray Buntine, John Moody and Chris Bishop have also helped to shape my treatment. I was introduced to the machine-learning literature and its distinctive goals by Donald Michie. Several people have read and commented on chapters, notably Phil Dawid, Francis Marriott and Ruth Ripley. I am grateful to Lionel Tarassenko and his co-authors for the cover picture of outlier detection in a mammogram (from Tarassenko *et al.*, 1995).

Parts of this book have been used as source material for graduate lectures and seminar courses at Oxford, and I am grateful to my students and colleagues for feedback; present readers will appreciate the results of their insistence on more details in the mathematics.

Some of the software used is supplied with Venables & Ripley (1994).

The examples were computed within the statistical system S-Plus of MathSoft Inc., using software developed by the author and other contributors to the library of software for that system (notably Trevor Hastie and Rob Tibshirani).

It has been a pleasure to work with CUP staff on the design and production of this volume; especial thanks go to David Tranah, the editor for this project who also contributed many aspects of the design.

B. D. Ripley
Oxford, June 1995

Errata and clarifications for this printing

page 18, line -14: The unconditional misclassification probability should be $\Pr\{\hat{c}(X) \neq C \text{ or } \mathcal{D}\}$

page 143, line -3: Clarification: $f \equiv 1$ means f is the identity function, not the constant

page 248, line 4: **disjoint** subsets A, B, C

page 266, line -1: Clarification: the 'subgraph of the ancestors' is the 'smallest ancestral set'

page 354, **uniform convergence**: Purists who know what this means should read 'max' as 'sup' (but then they should know this definition anyway)

Notation

The notation used generally follows the standard conventions of mathematics and statistics. Random variables are usually denoted by capital letters; if X is a random variable then x denotes its value. Often bold letters denote vectors, so $\mathbf{x} = (x_i)$ is a vector with components $x_i, i = 1, \ldots, m$, with m being deduced from the context.

$\mathcal{D}$	is the 'doubt' report.
E	denotes expectation. A suffix denotes the random variable or distribution over which the averaging takes place.
$I(A)$	is the indicator function of event A, one if A happens, otherwise zero.
$N_p\{\mu, \Sigma\}$	denotes a normal distribution in p dimensions.
$O(g(n))$	$f(n) = O(g(n))$ means $\|f(n)/g(n)\|$ is bounded as $n \to \infty$.
$O_p(g(n))$	$X_n = O_p(g(n))$ means given $\epsilon > 0$ there is a constant B such that $\Pr\{\|X_n/g(n)\| > B\} < \epsilon$ for all n.
$\mathcal{O}$	is the outlier report.
$p(x)$	denotes a probability density function.
$\Pr\{A\}$	denotes the probability of an event A.
$\Pr\{A \mid B\}$	denotes the conditional probability of A given B.
$\mathbb{R}^m$	m-dimensional Euclidean space.
X^T	denotes the transpose of a matrix X.
θ	a parameter or vector of parameters.
$\widehat{\theta}, \widetilde{\theta}$	a parameter estimate.
$[\,]_+$	the positive part, the maximum of the expression and zero.
$\lfloor\,\rfloor$	the integer part (rounding down). The *floor* function.
$\lceil\,\rceil$	the nearest integer (rounding up). The *ceiling* function.

1

Introduction and Examples

This book is primarily about *pattern recognition*, which covers a wide range of activities from many walks of life. It is something which we humans are particularly good at; we receive data from our senses and are often able, immediately and without conscious effort, to identify the source of the data. For example, many of us can

recognize faces we have not seen for many years, even in disguise,
recognize voices over a poor telephone line,
as babies recognize our mothers by smell,
distinguish the grapes used to make a wine, and sometimes
    even recognize the vineyard and year,
identify thousands of species of flowers and
spot an approaching storm.

Science, technology and business has brought to us many similar tasks, including

diagnosing diseases,
detecting abnormal cells in cervical smears,
recognizing dangerous driving conditions,
identifying types of car, aeroplane, ...,
identifying suspected criminals by fingerprints and DNA profiles,
reading Zip codes (US postal codes) on envelopes,
reading hand-written symbols (on a penpad computer),
reading maps and circuit diagrams,
classifying galaxies by shape,
picking an optimal move or strategy in a game such as chess,
identifying incoming missiles from radar or sonar signals,
detecting shoals of fish by sonar,
checking packets of frozen peas for 'foreign bodies',
spotting fake 'antique' furniture,

deciding which customers will be good credit risks and
spotting good opportunities on the financial markets.

Humans can (and do) do some of the tasks quite well, but the technological pressure is to build machines which can perform such tasks more
accurately or faster or more cheaply than humans, or even to release
humans from drudgery. There are also purely technological tasks such
as reading bar codes at which humans are poor. *Pattern recognition* is
the discipline of building such machines:

> 'It is felt that the decision-making processes of a human being are
> somewhat related to the recognition of patterns; for example the next
> move in a chess game is based upon the present position on the
> board, and buying or selling stocks is decided by a complex pattern
> of information. The goal of pattern recognition research is to clarify
> these complicated mechanisms of decision-making processes and to
> automate these functions using computers. However, because of the
> complex nature of the problem, most pattern recognition research has
> been concentrated on more realistic problems, such as the recognition
> of Latin characters and the classification of waveforms.'
> (Fukunaga, 1990, p. 1)

Since the best humans can perform many of these tasks very well,
even better than the best machines, it has been of great interest to
understand how we do so, and this is of independent scientific interest.
So there has for many years been an interchange of ideas between
engineers building pattern recognition systems and psychologists and
physiologists studying human and animal brains. Twice this has led to
great enthusiasm about machines influenced by ideas from psychology
and biology. The first was in the late 1950s with the *perceptron*, the
second in the mid 1980s over *neural networks*. Both rapidly left their
biological roots, and were studied by mathematical techniques against
engineering performance goals as pattern recognizers. This book is
not about the impact of the study of neural networks as models of
animal brains, but discusses what are more accurately (but rarely) called
artificial neural networks which have been developed by a community
which was originally biologically motivated (although many 'neural
network' methods were not). Thus for the purposes of this book, a
neural network is a method which arose or was popularized by the
neural network community and has been or could be used for pattern
recognition. Many of the originators of the current wave of interest
were more careful in their terminology; whereas Hopfield (1982) did
talk about neural networks, Rumelhart & McClelland (1986) used the
term 'parallel distributed processing', and 'connectionist' has also been
popular (for example, see Hinton, 1989a).

Marginal notes such as
this replace footnotes
and offer explanation,
sidelines, and opinion.

Many of the ideas had
arisen earlier in the
pattern recognition
context, but without the
seductive titles had
made little impact.

One characteristic of human pattern recognition is that it is mainly *learnt*. We cannot describe the rules we use to recognize a particular face, and will probably be unable to describe it well enough for anyone else to use the description for recognition. On the other hand, botanists can give the rules they use to identify flowering plants.

Most learning involves a *teacher*. If we try enough different wines from unlabelled bottles, we may well discover that there are common groupings, and that one group has the aroma of gooseberries (if the latter have been experienced). But we will need a teacher to tell us that the common factor is that they were made (in part) from the *sauvignon blanc* grape. The discovery of new groupings is called *unsupervised* pattern recognition. A more common mode of learning both for us and for machines is to be given a collection of labelled examples, known as the *training set*, and from these to distil the essence of the grouping. This is *supervised* pattern recognition and is used to classify future examples into one of the same set of classes (or say it is none of these).

There is a subject known as *machine learning* which has emerged from the artificial intelligence and computer science communities. It too is concerned with distilling structure from labelled examples, although the labels are usually 'true' and 'false'.

> 'Machine Learning is generally taken to encompass automatic learning procedures based on logical or binary operations, that learn a task from a series of examples.'

> 'Machine Learning aims to generate classifying expressions simple enough to be understood easily by humans. They must mimic human reasoning sufficiently well to provide insight into the decision process. Like statistical approaches, background knowledge may be exploited in development, but operation is assumed without human intervention.'
> (Michie *et al.*, 1994, p. 2)

This stresses the need for a comprehensible explanation, which is needed in some but not all pattern recognition tasks. We have already noted that we cannot explain our identification of faces, and to recognize Zip codes no explanation is needed, just speed and accuracy.

This quotation mentions statistical approaches, and statistics is the oldest of the disciplines concerned with automatically finding structure in examples. As in the quotation, statistics is often thought of as being less automatic than the other disciplines, but this is largely an artefact of its greater age; its current research frontiers are very much concerned with replacing the human choice of methods by computation. Furthermore, statistics encompasses what the community of statisticians do, of whom your author is one!

Gooseberries are the fruits of the species *Ribes grossularia*.

We should never underestimate the power of simply remembering some or all of the examples and comparing test examples with our memory.

1.1   How do neural methods differ?

Assertions are often made that neural networks provide a new approach
to computing, involving analog (real-valued) rather than digital signals
and massively parallel computation. For example, Haykin (1994, p. 2)
offers a definition of a neural network adapted from Aleksander &
Morton (1990):

> 'A neural network is a massively parallel distributed processor that has
> a natural propensity for storing experiential knowledge and making it
> available for use. It resembles the brain in two respects:
>
> 1.   Knowledge is acquired by the network through a learning process.
>
> 2.   Interneuron connection strengths known as synaptic weights are
>   used to store the knowledge.'

Many neural networks
are excluded by this
definition, including
those of Kohonen. One
could ask how a
machine comes to have
'natural' properties.

In practice the vast majority of neural network applications are run on
single-processor digital computers, although specialist parallel hardware
is being developed (if not yet massively parallel). However, all the
other methods we consider use real signals and can be parallelized to a
considerable extent; it is far from clear that neural network methods will
have an advantage as parallel computation becomes common, although
they are frequently so slow that they need a speed-up. (Parallelization
on real hardware has proved to be non-trivial; see Pitas, 1993 and
Przytula & Prasanna, 1993.) We will argue that a large speed-up can
be achieved by designing better learning algorithms using experience
borrowed from other fields.

The traditional methods of statistics and pattern recognition are
either *parametric* based on a family of models with a small number
of parameters, or *non-parametric* in which the models used are totally
flexible. One of the impacts of neural network methods on pattern
recognition has been to emphasize the need in large-scale practical
problems for something in between, families of models with large but
not unlimited flexibility given by a large number of parameters. The two
most widely used neural network architectures, *multi-layer perceptrons*
and *radial basis functions* (RBFs), provide two such families (and several
others already existed in statistics).

The name 'multi-layer
perceptrons' is
confusing; they are not
multiple layers of
perceptrons. We call
them feed-forward
neural nets.

Another difference in emphasis is on *'on-line'* methods, in which the
data are not stored except through the changes the learning algorithm
has made. The theory of such algorithms is studied for a very long
stream of examples, but the practical distinction is less clear, as this
stream is made up either by repeatedly cycling through the training set
or by sampling the training examples (with replacement). In contrast,
methods which use all the examples together are called *'batch'* methods.

It is often forgotten that there are intermediate positions, such as using small batches chosen from the training set.

1.2 The pattern recognition task

Except in Chapter 9 we will be exclusively concerned with supervised pattern recognition. Thus we are given a set of K pre-determined classes, and assume (in theory) the existence of an oracle that could correctly label each example which might be presented to us. When we receive an example, some measurements are made, known as *features*, and these data are fed into the pattern recognition machine, known as the *classifier*. This is allowed to report

> 'this example is from class ℓ' or
> 'this example is from none of these classes' or
> 'this example is too hard for me'.

The second category are called *outliers* and the third *rejects* or *'doubt'* reports. Both can have great importance in applications. Suppose we have a medical diagnosis aid. We would want it to report any patient who apparently had an unknown disease, and we would also want it to ask the opinion of a senior doctor if there was real doubt. Often rejects are referred to a more expensive second tier of classification, perhaps a human or (as in Zip code recognition) a slower but more powerful method or even (as in analytical chemistry) for more expensive measurements to be made. Many pattern recognition systems always make a firm classification, but this seems to us more often to be bad design than a conscious decision that a firm decision was necessary.

The primary assessment of a system will be by its performance; a Zip code recognition system might be required to reject less than 2% of the examples and mis-read less than 0.5% of the remainder. In medical diagnosis we will be more interested in some errors than others, in particular in missing a disease, so the errors will need to be weighted. There may be a cost trade-off between rejection and error rate.

The other aspect of performance stressed in the quote from Michie *et al.* (1994, p. 2) is the power of explanation. Users need to have confidence in the system before it will be adopted. No one really cares if an odd letter is mis-routed, but patients do care if they are mis-diagnosed, and when a civilian airliner is mistaken for an enemy aircraft, questions are raised. So for some tasks 'black boxes' are unacceptable whatever their performance advantage (possibly even if they appear perfect on test). The methods of Chapters 7 and 8 are often found to be more acceptable for such tasks.

Someone else may have made the measurements for us.

It may help to know which classes are plausible.

This might be unrealistic for hand-written addresses, and is well beyond current performance levels.

Some tasks are slightly different. We (and medics) often think of medical diagnosis as deciding which disease a patient has, but this ignores the possibility of two or more concurrent diseases; what we should really be asking is whether the patient has this disease for each of a range of diseases. This can be thought of as a compound decision, the classes each being a subset of the diseases, but it is normally helpful to make use of special structure within the classes.

Design issues

Although most of this book is about designing the pattern recognition machine, often the most important aspect of design is to choose the right features. If the wrong things are measured (or, more often these days with digital data, if the data are condensed too much) the task may be unachievable. Much of the enhanced success of Zip-code recognition systems has come from better features (for example, Simard *et al.*, 1993) rather than through more complicated classifiers. Sometimes good features can be found by training a classifier on a large number of features and extracting good ones (for example, by the methods of Chapters 9 and 10), but most often problem-specific insights are used.

In a few problem domains very specific rules are known which can be used to design a classifier; as an extreme example compilers can classify C programs as correct or invalid without needing to see any previous programs. Such information is often in the form of a formal *grammar*, and systems based on specifying such grammars are often called *syntactic* pattern recognition systems (Fu, 1982; Gonzalez & Thomason, 1978), but are of very restricted application. Allowing stochastic grammars in which the structure is given but the probabilities are learnt allows a little more flexibility. Chou (1989) gives an example of recognizing typeset mathematical expressions using a stochastic grammar.

In the vast majority of applications no structural assumptions are made, all the structure in the classifier being learnt from data. In the pattern recognition literature this is known as *statistical pattern recognition*. The training set is regarded as a sample from a population of possible examples, and the statistical similarities of each class extracted, or more precisely the significant differences between classes are found. A parametric or non-parametric model is constructed for the distribution of features for examples from each class, and statistical decision theory used to find an optimal classification. This is sometimes known (Dawid, 1976) as the *sampling paradigm*.

Another view, the *diagnostic paradigm*, goes back in the statistical literature at least to Cox (1958), and was developed in medical applications by Jerome Cornfield. This said that we were not interested in what the classes looked like, but only given an example in what the distribution over classes is *for similar examples*. The main method of this approach became known as *logistic discrimination* (Anderson, 1982), but was never widely known even in statistics and (as far as we could ascertain) appears in no pattern recognition text. This is the main approach of the neural network school.

When humans are learning concepts, we are often able to ask questions or to seek the classifications of examples which we synthesize (this being a paradigm of experimental science). Alternatively, we may describe our understanding to an expert, who will then supply a counter-example. Can we allow our machines to do the same? The idea has occurred in machine learning (Angluin, 1987, 1988, 1993), but apparently only for learning logical concepts.

We will sometimes have qualitative knowledge about the task in hand; we might know that only the sign of one of the features was material, or that the probability of a positive outcome was increasing in some continuous feature. Of course we should design the classifier to agree with such information, which Abu-Mostafa (1990, 1993, 1995a, b, c) calls 'hints'. Sometimes this is easy (just use the sign of the feature) but it can be very difficult (as in monotonicity). Generally hints (if true) help to avoid over-fitting to the training set, and this seems to be the real explanation of the gains in exchange-rate performance observed by Abu-Mostafa (1995a).

Method tuning and checking

All methods have some knobs which can be tweaked. Sometimes taking the class of the nearest training-set example is regarded as a fully automatic method, but we need to specify the metric used to find the nearest. (If the answer is 'use Euclidean distance' we still have to specify the units of measurement.)

Note that this is not the procedure called *cross-validation*, despite the misuse of that term in the neural networks literature.

How should those knobs be set? The most obvious way is to choose them to maximize performance. One thing we should *not* do is to evaluate the performance on a *test set* and choose the best-performing classifier, since we will then have no way to measure the true performance. We can keep back another test set, called a *validation set*, and use the performance on that to set the knobs. However, to obtain a sensitive measure of the performance, the validation set will need to be very large, and this is data which could otherwise be used for learning.

This problem has been ignored for a long time, but now methods to use the training set for both learning and knob-setting are beginning to be used. These are discussed in Chapter 2 and illustrated on the quite small running examples that we chose.

To see why this is a real issue, consider Figure 1.1. Without knowing the true curve, it is hard to tell which of plots (b) and (c) is closer to the truth.

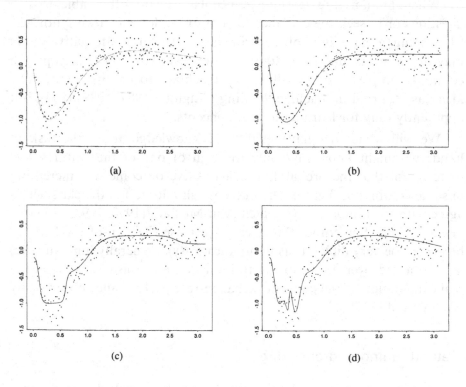

Figure 1.1: An illustration of model selection. Plot (a) shows 250 points generated by from the curve shown plus random noise, and plots (b–d) show fits by a single-hidden-layer neural network with 2, 4 and 8 hidden units.

Performance assessment

We will often want to choose between different candidate classifiers, and it will be usual to check that the performance targets are likely to be met. This needs an experimental test of the classifiers on some unseen examples. Such experiments are often (usually?) very poorly designed, and slanted towards a favourite method. The reader is urged to consult a good book on experimental design (such as Box *et al.*, 1978) before conducting such experiments.

Many of the experiments reported in the literature are designed to compare methods, when there is even more scope for confusion. In medicine, methods (treatments) are compared in *double-blind* trials so there can be no preferential treatment, and in pure science experiments must be repeatable. (The large-scale trial of the StatLog project reported in Michie *et al.*, 1994, was designed to be run in these ways.) One source of confusion is that such trials may confuse the merits of the methods with the expertise of the experimenter in using them; this is a particular difficulty when the experimenter's own invention is in the trial. Two cases are of interest. One is where every method is used by a real expert and so assesses the best attainable performance. The other is when all methods are used by typical (or even new) users, which might provide a basis for recommendations to such users.

Prechelt (1994) surveyed two leading neural network journals for 1993 and half of 1994. He deemed an evaluation of an algorithm acceptable if it used two or more realistic or real problems and compared at least one alternative algorithm. Only 18% passed—in his words 'sad, but true'. Note that this book is not about evaluating algorithms, but we have used real examples to explore the merits and limitations of the methods. Amazingly, almost all books on pattern recognition or neural networks include no real or realistic examples.

This test does not consider experimental biases nor if an evaluation of the significance of the results was made.

1.3 Overview of the remaining chapters

Our approach to building a classifier will be based on statistical decision theory. In Chapter 2 we consider the Bayes rule, the best possible classifier if we knew everything about the population of examples, and then various approximations we can make if we have to learn from a training set. This includes several ways to use parametric models (which we assume to be false but perhaps convenient approximations); these sections include the classic methods based on the multidimensional normal distribution but also some improvements which are much less well known.

The next questions are: how complicated do our models need to be, and how well do they perform? These are discussed in Sections 2.6 and 2.7. There is a trade-off between adapting well to complexity of the real structure in the examples and fitting the structure of our particular training set (Figure 1.1). This explains why we are not interested in the usual asymptotics of mathematical statistics; as we receive more data we will want to choose more complicated models, and only limit the model complexity to avoid over-fitting the current training set. Another view of the effect of model flexibility on over-fitting is the study of *generalization* in Section 2.8.

Chapters 3 to 5 make weaker assumptions than standard parametric models. In Chapter 3 we study how we could use linear methods. Both Chapters 4 and 5 discuss how to apply flexible families of functions from the feature space $\mathcal{X}$ to d-dimensional Euclidean space $\mathbb{R}^d$, building on the linear methods, and consider the commonest such families, neural networks and radial basis functions, as well as splines and their generalizations.

The sixth chapter is on (nearly) non-parametric methods, where minimal assumptions are made about the classes. Most of these methods are based on looking at the classes of nearby examples, in some methods after designing a set of representative examples to replace the training set. That chapter also includes the use of mixtures of densities to model very general distributions.

Chapter 7 is about a rather different class of methods that partition the feature space $\mathcal{X}$ into regions and assign a class to each. This is done by splitting along a feature at a time, and then subdividing each subregion recursively. *Classification trees* have been considered in both statistics and machine learning; they are often easy to interpret but not amongst the highest performers.

Belief networks, also known as causal probability networks and Bayes networks, are not primarily designed for classification, but to explain the relationships between all of the observations. They are the subject of Chapter 8. They are very good for explanation, but may be less good for classification (as the finite amount of training data has to be used to learn more structure than just the relationship of the class to their features). Their strength is that they can incorporate qualitative knowledge about causal relationships amongst the features (an earlier and more sophisticated use of 'hints'). Also included in that chapter are the methods of Boltzmann machines and hierarchical mixtures of experts which can be considered within the framework of belief nets.

... and many more names beside

Chapters 9 and 10 are concerned with finding good features and choosing which features to use.

The appendix discusses a number of complements; some are statistical background and some explore issues a little further than is needed for pattern recognition.

1.4   Examples

The examples have been chosen to illustrate the properties of the methods we describe; not every method is used on each.

Cushing's syndrome

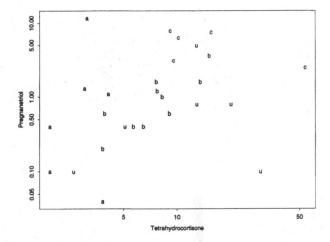

Figure 1.2: Results of two diagnostic tests on patients with Cushing's syndrome.

These data are taken from Aitchison & Dunsmore (1975, Tables 11.1–3) on diagnostic tests on patients with Cushing's syndrome, a hypersensitive disorder associated with over-secretion of cortisol by the adrenal gland. This dataset has three recognized types of the syndrome represented as a, b, c. (These encode 'adenoma', 'bilateral hyperplasia' and 'carcinoma', and represent the underlying cause of over-secretion. This can only be determined histopathologically.) The observations are urinary excretion rates (mg/24h) of the steroid metabolites tetrahydrocortisone and pregnanetriol, and are considered on log scale.

One of the patients of unknown type (marked u) was later found to be of a fourth type, and another was measured faultily.

Titterington (1976) discusses a different dataset which had 87 patients, five types, and fifteen measurements per patient, which suggests the current dataset is an abstraction of the full problem.

Synthetic two-class problem

This is a 'realistic' problem from Ripley (1994a), used there (and here) to illustrate how methods work. There are two features and two classes; each class has a bimodal distribution as should be clear from Figure 1.3. The class distributions were chosen to allow a best-possible error rate of about 8%, and are in fact equal mixtures of two normal distributions. The component normal distributions have a common covariance matrix.

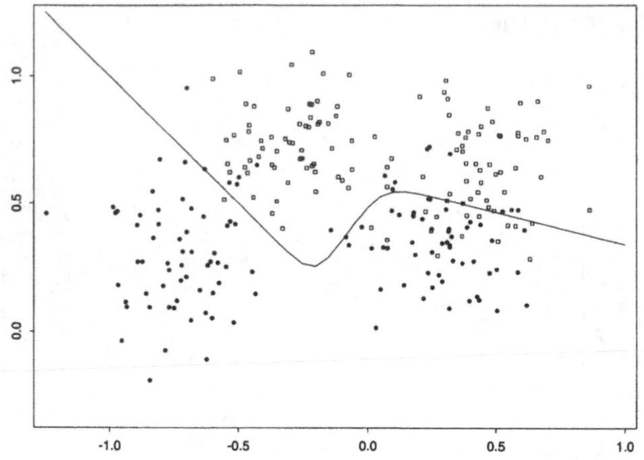

Figure 1.3: Two-class synthetic data from Ripley (1994a). The two classes are shown by solid circles and open squares: there are 125 points in each class.

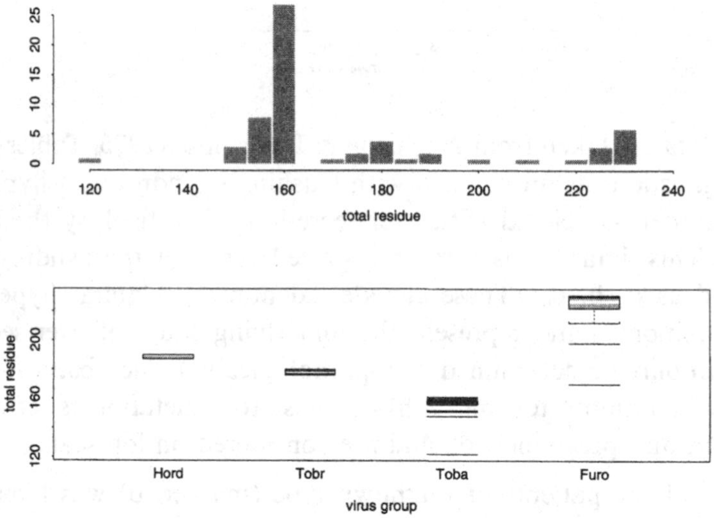

Figure 1.4: Histogram and boxplot by group of the viruses dataset. A boxplot is a representation of the distribution; the central grey box shows the middle 50% of the data, with median as a white bar. 'Whiskers' go out to plausible extremes, with outliers marked by bars.

Viruses

This is a dataset on 61 viruses with rod-shaped particles affecting various crops (tobacco, tomato, cucumber and others) described by Fauquet *et al.* (1988) and analysed by Eslava-Gómez (1989). There are 18 measurements on each virus, the number of amino acid residues per molecule of coat protein; the data come from a total of 26 sources. There is an existing classification by the number of RNA molecules and mode of transmission, into

No experimental details are provided in the source.

39 *Tobamoviruses* with monopartite genomes spread by contact,
6 *Tobraviruses* with bipartite genomes spread by nematodes,
3 *Hordeiviruses* with tripartite genomes, transmission mode unknown and
13 'furoviruses', 12 of which are known to be spread fungally.

The question of interest to Fauquet *et al.* was whether the furoviruses form a distinct group, and they performed various multivariate analyses.

One initial question with this dataset is whether the numbers of residues are absolute or relative. The data are counts from 0 to 32, with the totals per virus varying from 122 to 231. The average numbers for each amino acid range from 1.4 to 20.3. As a classification problem, this is very easy as Figure 1.4 shows. The histogram shows a multimodal distribution, and the boxplots show an almost complete separation by virus type. The only exceptional value is one virus in the furovirus group with a total of 170; this is the only virus in that group whose mode of transmission is unknown and Fauquet *et al.* (1988) suggest it has been tentatively classified as a *Tobamovirus*. The other outlier in that group (with a total of 198) is the only beet virus. The conclusions of Fauquet *et al.* may be drawn from the totals alone.

It is interesting to see if there are subgroups within the groups, so we will only use this dataset in Chapter 9, principally to investigate further the largest group (the *Tobamoviruses*). There are two viruses with identical scores, of which only one is included in the analyses. (No analysis of these data could differentiate between the two.)

Leptograpsus crabs

Campbell & Mahon (1974) studied rock crabs of the genus *Leptograpsus*. One species, *L. variegatus*, had been split into two new species, previously grouped by colour form, orange and blue. Preserved specimens lose their colour, so it was hoped that morphological differences would enable museum material to be classified.

Data are available on 50 specimens of each sex of each species, collected on sight at Fremantle, Western Australia. Each specimen has measurements on the width of the frontal lip FL, the rear width RW, and length along the midline CL and the maximum width CW of the carapace, and the body depth BD in mm.

Forensic glass

Our next example comes from forensic testing of glass collected by B. German on 214 fragments of glass, and taken from Murphy & Aha (1995). Each case has a measured refractive index and composition (weight percent of oxides of Na, Mg, Al, Si, K, Ca, Ba and Fe). The fragments were originally classed as seven types, one of which was absent in this dataset. The categories which occur are window float glass (70), window non-float glass (76), vehicle window glass (17), containers

(13), tableware (9) and vehicle headlamps (29). The composition sums
to around 100%; what is not anything else is sand.

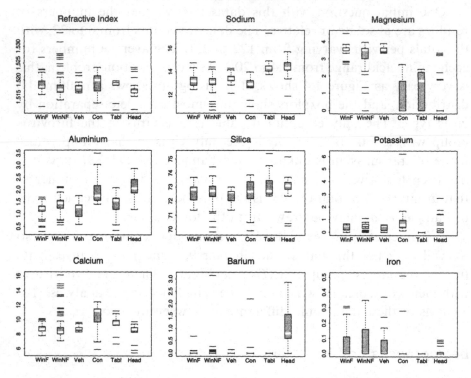

Figure 1.5: Boxplots of the features of the forensic glass data.

Figure 1.5 shows boxplots of the features. Some discrimination
between glass types is apparent even from single features; for example
headlamp glass is high in barium (although some examples have none),
high in sodium and aluminium and low in iron. The three types of
window glass appear similar, with one exceptional fragment of window
non-float glass having a high refractive index, high barium and calcium
and low magnesium and sodium. The containers group also contains
a couple of exceptions. Characterizing populations with exceptions
(especially 2 out of 13) can be difficult, and it may be easier to remove
the exceptions in the training phase.

This example is really too small to divide, so methods have been
assessed by 10-fold cross-validation using the same random partition
for each method. The best methods have an estimated error rate of
about 24%.

This is discussed in Section 2.7.

Diabetes in Pima Indians

A population of women who were at least 21 years old, of Pima Indian
heritage and living near Phoenix, Arizona, was tested for diabetes according to World Health Organization criteria. The data were collected

by the US National Institute of Diabetes and Digestive and Kidney Diseases, and are available from Murphy & Aha (1995). A previous report by Smith *et al.* (1988) found an error rate of about 24%. The reported variables are

  number of pregnancies
  plasma glucose concentration in an oral glucose tolerance test
  diastolic blood pressure (mm Hg)
  triceps skin fold thickness (mm)
  serum insulin (μU/ml)
  body mass index (weight in kg/(height in m)^2)
  diabetes pedigree function
  age in years

Many of these had zero values where these were impossible, so are taken to be missing values. Of the 768 records, 376 were incomplete (most prevalently in serum insulin). Most of our illustrations omit serum insulin and use the 532 complete records on the remaining variables. These were randomly split into a training set of size 200 and a test set of size 332. Methods which can deal with missing values were given 100 of the incomplete cases as part of the training set.

Note that 33% of the population were reported to have diabetes, so an error rate of 33% can be achieved by declaring all test cases to be non-diabetic. Our best methods reduce this to about 20%.

Some aspects of this dataset were considered by Wahba *et al.* (1995).

Data availability

All these datasets are available by anonymous ftp from the Internet site

  `ftp.stats.ox.ac.uk`   IP address 163.1.20.1

in directory /pub/PRNN. The datasets and other material are available by pointing your World Wide Web browser at

  `http://www.stats.ox.ac.uk/~ripley/PRbook/`

1.5   Literature

The classic books on pattern recognition are Duda & Hart (1973), Devijver & Kittler (1982) and Fukunaga (2nd edn 1990), all of which pre-date the impact of neural networks on the subject. There are a small number of introductory texts (James, 1988; Therrien, 1989; Schalkoff, 1992) and two specialist monographs on kernel methods

(Hand, 1982; Coomans & Broeckaert, 1986). Some conference proceedings, for example Devijver & Kittler (1987), provide a good overview of applications.

Classical statistical techniques are discussed in most texts on *multivariate analysis* such as T. W. Anderson (1984) and Mardia *et al.* (1979) and in slightly more specialized books by Lachenbruch (1975), Goldstein & Dillon (1978), Hand (1981) and McLachlan (1992).

There are now very many books on neural networks, particularly on parts of the subject not discussed here. Approaches to modelling memory from the point of view of statistical physics are covered by Amit (1989), Peretto (1992) and Hertz *et al.* (1991). Haykin (1994) is modern, comprehensive but unselective (and untroubled by real applications). Amari (1993) and Ripley (1993) give two statistical views of the neural network field, and Bishop (1995a) is slanted towards pattern recognition. Arbib (1995) provides many short sketches of topics over a very wide range of neural networks. One important area of neural network methods which we do not consider is the prediction of time series, the subject of a competition analysed by Weigend & Gershenfeld (1993), including expository papers.

There is now one text on general machine learning, Langley (1996), and it appears in some artificial intelligence texts (for example, Winston, 1992; Russell & Norvig, 1995). There are many more aspects than we shall consider, including incorporating domain knowledge as illustrated by King *et al.* (1992). Langley & Simon (1995) and Bratko & Muggleton (1995) discuss applications of machine learning with claimed real-world benefits.

Books which cover more than one of these three areas are rare. Krishnaiah & Kanal (1982) was a very good overview at its time; the recent edited volumes by Cherkassky *et al.* (1994) and Michie *et al.* (1994) contain several good overviews.

Face recognition is a popular application of pattern recognition surveyed by Samal & Iyengar (1992). Golomb *et al.* (1991) and Flocchini *et al.* (1992) give two example systems.

There is a very large literature on character recognition, and non-European alphabets with at least hundreds of classes provide a severe test of pattern recognition methods. The articles by Baird (1993), Cohen *et al.* (1991), Le Cun *et al.* (1989, 1990a), Gader *et al.* (1991), Guyon *et al.* (1992), Impedovo *et al.* (1991), Knerr *et al.* (1991), Lee (1991), Martin & Pitman (1990, 1991), Pavlidis (1993), Simard *et al.* (1993), Singer & Tishby (1994), Suen *et al.* (1992, 1993) and Wakahara (1993) provide some flavours.

2

Statistical Decision Theory

This chapter presents basic statistical decision theory for classification problems with predefined classes.

The framework in its simplest form is as follows. Certain objects are to be classified as coming from one of a fixed number of types, or classes, say $1, \ldots, K$. Each object gives rise to certain measurements which together form the *feature vector* X, belonging to a suitable feature space $\mathscr{X}$. This is typically a subset of $\mathbb{R}^p$ or perhaps of the type $\mathscr{X}_1 \times \cdots \times \mathscr{X}_p$ with each $\mathscr{X}_j$ either a given finite set or $\mathbb{R}$. The proportion of class k cases in the population under study is some known or unknown π_k. Feature vectors from class k are distributed according to the density $p_k(x)$. The task is to classify an object, which means reaching one of $K + 2$ possible decisions $1, \ldots, K, \mathscr{D}, \mathscr{O}$ on the basis of the observed value $X = x$; decision k corresponds to claiming 'X is from class k', whereas $\mathscr{D}$ means 'being in doubt', possibly postponing the decision until further measurements have been extracted, and $\mathscr{O}$ signifies an outlier, an object definitely not belonging to any of the K predefined classes.

We regard probability mass functions of discrete distributions as densities.

Section 2.1 treats the idealized case when class densities $p_k(x)$ as well as class prior probabilities (π_k) are known. This gives valuable insight and also provides limits for the performance of real-life classifiers that in some way must estimate class densities. In Section 2.2 some of the most important parametric models for classification are studied. The parameters are typically estimated using maximum likelihood, but alternatives are discussed in Section 2.3 (assuming less of the model), Section 2.4 (taking the variability of the parameters into account) and Section 2.5 (bias correction).

Sections 2.6 and 2.7 discuss how we assess the adequacy of a parametric model and estimate the performance of a classification rule. The final section considers 'generalization', a more abstract way to find bounds on the expected performance of a class of classifiers.

Not all of the material here is essential for the later applications. The most important sections are Section 2.1, Section 2.2 omitting 'Bayes risk consistency' and 'Fitting parametric families when they are wrong', Sections 2.3, 2.4, 2.6 and 2.7.

Marginal notes point out the less important material.

2.1 Bayes rules for known distributions

In this present section we assume that the class densities p_k and the prior probabilities π_k are known. This makes it possible to construct classification procedures with well understood optimal properties. Such results are not directly applicable since the class densities, at least, are unknown in practice, but they will serve as guidelines for the estimated rules of Sections 2.2 and 2.3, and have intrinsic theoretical interest.

Let C denote the class label of a random feature vector X, in particular $C = k$ with probability π_k. The classification task is to estimate the true C after having observed X. Let $\widehat{c}: \mathscr{X} \to \{1, \ldots, K, \mathscr{D}\}$ be a classification procedure (also known as a *classifier*). (We will deal with outlier decisions in a later subsection.) To determine whether such a procedure is 'good' or not one has to agree on reasonable overall criteria, for example involving the misclassification probabilities

$$\mathsf{pmc}(k) = \Pr\{\widehat{c}(X) \neq k, \widehat{c}(X) \in \{1, \ldots, K\} \mid C = k\} \qquad (2.1)$$

and the reject or *doubt probabilities* $\mathsf{pd}(k) = \Pr\{\widehat{c}(X) = \mathscr{D} \mid C = k\}$. The quantities pmc and pd denote the *un*conditional misclassification probability $\Pr\{c(X) \neq C\}$ and doubt probability $\Pr\{c(X) = \mathscr{D}\}$ respectively.

Minimizing the expected error rate

The usual way of formalizing a goodness criterion is by means of a *loss function*. Let $L(k, l)$ be the loss incurred by making decision l if the true class is $C = k$. One should have $L(k, k) = 0$ and maybe $L(k, \mathscr{D}) = d$ for all k, whereas the other $L(k, l)$'s could in principle be any set of positive numbers. If every misclassification is equally serious, then

$$L(k, l) = \begin{cases} 0 & \text{if } l = k \text{ (correct decision)}, \\ 1 & \text{if } l \neq k \text{ and } l \in \{1, \ldots, K\} \text{ (wrong decision)}, \\ d & \text{if } l = \mathscr{D} \text{ (being in doubt)}, \end{cases} \qquad (2.2)$$

Loss (2.2) is used unless otherwise stated.

for $k = 1, \ldots, K$ and $l = 1, \ldots, K, \mathscr{D}$, is a reasonable choice. In what follows we will often employ the loss function (2.2) to illustrate

A medical example with assessed costs is Table 4 of Titterington *et al.* (1981, p. 154).

important concepts; it is often used when there is no way to assign more accurate costs. However, we should warn that its use when inappropriate can cause difficulties or even be dangerous; the costs of failing to spot a disease are usually very much higher than those of a false positive in a series of screening tests.

The *risk function* for classifier $\hat{c}$ is the expected loss when using it, as a function of the unknown class k:

$$R(\hat{c}, k) = E[L(k, \hat{c}(X)) \mid C = k]$$

$$= \sum_{l=1}^{K} L(k, l) \Pr\{\hat{c}(X) = l \mid C = k\} + L(k, \mathcal{D}) \Pr\{\hat{c}(X) = \mathcal{D} \mid C = k\}$$

$$= \mathsf{pmc}(k) + d\,\mathsf{pd}(k).$$

The *total risk* is the total expected loss, viewing both the class C and the vector X as random;

$$R(\hat{c}) = ER(\hat{c}, C) = \sum_{k=1}^{K} \pi_k \,\mathsf{pmc}(k) + d \sum_{k=1}^{K} \pi_k \,\mathsf{pd}(k). \qquad (2.3)$$

This is seen to be the overall misclassification probability plus d times the overall doubt (or reject) probability. It is also the long-term average loss, the limit of $n^{-1} \sum_{j=1}^{n} L(C_j, \hat{c}(X_j))$, where $\{(C_j, X_j)\}$ is a random sample of size n.

For our first main result, let

$$p(k \mid x) = \Pr\{C = k \mid X = x\} = \frac{\pi_k\, p_k(x)}{\sum_{l=1}^{K} \pi_l\, p_l(x)} \qquad (2.4)$$

be the *posterior probability* of class k given $X = x$. Then the following holds.

Proposition 2.1 *The classification rule which minimizes the total risk under loss (2.2) is*

$$c(x) = \begin{cases} k & \text{if } p(k \mid x) = \max_{l \leqslant K} p(l \mid x) \text{ and this exceeds } 1 - d, \\ \mathcal{D} & \text{if each } p(k \mid x) \leqslant 1 - d, \end{cases} \qquad (2.5)$$

and for a general loss function is

$$c(x) = \begin{cases} k & \text{if this attains } \min_{l \leqslant K} \sum_{j} L(j, l) p(j \mid x) < d, \\ \mathcal{D} & \text{otherwise.} \end{cases} \qquad (2.6)$$

Proof: We have

$$R(\widehat{c}) = \mathsf{E}\big[\mathsf{E}[L(C, \widehat{c}(X)) \mid X]\big]$$

$$= \int_{\mathscr{X}} \mathsf{E}[L(C, \widehat{c}(x)) \mid X = x]\, p(x)\, \mathrm{d}x$$

where $p(x) = \sum_{k=1}^{K} \pi_k p_k(x)$ is the marginal density for X. It suffices to minimize the conditional expectation, which we can write as $\sum_{k=1}^{K} L(k, c)\, p(k \mid x)$, with respect to c, for each x. For $c = \mathscr{D}$ we have $\sum_{k=1}^{K} L(k, \mathscr{D})\, p(k \mid x) = d$. Under loss (2.2) the minimand becomes

$$1 - p(1 \mid x), \ldots, 1 - p(K \mid x), d$$

when $\widehat{c} = 1, \ldots, K, \mathscr{D}$ respectively. Thus we look for the maximum of $p(1 \mid x), \ldots, p(K \mid x), 1 - d$, and find the solution given. $\square$

Under loss (2.2) another way to write the optimal rule is to choose the class with the highest $\pi_k p_k(x)$ provided this exceeds $(1 - d) p(x)$, from (2.4).

This optimal classifier is also referred to as the *Bayes rule*. When two or more classes attain the maximal $p(k \mid x)$ the tie can be broken arbitrarily. The value $R(\widehat{c})$ of the total risk (2.3) for the Bayes rule is called the *Bayes risk*. This value is the best one can achieve if the π_k's and p_k's are known, and provides a benchmark for all other procedures. For two classes and without the doubt option the Bayes risk is $\mathsf{E}\min[p(1 \mid x), p(2 \mid x)]$, for any number of classes it is $\mathsf{E}[1 - \max p(k \mid x)]$. Let $r(x) = 1 - \max_k p(k \mid x)$. Then with the doubt option

$$\mathsf{pmc} = \mathsf{E}\{r(X) I[r(X) \leqslant d]\}, \qquad \mathsf{pd} = \mathsf{Pr}\{r(X) > d\},$$

and the Bayes risk is $R = \mathsf{pmc} + d\mathsf{pd}$. The *error-reject* curve plots pmc against pd for varying d. Note (Chow, 1970) that

$$\mathsf{pmc}(d) = -\int_0^d \zeta\, \mathrm{d}\mathsf{pd}(\zeta)$$

so $\mathsf{pd}(d)$ as a function of d determines all the performance quantities.

Proposition 2.1 highlights the central role of the posterior probabilities. Most of the rest of the theory presented here can be regarded as ways to estimate or approximate the posterior probabilities from the training set.

This definition follows Lehmann (1986), Devijver & Kittler (1982) and many others; another school, represented by Berger (1985), calls the total risk the Bayes risk, and the Bayes risk the minimum Bayes risk. Fukunaga (1990) calls it the *Bayes error*.

Example: Normal classes with common covariance matrix

The most important distribution in statistical theory is the *normal distribution*, with the familiar density $(\sqrt{2\pi}\sigma)^{-1}\exp[-\frac{1}{2}(x-\mu)^2/\sigma^2]$ in the one-dimensional case. We write $X \sim \mathsf{N}\{\mu,\sigma^2\}$ to signify that X has this distribution, with mean parameter μ and variance parameter σ^2, and will also say that 'X is normal (μ,σ^2)'. In p dimensions the density is

$$p(x) = (2\pi)^{-p/2}|\Sigma|^{-1/2}\exp\left[-\tfrac{1}{2}(x-\mu)^T\Sigma^{-1}(x-\mu)\right] \quad \text{for } x \in \mathbb{R}^p. \quad (2.7)$$

We write $X \sim \mathsf{N}_p\{\mu,\Sigma\}$ or say that X is normal (μ,Σ) when X is a p-dimensional vector with this distribution. (Thus we omit the qualifying 'multi' or 'multivariate' that often is included.) The expected value is μ and the covariance matrix is Σ (Mardia *et al.*, 1979, p. 37).

Suppose the feature vectors from class k are $\mathsf{N}_p\{\mu_k,\Sigma\}$. If we disregard the doubt option a new feature vector x is allocated to the class k with smallest value of $\delta(x,\mu_k)^2 - 2\log\pi_k$, where

$$\delta(x,\mu_k) = [(x-\mu_k)^T\Sigma^{-1}(x-\mu_k)]^{1/2}$$

is (the definition of) the Mahalanobis (1936) distance from x to the centre of class k. Since the quadratic term $x^T\Sigma^{-1}x$ is common to each class the optimal rule can be written

$$\text{minimize } -2\mu_k^T\Sigma^{-1}x + \mu_k^T\Sigma^{-1}\mu_k - 2\log\pi_k \quad \text{over } k=1,\ldots,K. \quad (2.8)$$

This is called (the population version of, or the theoretical version of) *linear discriminant analysis*. If the classes are equally likely *a priori* then x is classified as coming from the nearest class, in the sense of having the smallest Mahalanobis distance to its mean. If in addition Σ is proportional to the identity matrix then distance can be Euclidean distance.

The error rate for the optimal rule can be computed explicitly in the two-class case. One should allocate to class 1 whenever

$$\log\pi_1 - \tfrac{1}{2}\delta(x,\mu_1)^2 > \log\pi_2 - \tfrac{1}{2}\delta(x,\mu_2)^2.$$

This can be reorganized as

$$A = (\mu_1-\mu_2)^T\Sigma^{-1}(x-\bar{\mu}) > \log(\pi_2/\pi_1), \quad (2.9)$$

where $\bar{\mu} = \frac{1}{2}(\mu_1+\mu_2)$. If X comes from class 1 then $A \sim \mathsf{N}\{\frac{1}{2}\delta^2,\delta^2\}$, in terms of the Mahalanobis distance

$$\delta = \left\{(\mu_1-\mu_2)^T\Sigma^{-1}(\mu_1-\mu_2)\right\}^{1/2} \quad (2.10)$$

between the two classes. Similarly, if X comes from class 2, then $A \sim N\{(-\frac{1}{2}\delta^2, \delta^2\}$. Accordingly

$$
\begin{aligned}
\mathsf{pmc} &= \pi_1 \mathsf{Pr}\big[N\{\tfrac{1}{2}\delta^2, \delta^2\} \leqslant \log(\pi_2/\pi_1)\big] \\
&\quad + \pi_2 \mathsf{Pr}\big[N\{-\tfrac{1}{2}\delta^2, \delta^2\} > \log(\pi_2/\pi_1)\big] \\
&= \pi_1 \Phi\Big(-\tfrac{1}{2}\delta + \tfrac{1}{\delta}\log(\pi_2/\pi_1)\Big) + \pi_2 \Phi\Big(-\tfrac{1}{2}\delta - \tfrac{1}{\delta}\log(\pi_2/\pi_1)\Big),
\end{aligned}
$$

where $\Phi(\cdot)$ is the cumulative distribution function for the standard normal. Note that the error rate is expressed in terms of the one-dimensional normal distribution even when the class distributions are p-dimensional normal. In the symmetric case with equal prior probabilities both class-wise error rates are equal, and the minimum attainable misclassification rate is $\mathsf{pmc} = \Phi(-\frac{1}{2}\delta)$.

Example: three Poisson groups

Suppose there are three equally likely groups of Poisson data, with mean parameters $\lambda_1 = 10$, $\lambda_2 = 15$, $\lambda_3 = 20$. Then the optimal rule is to allocate to class 1 if $X \leqslant 12$, to class 2 if $13 \leqslant X \leqslant 17$, and to class 3 if $X \geqslant 18$. The class-wise success rates, or probabilities of correct classification, are

$$
\begin{aligned}
\mathsf{pcc}(1) &= \mathsf{Pr}\{X \leqslant 12 \mid C = 1\} = 0.792, \\
\mathsf{pcc}(2) &= \mathsf{Pr}\{13 \leqslant X \leqslant 17 \mid C = 2\} = 0.481, \\
\mathsf{pcc}(3) &= \mathsf{Pr}\{X \geqslant 18 \mid C = 3\} = 0.703.
\end{aligned}
$$

The overall error rate is 0.341.

Suppose next that one can obtain two independent measurements X_1 and X_2 from the object to be classified. How do the allocation rules and the error rates change? Some easy calculations show that one should allocate to class 1 if $\overline{X} \leqslant 12.0$, to class 2 if $12.5 \leqslant \overline{X} \leqslant 17.0$, and to class 3 if $\overline{X} \geqslant 17.5$, where $\overline{X} = (X_1 + X_2)/2$. The revised class-wise success rates are

$$
\begin{aligned}
\mathsf{pcc}(1) &= \mathsf{Pr}\{\overline{X} \leqslant 12.0 \mid C = 1\} = 0.843, \\
\mathsf{pcc}(2) &= \mathsf{Pr}\{12.5 \leqslant \overline{X} \leqslant 17.0 \mid C = 2\} = 0.640, \\
\mathsf{pcc}(3) &= \mathsf{Pr}\{\overline{X} \geqslant 17.5 \mid C = 3\} = 0.806.
\end{aligned}
$$

The overall error rate has been reduced to 0.237.

Remarks

The constant d in (2.2) acts as a safety threshold, and should in principle be specified by the user of the resulting classifier. The inconveniences caused by a reject have to be judged against the consequences of a misclassification. In a serious application where the classifier is meant to work routinely on future examples one would typically try several d values on a training set of vectors with known classes, and obtain estimates of misclassification and doubt rates (see Section 2.7) before a 'final value' is chosen. Plotting misclassification rate against d is useful (see Figure 3.5 on page 114). If d is near zero then 'doubt' is inexpensive. This will lead to low error rates but on few classified vectors and a high doubt rate. If on the other hand $d \geqslant 1 - 1/K$ then decision $\mathcal{D}$ is so expensive that it never will be used.

There are no restrictions on the type of densities $p_1, \ldots, p_K$; in particular they need not be densities with respect to Lebesgue measure. (For professional probabilists: as long as $p_k = \mathrm{d}P_k/\mathrm{d}\mu$ for some σ-finite measure μ dominating the class distributions $P_1, \ldots, P_k$ both (2.4) and (2.5) continue to hold. We may in fact take $\mu = \sum_{k=1}^{K} P_k$.) Thus some or all of the P_k's may have discrete components, they may represent normal distributions with singular covariance matrices, and so on.

The small piece of theory presented here is fairly standard, although the rigorous derivation of the optimal reject ('doubt') region, by means of the loss function, is less known. The most popular special cases of the optimal rule are the normal distribution cases with common or different covariance matrices; see the example above and those discussed in Section 2.2. Indeed, discriminant or classification analysis started with a sample version of (2.9), in Fisher (1936). He derived the best linear rule in the two-class case but from a different perspective; see Section 3.1.

Missing values

Some problems (such as the Pima Indians data) have examples with missing values for some of the features. In principle these are easily accommodated; just compute the posterior probabilities $p(c \mid x^*)$ using the observed features x^*. However, these may be difficult to calculate. One technique is to simulate the missing features from $p(x \mid x^*)$ and average $p(c \mid x)$ over the simulated values. For this to be possible, the marginal density $p(x)$ must be known. If there are several missing features, the Gibbs sampler (Section A.3) may be used to allow them to be sampled one at a time.

This technique is known as *multiple 'hot deck' imputation* in survey sampling.

For normal classes both procedures are easy, as the distribution of some of the features is again joint normal, so we find a modified linear rule in the observed features. The density $p(x \mid x^*)$ is a mixture of normal distributions (one for each class) and so is easy to sample from.

Simpler procedures are often used, such as replacing missing values by 'typical' values, for example by the average over observed values. This is potentially dangerous, as the conditional density $p(y \mid x^*)$ of a feature y may have a very different mean from the unconditional density.

Missing values have been largely ignored in the pattern recognition literature. They are common in medical diagnosis, but rare in domains where data are collected automatically. It is a subject which has been treated most extensively in the literature on sampling surveys (Little & Rubin, 1987). There the problem may be that 'missing' actually indicates a refusal to respond, and so is informative about the features. This can also occur in medical diagnosis, where the medical practitioner may not order a test whose outcome appears certain or not relevant to the diagnosis. It could also be that a feature is missing because it proved to be too difficult to measure. Note that informative missingness of y is only a problem if it indicates a departure from the distribution $p(y \mid x^*)$. Thus a missing test whose outcome could be predicted from the remaining features would not be a difficulty (although the medic may be predicting from qualitative data which are not recorded). On the other hand, the refusal to answer a test may well be unpredictable and so informative. Where this is suspected, often the only possible action is to code 'missing' as a value of the feature, and somehow to find the densities required using the expanded feature(s).

Outliers

The concept of outliers does not fit cleanly into the decision-theory framework; one is supposed to have described the whole problem, and 'outliers' suggest incorrect specification. So one way forward is to anticipate outliers and build them into the specification as a separate class, with a specified $\pi_\mathcal{O}$ and class density $p_\mathcal{O}(x)$. Where might these come from? As outliers express surprise, the class density should perhaps reflect ignorance, and so be a suitable uniform distribution over $\mathcal{X}$. This is likely to cause difficulties, as for many feature spaces the uniform distribution is not normalizable to a probability distribution. These can be circumvented; for example for $\mathcal{X} = \mathbb{R}^p$ we could take a normal distribution with a very large variance. However, the difficulties persist, as both the 'shape' of the variance (for $p > 1$) and the scale of the variance can affect dramatically the reporting of outliers.

An alternative to assuming ignorance for the class of outliers is to follow the procedure we will use for all the other classes, and estimate $\pi_{\mathcal{O}}$ and $p_{\mathcal{O}}(x)$ from the training set. Sometimes this *is* feasible; for instance in reading Zip codes and in object recognition, data are sufficiently plentiful to enable a representative sampling of outliers. But such training sets are not commonplace, and often training sets are collected under carefully controlled circumstances where outliers are less common than usual (or even removed entirely).

Once the outlier distribution is given, under loss (2.2) outliers are declared if

$$\pi_{\mathcal{O}} \, p_{\mathcal{O}}(x) \geqslant (1 - d) \, p(x), \; \max_k \pi_k \, p_k(x)$$

If $p_{\mathcal{O}}$ is 'uniform' this classifies as an outlier when both $p(x) = \sum \pi_k \, p_k(x)$ and each component is small.

Another way to view an outlier would be as an observation x which was implausible under each of the class densities p_k or under all classes, that is under $p(x) = \sum \pi_k \, p_k(x)$. Note that these two concepts can be very different if the classes have very different prior probabilities; the second seems preferable as we would want to report as an outlier a mildly-unusual observation for a very rare class. Thus in this approach outliers are detected by first screening observations x and declaring those with small $p(x)$. How small? This is the same scenario as a pure significance test in statistical hypothesis testing (Cox & Hinkley, 1974; Lehmann, 1986) and the same ideas apply. Typically we will fix a level α of acceptable false detections of outliers, and fix a level p_c so that

$$\Pr\{p(X) < p_c\} \leqslant \alpha.$$

However, the integration needed here will often be intractable, and in the examples we relate $p(x)$ to its average value on the training set.

The two routes lead to the same practical conclusion; declare an outlier when $p(x)$ is small. Note that this is one place where knowledge of the posterior probabilities is not sufficient. We have to be very careful to ensure that 'uniformity' is an acceptable assumption for $p_{\mathcal{O}}$; as this is a density it will depend on the particular transformation of the features used. Often structural constraints on the features will rule out uniformity, and some other plausible guess at $p_{\mathcal{O}}$ will be needed.

The data on Cushing's syndrome shown in Figure 1.2 on page 11 provide an illustration of the difficulties of outliers in even a small number of dimensions. (Typically there are many more points in many more dimensions, so the data may be equally 'sparse'.) One of the unknown results seems a clear outlier, both for all three types individually and from the whole distribution, but so does one of the results of type c, and the latter is believed to be genuine. (However, there are only five patients of type c with known results.) Another of the unknown results looks like a marginal outlier. In this problem we will assume that a uniform distribution over log excretion rates is plausible, even though there must be an effective maximum and minimum, and we might perhaps expect the two rates to be correlated.

Ignoring the possibility of outliers can lead to misleading results. In the early 1950s anthropologists were discussing recently discovered hominoid fossils, and in particular whether *Australopithecus africanus* should be classified as an ape or a human. Bronowski & Long (1951) considered a linear discriminant analysis of teeth between chimpanzees and *Homo sapiens* and found agreement with *Homo* but not chimpanzees; Rao (1960) pointed out that they thereby overlooked the fact that on the full set of variables the sample tooth of *A. africanus* was implausible for either population.

Sometimes outliers are the main interest in a classification problem, in what is known in signal processing as *novelty detection*. For example in detecting tumorous tissue in mammograms, the tumours are so rare that what is required is to highlight unusual tissue for further inspection (Tarassenko *et al.*, 1995).

See the cover for an example.

2.2   Parametric models

We have seen in Proposition 2.1 the central role of the posterior probabilities $p(k \mid x)$, although the consideration of outliers showed that this is not universal. Since the posterior probabilities are in general unknown, we have to estimate them from the data, and to do so we use models. The difference between the parametric models we consider here and the non-parametric models we consider in Chapter 6 is less clear-cut than the terms would suggest: the real distinction is between families of probabilities which are quite constrained by having only a few parameters, and those which are so flexible that they can approximate (almost) any posterior probabilities.

We first give some general comments about the use of parametric models in classification, including discussion of what the methods actually do when the underlying assumed models are incorrect. We then present classification rules based on some of the most important parametric models.

The most theoretically satisfying approach comes in Section 2.4.

Figure 2.1: It is not always necessary to model the class-conditional densities (upper figure) accurately, as the posterior probabilities in the lower figure are effectively unchanged by most aspects of modelling the right peak of the class-conditional density shown dashed. Only the densities in the interval $[1,2]$ matter.

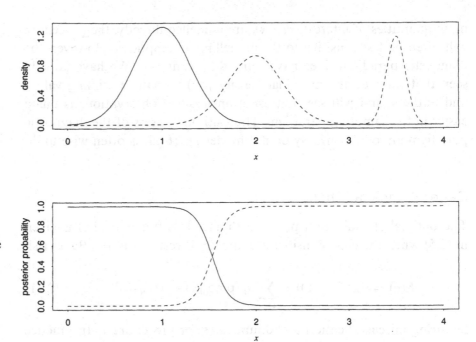

Theoretical and practical issues related to debiasing of maximum likelihood density estimates, predictive classifiers and robust estimation are addressed in later sections. Our first approach is that of classical statistics, to model either the class densities (this section) or the conditional probabilities (discussed in Section 2.3). It will be helpful to distinguish clearly these two tasks, which Dawid (1976) calls the *sampling* and *diagnostic* paradigms. Both give a parametric model of the joint density $p(x,c;\theta)$ of a random sample (X,C) of a set of features and its (reported) classification. In the sampling paradigm, interest centres on $p_k(x;\theta)$, and we have $p(x,c;\theta) = \pi_c\, p_c(x;\theta)$, with the prior probabilities (π_k) for the classes assumed to be either known or completely unknown. In the diagnostic paradigm, interest centres on the posterior probabilities $p(c\,|\,x;\theta)$, with $p(x,c;\theta) = p(c\,|\,x;\theta)\,p(x;\theta)$, but any information about θ in the unconditional density $p(x;\theta)$ is normally discarded by conditioning on the observed x's.

In later chapters we will concentrate on the diagnostic paradigm, which is illustrated in Section 2.3. The sampling paradigm is considered in this section, Sections 2.4 and 2.5 and Chapter 6. In Chapter 8 (X,C) is modelled simultaneously without stressing the importance of the class C.

Each of these approaches has strengths and weaknesses. As Figure 2.1 shows, direct modelling of the posterior probabilities may need fewer parameters than modelling via the class-conditional densities, and as the main quantities of interest, $p(c \mid x)$, are modelled directly, the procedure will often be less sensitive to the modelling assumptions. However, the diagnostic paradigm does have some disadvantages. We have already seen that we need the marginal density $p(x)$ to handle missing values and outliers, and will see that using unclassified observations is much easier in the sampling paradigm. Thus although users of the diagnostic paradigm almost invariably do not model $p(x; \theta)$, it is often wise to do so.

General considerations

The optimal classification procedure under loss function (2.2) is given in (2.5) when the class densities are known. It resulted in the Bayes risk

$$R(c) = \mathsf{E}L(C, c(X)) = \sum_{k=1}^{K} \pi_k [\mathrm{pmc}_0(k) + d\, \mathrm{pd}_0(k)], \qquad (2.11)$$

featuring misclassification and doubt rates for procedure c. In practice the p_k's are at least partly unknown, and the statistical task becomes one of providing good alternative procedures with Bayes risk as close to $R(c)$ as possible.

It is assumed in this section that the prior probabilities π_k are known and that the class densities are modelled parametrically, say

> We would estimate π_k by $\widehat{\pi}_k = n_k / \sum n_j$.

$$p_k(x) = p_k(x; \theta) \quad \text{for } k = 1, \dots, K,$$

where $\theta \in \Theta$ is the vector of unknown parameters needed to describe the K class densities. Suppose a training set of the form

$$\mathscr{T} = \{X_{k,1}, \dots, X_{k,n_k}, k = 1, \dots, K\} \qquad (2.12)$$

is available, with the n_k $X_{k,j}$'s coming from class k. These give rise to an estimate $\widehat{\theta}_k$ of θ_k. A natural proposal is then the classification rule

$$\widehat{c}(x) = \begin{cases} k & \text{if } \widehat{p}(k \mid x) = \max_l \widehat{p}(l \mid x) \text{ and this exceeds } 1 - d, \\ \mathscr{D} & \text{if each } \widehat{p}(k \mid x) \leqslant 1 - d, \end{cases} \qquad (2.13)$$

where parameter estimates are inserted in class densities to produce approximate posterior probabilities

$$\widehat{p}(k \mid x) = \frac{\pi_k\, p_k(x; \widehat{\theta})}{\sum_{l=1}^{K} \pi_l\, p_l(x; \widehat{\theta})}. \qquad (2.14)$$

The rule (2.13) is called the *plug-in classifier*. Some of the most widely used classification methods are of this form, as shown in the examples below.

It remains to decide exactly which estimator should be plugged in. The maximum likelihood (ML) estimator has been the most popular choice in statistical practice, together with modifications to reduce its bias. (It is defined and discussed below.) The widespread use of the plug-in rule with the ML estimator has been caused by the good general reputation the ML method enjoys and the fact that several pioneers in statistical classification theory have directly or implicitly recommended it. The use has been rather uncritical, though. Although $\widehat{\theta}$ may be excellent as an estimator of θ there is no guarantee that $p_k(x;\widehat{\theta})$ is a good guess for $p_k(x;\theta)$, nor is $\widehat{c}(x)$ necessarily a good approximation to $c(x)$. The performance of plug-in rules and other procedures should really be judged by the criterion of total risk, $R(\widehat{c})$ defined in (2.3), if (2.2) is still considered to be the appropriate loss function, or by other criteria more tied to classification accuracy than to the behaviour of $\widehat{\theta}$ as an estimator for θ.

For example, to apply Proposition 2.1 we only need to know which of the posterior probabilities is the largest (or which of a weighted sum is the smallest), which requires high accuracy of modelling only for some parts of the feature space. (If one posterior probability dominates, it does not matter if it is fitted as 0.999 when it is really 0.85.) We know of no work aimed at this aspect of the problem, although some approaches are closer than others to its goals.

These questions and related problems are returned to later, but first we give some general comments pertaining to the use of parametric models in discriminant analysis.

Bayes risk consistency

A reasonably simple observation that has been taken as support for the use of plug-in parametric rules is the following: As the training set increases, that is each of $n_1,\ldots,n_K$ grows, then provided only each $\widehat{\theta}_k$ is consistent and the class densities are continuous in their parameters, $\widehat{c}$ of (2.13) becomes identical to the optimal c and its total risk $R(\widehat{c})$ converges to Bayes risk $R(c)$. Many plug-in rules, corresponding to a large class of possible estimators $\widehat{\theta}$, have this property; see Van Ryzin (1966) and Glick (1972, 1976).

Here consistency means almost sure convergence to the 'true' value.

There is an important assumption behind this argument, that the class densities $p_1,\ldots,p_K$ in fact obey the parametric structure in question. As statisticians sometimes admit, their parametric models are only approximations to reality, implying in the present context that even when the size of the training set increases beyond bounds, $\widehat{c}$ of (2.13) will become close to only an approximation to c of (2.5), and the total risk $R(\widehat{c})$ will converge to a number greater than $R(c)$. Expressions for this limit can be found using the theory presented below. It is often possible to construct procedures that are *Bayes risk consistent* in the sense that the sequence of total risks converges to the Bayes risk $R(c)$ when the training sets grow. Unless one firmly believes in a certain parametric model the Bayes risk consistent rules will necessarily involve non-parametric or very flexible parametric methods, a topic returned to in Chapters 4 and 6.

These comments are not meant to imply that parametric models are useless; they may indeed constitute good and compact approximations to more complicated models. Classifiers built on parametric assumptions may work excellently. Non-parametric methods often demand for their successful application far larger training sets than parametric alternatives. Thus there is a trade-off between perhaps simple, easily implementable algorithms that work well even for moderately sized training sets, and non-parametric ones that may behave awkwardly for small to moderate training sets. The non-parametric ones will nevertheless (nearly always) win if sufficient training data are available.

Likelihoods and unclassified observations

The likelihood for the training set $\mathcal{T}$ is

$$\ell(\theta;\mathcal{T}) = \prod_{k=1}^{K}\prod_{j=1}^{n_k} p_k(x_{k,j};\theta)\pi_k(\theta)$$

and this applies whether the n_k were fixed in advance or resulted from a random sample taken from the whole population. We will use $L(\theta;\mathcal{T}) = \log\ell(\theta;\mathcal{T})$ for most of our calculations. Conventionally likelihoods are only defined up to a factor which does not depend on θ and hence log-likelihoods up to an additive constant.

Here we assume that either $\pi_k(\theta)$ is known, hence does not depend on θ and can be dropped from the likelihood, or completely unknown and forms part of the parameter vector, which is then really $\psi = (\theta, \pi_1, \ldots, \pi_K)$. The maximum likelihood estimator of ψ is the maximizer of the (log-)likelihood. We have

$$L(\theta,(\pi_k);\mathcal{T}) = \sum_{k}\sum_{j}\log p_k(x_{k,j};\theta) + \sum_{k} n_k \log \pi_k.$$

We can maximize first over the second term; after introducing a Lagrange multiplier for the condition $\sum \pi_k = 1$ we find $\widehat{\pi}_k = n_k / \sum n_i$.

Plugging this in gives the log *profile likelihood*

$$L(\theta, (\widehat{\pi}_k); \mathscr{T}) = \sum_k \sum_j \log p_k(x_{k,j}; \theta) + \text{const}$$

which is the same as the log-likelihood knowing (π_k).

In most cases we have to maximize over θ directly (numerically or analytically). Sometimes the parameter θ divides into separate parts for each class, in which case we can fit $p_k(x; \theta)$ to each class separately by maximum likelihood.

If some of the features are missing, we replace $p_k(x_{k,j}; \theta)$ by $p_k(x_{k,j}^*; \theta)$ for the observed features $x_{k,j}^*$.

There are some problems in which observations X are cheap but classifications C are expensive, so we can envisage having a set of unclassified observations $\mathscr{U} = \{x_j'\}$ in addition to the training set. These must be regarded as independent samples from the mixture distribution $p(x; \theta) = \sum_k \pi_k p_k(x; \theta)$, and the log-likelihood (or profile likelihood) becomes

$$L(\theta, (\pi_k); \mathscr{T}) = \sum_k \sum_j \log p_k(x_{k,j}; \theta) + \sum_j \log p(x_j'; \theta) \qquad (2.15)$$

which will couple the class densities even if they could previously be separated. Note that the extra observations may carry much useful information; consider classes with $N_p\{\mu_k, \Sigma_k\}$ distributions. Given enough unclassified data, we could estimate all the parameters μ_k, Σ_k, π_k as precisely as desired, *except* we would be unable to say which group applied to which class. The classified observations provide the information on this matching.

Fitting parametric families when they are wrong

We will give a brief discussion of the behaviour of ML estimates when the underlying parametric model is not necessarily true. Assume that $X_1, \ldots, X_n$ are independent and identically distributed with a density $p(x)$, and that the parametric model $p(x; \theta) = p_\theta(x)$ is forced on the data, θ being a q-dimensional parameter belonging to some open parameter set. The ML estimator $\widehat{\theta}$ maximizes the log-likelihood function

$$L_n(\theta) = \sum_{i=1}^n \log p_\theta(X_i) \qquad (2.16)$$

with respect to θ. By the law of large numbers $n^{-1} L_n(\theta)$ tends to $\int p \log p_\theta \, dx$, the mean of $\log p_\theta(X_i)$, with probability 1 (often termed 'almost surely').

For many important parametric models this function has a unique maximum at a parameter value $\theta = \theta_0$. This θ_0 is not necessarily the 'true value' because we have not assumed that p belongs to the family of p_θ's. In a sense θ_0 is the value of θ making p_θ closest to the true p, in that it minimizes the *Kullback–Leibler divergence*

$$d(p, p_\theta) = \int p(x) \log \frac{p(x)}{p_\theta(x)}\, \mathrm{d}x. \qquad (2.17)$$

This measure is not symmetric in its arguments and therefore not a distance in the usual sense. It is rather a 'directed' distance from the true density to the modelled density, and we think of θ_0 as the *'least false'* parameter value. Under weak regularity conditions $\widehat{\theta} \to \theta_0$ almost surely (see for example: Huber, 1967; White, 1982) thus generalizing the classical consistency result for ML estimators. If the true density is in the parametric family, $p(x) = p(x; \theta_0)$ and $d(p, p_{\theta_0}) = 0$.

Applying this result to each of the parametrically estimated class densities we see that the ML plug-in rule $\widehat{c}$ defined in (2.13) and (2.14) converges pointwise to a rule c^* defined analogously to (2.5) but with posterior probabilities of the form

$$p(k\,|\,x) = \frac{\pi_k\, p_k(x; \theta_0)}{\sum_{l=1}^{K} \pi_l\, p_l(x; \theta_0)}.$$

Furthermore,

$$R(\widehat{c}) \to R(c^*) \text{ almost surely, and } R(c^*) > R(c). \qquad (2.18)$$

The classical result on the limiting distribution of $\sqrt{n}(\widehat{\theta} - \theta_0)$ may also be generalized to the present agnostic state of affairs where the parametric family does not necessarily contain the true p (Huber, 1967, p. 231; White, 1982, Theorem 3.2).

Proposition 2.2 *Under mild regularity conditions*

$$\sqrt{n}(\widehat{\theta} - \theta_0) \to_d \mathsf{N}_q\{0, J^{-1}KJ^{-1}\} \qquad (2.19)$$

where $\to_d$ *denotes convergence in distribution and*

$$J = -\mathsf{E}_p \frac{\partial^2 \log p(X_i; \theta_0)}{\partial\theta\, \partial\theta^T} \quad \text{and} \quad K = \mathsf{Var}_p \frac{\partial \log p(X_i; \theta_0)}{\partial\theta}.$$

If the true density belongs to the parametric family, $J = K$.

Proof: The ML estimator solves the vector equation $U_n(\widehat{\theta}) = 0$, where $U_n(\theta) = \sum_{i=1}^{n} \frac{\partial}{\partial\theta} \log p(X_i; \theta)$. A Taylor expansion shows that

$$0 = U_n(\widehat{\theta}) = U_n(\theta_0) + I_n(\widetilde{\theta})(\widehat{\theta} - \theta_0),$$

where $I_n(\theta)$ is the Hessian of the log-likelihood function and $\widetilde{\theta}$ lies on the vector between the ML estimator and θ_0. This implies

$$\sqrt{n}(\widehat{\theta} - \theta_0) = \left[-n^{-1}I_n(\widetilde{\theta})\right]^{-1} n^{-1/2} U_n(\theta_0)$$
$$\to_d J^{-1} N_q\{0, K\} = N_q\{0, J^{-1}KJ^{-1}\}.$$

If the family contains the true model that $J = K$ is well known (Cox & Hinkley, 1974, p. 108; Lehmann, 1983, p. 118). □

The usual definition of the Fisher information matrix is K. The regularity conditions needed imply that J and K are positive definite.

We also need to consider the effect of approximating the log-likelihood (M. Stone, 1977b; Murata *et al.*, 1991, 1993, 1994).

The *deviance* is defined in the glossary. Here the reference model is the true distribution.

Proposition 2.3 *Let* $D = 2\,\mathsf{E}\left[\log p(X) - \log p(X; \widehat{\theta})\right]$, *the expected deviance on a single test example. Then*

$$n \times D = \mathsf{E}\,\text{deviance} + 2q^* + O(1/\sqrt{n}) \tag{2.20}$$

where $q^* = \text{trace}\left[KJ^{-1}\right]$. *If the parametric family contains the true density,* $q^* = q$, *the number of parameters.*

Proof: Let $i(x, \theta)$ be the Hessian of the log-likelihood for just one sample. We approximate D via the Taylor expansion about θ_0

$$2\log p(x; \widehat{\theta}) \approx 2\log p(x; \theta_0) + 2(\widehat{\theta} - \theta_0)^T \partial \log p(x; \theta_0)/\partial\theta$$
$$+ (\widehat{\theta} - \theta_0)^T i(x, \theta_0)(\widehat{\theta} - \theta_0)$$
$$= 2\log p(x; \theta_0) + 2(\widehat{\theta} - \theta_0)^T \partial \log p(x; \theta_0)/\partial\theta$$
$$+ \text{trace}\left[i(x, \theta_0)(\widehat{\theta} - \theta_0)(\widehat{\theta} - \theta_0)^T\right].$$

We assume $\partial d(p, p_\theta)/\partial\theta = \mathsf{E}\,\partial \log p(X; \theta)/\partial\theta = 0$ at θ_0, so

The second step uses the independence of X and $\widehat{\theta}$.

$$D \approx 2\,d(p, p_{\theta_0}) - \mathsf{E}\,\text{trace}\left[i(X, \theta_0)(\widehat{\theta} - \theta_0)(\widehat{\theta} - \theta_0)^T\right]$$
$$= 2\,d(p, p_{\theta_0}) + \text{trace}\left[J\,\text{Var}(\widehat{\theta})\right] = 2\,d(p, p_{\theta_0}) + \frac{1}{n}\text{trace}\left[JJ^{-1}KJ^{-1}\right]$$
$$= 2\,d(p, p_{\theta_0}) + \frac{1}{n}\text{trace}\left[KJ^{-1}\right]. \tag{2.21}$$

For the training set we expand about $\widehat{\theta}$:

$$2\sum_i \log \frac{p(X_i)}{p(X_i;\theta_0)}$$

$$\approx 2\sum_i \log \frac{p(X_i)}{p(X_i;\widehat{\theta})} - \sum_i \text{trace}\left[i(X_i,\widehat{\theta})(\widehat{\theta}-\theta_0)(\widehat{\theta}-\theta_0)^T\right]$$

$$\approx 2\sum \log \frac{p(X_i)}{p(X_i;\widehat{\theta})} + q^* = \text{deviance} + q^*.$$

Checking the error terms shows the error to be $O(1/\sqrt{n})$.   □

This is the basis of Akaike's (1973, 1974) AIC and Murata *et al.*'s (1991, 1993, 1994) NIC criteria for model selection, which are of the form (deviance$+2\,q^*$), with q^* replaced by q for AIC. M. Stone (1977b) derived NIC while considering cross-validation and AIC, but did not comment that it might provide a better approximation. Moody's (1991, 1992) p_{eff} is a more general version which we discuss in Section 4.3.

To use (2.20) we replace the expectation of the deviance by the observed value. The main error comes in the fluctuations of the deviance at θ_0, $2\sum \log p(X_i)/p(X_i;\theta_0)$, about its mean $2n\,d(p,p_{\theta_0})$, which by the central limit theorem (assumed applicable) will be of order $O_p(\sqrt{n})$, and we have

$$n \times D = \text{NIC} + O(1/\sqrt{n}) + O_p(\sqrt{n}). \qquad (2.22)$$

AIC was named by Akaike (1974) as 'An Information Criterion', although it seems commonly believed that the A stands for Akaike. NIC is an abbreviation of 'Network Information Criterion'. Some definitions of AIC and the definition of NIC divide by n, which is fixed.

(The notation $O_p()$ is explained on page xii.)

Now consider comparing several models via their values of NIC, and choosing the model with the smallest. Equation (2.22) shows that for large enough n we will choose one of the models with smallest D. Of course, there may be many such models if we have a nested set, so NIC will there choose a model which includes the smallest true model, but not necessarily the smallest such model.

One major source of the fluctuations in (2.22) is the variability of the training set (X_i), and this is common to all models. However, the claim by Murata *et al.* (1994, §5) that for differences in NIC amongst nested models this fluctuation term in the differences is $O_p(1/\sqrt{n})$ is false. Suppose we have nested models with $q_1 > q_2$, $\Delta q = q_1 - q_2$ and the smaller model (and hence both) are true. Then as M. Stone (1977b) pointed out,

$$\text{AIC}_1 - \text{AIC}_2 = 2(\text{LR test of 1 } vs \text{ 2}) - 2\Delta q \sim \chi^2_{\Delta q} - 2\Delta q$$

for large samples, and the right-hand side has fluctuation $O_p(1)$. Even asymptotically we might find $\text{AIC}_1 < \text{AIC}_2$ and so choose the larger model. If the models are not nested and equally good we can have a fluctuation term in the difference of $O_p(\sqrt{n})$; consider the perverse example of two models which choose the true family for the odd numbered X_i, fixed θ for the even ones and *vice versa*, so the effective training sets are disjoint.

To make use of these results in practice, we have to be able to estimate J and K. Now J is the expectation of the *observed information*, the Hessian of the negative log-likelihood, with the observed information evaluated at θ_0 and the expectation over the true distribution of examples. Thus we can form a reasonable estimate by replacing the expectation by the average over a training (or test) set, and replacing θ_0 by $\widehat{\theta}$. The same argument suggests estimating K by the variance of $\partial \log p(X,\widehat{\theta})/\partial\theta$ over the training or test set. If q/n is not negligible, there is a danger of bias here, especially in estimating K, and hence of underestimating q^*. To see this, let $U(x,\theta) = \partial \log p(x;\theta)/\partial\theta$ denote the scores. Then $\mathsf{E}U(X,\theta_0) = 0$ from the definition of θ_0, so $K = \mathsf{E}U(X,\theta_0)U(X,\theta_0)^T$. For a training set $\sum_i U(X_i,\widehat{\theta}) = 0$, which imposes q constraints on the scores, and the divisor in the variance should perhaps be $n-q$. For a test set it is perhaps best to use the variance with divisor $n-1$.

Very little of the argument here depends on using a maximum likelihood estimator, and Huber's (1967) results hold much more generally. All we need is that $\widehat{\theta}$ maximizes $\sum \psi(X_i;\theta)$ for a suitably smooth function ψ playing the role of $\log p$, and that a unique θ_0 minimizes $\mathsf{E}\,\psi(X;\theta)$. Of course, the definitions of J and K change by replacing $\log p$ by ψ. (We use this freedom on page 140.)

Example

Consider the normal distribution $\mathsf{N}_q\{\mu, \Sigma\}$ as an approximation to a given density p on $\mathbb{R}^q$. The density is given at (2.7). Some analysis (Huber, 1985, Lemma 12.4) shows that the parameter values (μ_0, Σ_0) that provide the best approximation according to the Kullback–Leibler criterion (2.17) are

$$\mu_0 = \mathsf{E}_p X = \int x\, p(x)\, dx \qquad \text{and}$$

$$\Sigma_0 = \mathsf{Var}_p X = \int (x-\mu_0)(x-\mu_0)^T\, p(x)\, dx.$$

Thus when the normal model is used to describe data from a density p that perhaps is known *a priori* not to be normal and the ML estimators $\widehat{\mu}$, $\widehat{\Sigma}$ are computed, the theory shows that what they really estimate are μ_0, Σ_0, the population mean and variance. It is worth pointing out that this was proved without using any explicit expressions for the estimates themselves. These are derived below, after which another and more direct proof of $\widehat{\mu} \to \mu_0$ and $\widehat{\Sigma} \to \Sigma_0$ can be given.

The normal model and the best linear rule

Suppose p_k is the $N_p\{\mu_k, \Sigma\}$ density for $k = 1, \ldots, K$, as defined in (2.7). There we saw that the Bayes rule is a linear rule in the sense that we choose the maximum of K linear combinations, and for two classes we divide the linear combination $(\mu_2 - \mu_1)\Sigma^{-1}x$ (from (2.9)).

The total likelihood for a training data set of form (2.12) is

$$\prod_{k=1}^{K} \prod_{j=1}^{n_k} |\Sigma|^{-1/2} \exp[-\tfrac{1}{2}(X_{k,j} - \mu_k)^T \Sigma^{-1}(X_{k,j} - \mu_k)].$$

This is maximized by $\widehat{\mu}_k = \overline{X}_k = n_k^{-1} \sum_{j=1}^{n_k} X_{k,j}$ and by

$$\widehat{\Sigma} = \sum_{k=1}^{K} \frac{n_k}{N} \widehat{\Sigma}_k = \frac{1}{N} \sum_{k=1}^{K} \sum_{j=1}^{n_k} (X_{k,j} - \widehat{\mu}_k)(X_{k,j} - \widehat{\mu}_k)^T, \qquad (2.23)$$

where $N = \sum_{k=1}^{K} n_k$ is the total training set size (and the maximum will be infinity unless $N \geqslant p + K$). See, for example, Mardia *et al.* (1979, §4.2.2). The ML-estimated best linear rule takes the form (2.8) with these estimates plugged in:

$$\text{minimize } -2\widehat{\mu}_k^T \widehat{\Sigma}^{-1} x + \widehat{\mu}_k^T \widehat{\Sigma}^{-1} \widehat{\mu}_k - 2\log \pi_k \text{ over } k = 1, \ldots, K. \quad (2.24)$$

For two classes this is Fisher's (1936) linear discriminant, derived from another criterion; this approach stems from Rao (1948). Often the bias-corrected estimator of Σ with divisor $N - K$ is preferred (and $N - 1$ appears in at least one computer package). This makes no difference to the linear rule unless the prior probabilities differ, in which case the effect is to change the constant terms to reduce slightly the influence of the data term relative to the prior.

The best linear rule for the data on Cushing's syndrome on page 11 is shown in Figure 2.2. The equal-covariance normal model does not seem appropriate for this dataset.

The best quadratic rule

Now let the model for class k be $N_p\{\mu_k, \Sigma_k\}$. The ML estimators can be found from a likelihood expression as before, and since there are no

Figure 2.2: The decision regions of the best linear rule for the data on Cushing's syndrome, together with contours for $\widehat{p}(x)$ at negative powers of 10 of the average for the training set.

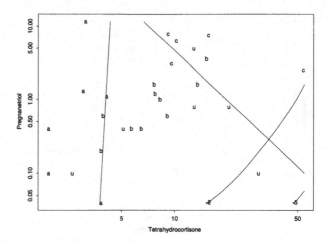

parameters common to more than one class, $\widehat{\mu}_k$ and $\widehat{\Sigma}_k$ are found by maximizing the likelihood for class k separately. The result is

$$\widehat{\mu}_k = \overline{X}_k \quad \text{and} \quad \widehat{\Sigma}_k = \frac{1}{n_k} \sum_{j=1}^{n_k} (X_{k,j} - \widehat{\mu}_k)(X_{k,j} - \widehat{\mu}_k)^T \qquad (2.25)$$

where we need $n_k \geqslant p + 1$ for each class for a finite maximum of the likelihood. This produces the plug-in version of the *best quadratic rule*;

$$\text{minimize} \ \tfrac{1}{2} \log |\widehat{\Sigma}_k| + \tfrac{1}{2}(x - \widehat{\mu}_k)^T \widehat{\Sigma}_k^{-1}(x - \widehat{\mu}_k) - \log \pi_k \qquad (2.26)$$

over $k = 1, \ldots, K$. The rule goes back to C. A. B. Smith (1947).

The number of estimated parameters has increased dramatically from $Kp + p(p + 1)/2$ for the best linear rule to $Kp + Kp(p + 1)/2$, so parameter estimates may be rather variable for the quadratic rule. Even though this method is guaranteed to outperform the linear rule for very large sample sizes, it can very well be outperformed by the linear rule for moderate sample sizes.

Since it may be preferable to use a linear rule, we can ask which linear rule produces the smallest error rate. This has been considered for two classes by Riffenburgh & Clunies-Ross (1960), Clunies-Ross & Riffenburgh (1960) and Anderson & Bahadur (1962). (See Anderson, 1984, §6.10.2.) The optimal linear rule is not that derived by pooling the covariance matrices and using (2.24) (for example, with $\widehat{\Sigma}$ the MLE or the average of $\widehat{\Sigma}_i$), although the linear combination used does derive from a convex combination of the two covariances. In practice it may be better to take some intermediate position, and compromise between the linear and quadratic rules. This is discussed in Section 3.4.

The data on Cushing's syndrome look suitable for quadratic discrimination, since although the numbers in the classes are very small,

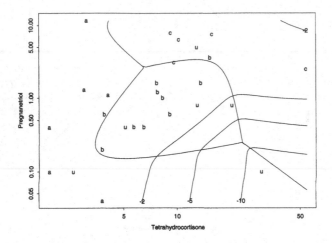

Figure 2.3: The decision regions of the best quadratic rule for the data on Cushing's syndrome, together with contours for $\widehat{p}(x)$ at negative powers of 10 of the average for the training set.

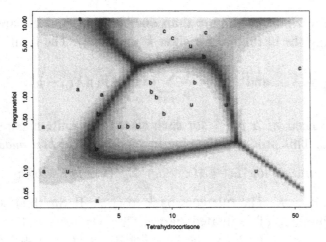

Figure 2.4: The uncertainty of the best quadratic rule for the data on Cushing's syndrome. The greyscales represent the maximum posterior probability of a class, with light grey as one and black as zero.

the covariance ellipsoids vary very considerably in orientation. The results are shown in Figures 2.3 and 2.4. The six unknown types are all given quite high posterior probabilities (the lowest is 70%, and the two apparent outliers have low values of $\widehat{p}(x)$, roughly 10^{-12} and 10^{-3} times the average for the training set. Thus both are rated as outliers (and they were medically, the more extreme being due to difficulties in the measurement procedure, and the less extreme to another type not represented in the training set).

It is possible that Σ_k is singular in one or more groups. (This happens in the forensic glass data—none of the samples of tableware contains any potassium, barium or iron.) A singular covariance matrix implies that the population for the class lies in a subspace of $\mathcal{X}$; equivalently it satisfies one or more linear constraints. Then a future example which does not satisfy those constraints does not come from the class, and one which does will come from this class (or any other that has the same constraints).

Multivariate t models

The univariate normal distribution is well known to have shorter tails than distributions which occur in applied problems, and a t distribution with a moderate number of degrees of freedom is often regarded as a better fit. The multivariate analogue of a t distribution is usually described by the analogue of the distribution of Student's t statistic: the multivariate t with location vector μ and scale matrix Σ is the distribution of $\mu + X/S$ where $X \sim N_p\{0, \Sigma\}$ and $vS^2 \sim \chi_v^2$ (Johnson & Kotz, 1972, §37.3; Mardia *et al.*, 1979, p. 57). (Unfortunately, several variant definitions exist in the literature, not all of which are actually densities!)

With this definition, for $v > 2$ the mean is μ and the covariance matrix is $v\Sigma/(v-2)$. The density

$$\frac{\Gamma(\frac{1}{2}(v+p))}{(v\pi)^{p/2}\,\Gamma(\frac{1}{2}v)}|\Sigma|^{-\frac{1}{2}}\left[1 + \tfrac{1}{v}(x-\mu)^T\Sigma^{-1}(x-\mu)\right]^{-\frac{1}{2}(v+p)} \qquad (2.27)$$

has elliptical contours with shape determined by Σ but which spread out more slowly than a normal distribution. The optimal classifier is

$$\text{minimize } \tfrac{(v+p)}{2}\log\left[1 + \tfrac{1}{v}(x-\mu_k)^T\Sigma_k^{-1}(x-\mu_k)\right] + \tfrac{1}{2}\log|\Sigma_k| - \log\pi_k. \qquad (2.28)$$

If the prior probabilities are equal and the scale matrix is common to all groups we again have the best linear rule.

The log-likelihood for the multivariate t is similar to that for the normal, except that the quadratic term $Q_i = (x_i - \mu)^T\Sigma^{-1}(x_i - \mu)$ is replaced by $(v + p)\log(1 + Q_i/v)$. Thus the maximum likelihood estimators of μ and Σ are weighted versions of the mean and scale matrix, with weights $w_i(\mu, \Sigma) = 1/(1 + Q_i/v)$:

These equations follow from Huber (1981, §8.4). They extend in the obvious way to a common scale matrix for all groups. See also Kent *et al.* (1994) and Lange *et al.* (1989).

$$\widehat{\mu} = \sum w_i(\widehat{\mu}, \widehat{\Sigma})x_i\Big/\sum w_i(\widehat{\mu}, \widehat{\Sigma}),$$

$$\widehat{\Sigma} = \frac{1}{n_k}\frac{v+p}{v}\sum w_i(\widehat{\mu}, \widehat{\Sigma})(x_i - \widehat{\mu})(x_i - \widehat{\mu})^T.$$

The effect of the longer tails of the t distribution is to down-weight observations which are far from the mean. The maximum likelihood estimators can be found by an iterative algorithm which updates the weights, although it would be wise to choose resistant estimates of the mean and covariance matrix (see Section 2.5) as starting points.

Kent *et al.* (1994) show that the solution is unique for $v > 1$ and $n \geqslant p + 1$.

Details of existence and convergence are a special case of arguments of Maronna (1976) and Huber (1981, §8.6). Note that as $\|x_i\| \to \infty$ its effect on the location estimate goes to zero, whereas on the scale estimate its effect remains bounded but does not vanish.

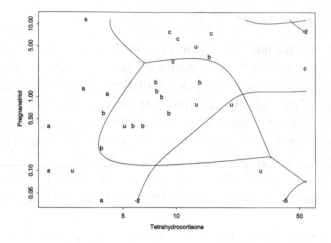

Figure 2.5: The decision regions of the rule based on the multivariate t_5 for the data on Cushing's syndrome, together with contours for $\hat{p}(x)$ at negative powers of 10 of the average for the training set.

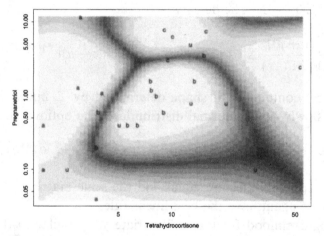

Figure 2.6: The uncertainty of the rule based on multivariate t_5 for the data on Cushing's syndrome. The greyscales represent the maximum posterior probability of a class, with light grey as one and black as zero.

One use of a multivariate t is as an agnostic model for distributions with elliptical densities with long tails, in the spirit of robust statistics (Huber, 1981; Lange *et al.*, 1989). In that setting it is interesting to consider what the least false parameters are, as they indicate what the parameters measure in the population. They are weighted versions of the mean and variance, weighted by $w(\mu_0, \Sigma_0) = 1/\left[1 + (X - \mu_0)^T \Sigma_0^{-1} (X - \mu_0)/v\right]$. Thus if the true density has elliptical contours, μ_0 will be the centre of the ellipses and Σ_0 will be proportional to the moment matrix of the ellipses (with the constant of proportionality depending on the true density).

The decision rule for multivariate t distributions on 5 degrees of freedom is shown in Figure 2.5 for the data on Cushing's syndrome. The number of degrees of freedom was chosen arbitrarily to give fairly 'fat' tails; despite this there is little difference from the best quadratic rule. Some of the difference is due to different mean and scale estimates, but the differences in the lower right reflect the tail behaviour. The much greater uncertainty shown in Figure 2.6 in the lower right also reflects the tail behaviour.

A mixed model for discrete and continuous components

Suppose $X = (A, Y_1, \ldots, Y_p)$ where A is discrete and takes values in $\{1, \ldots, m\}$ and $Y = (Y_1, \ldots, Y_p)^T$ has a continuous distribution. A simple and sometimes quite effective model for such feature vectors is to postulate $Y \mid \{A = a\} \sim \mathsf{N}_p\{\mu_{k,a}, \Sigma\}$ while $\mathsf{Pr}\{A = a \mid k\} = g_k(a)$ for class k (Olkin & Tate, 1961; Krzanowski, 1975). Many variations exist around this theme; the Σ matrix which is assumed common here can be taken to vary with either or both of a and k, for example. There is also a possibility of modelling $g_k(a)$ if the number of possible values for A is anything but small. (This is termed a conditional Gaussian distribution; see, for example, Edwards, 1995; Lauritzen, 1996, Chapter 6.) We shall be content here to illustrate the general principle with the simple model, sometimes called the 'location model'.

The class densities are

$$p_k\big((a, x)\big) = g_k(a)(2\pi)^{-p/2}|\Sigma|^{-1/2} \exp\big[-\tfrac{1}{2}(x - \mu_{k,a})^T \Sigma^{-1}(x - \mu_{k,a})\big]$$

and so from Proposition 2.1 we find the class k maximizing

$$\log \pi_k + \log g_k(a) - \tfrac{1}{2}(x - \mu_{k,a})^T \Sigma^{-1}(x - \mu_{k,a})$$

when (a, x) is observed. We need to find and plug in the maximum likelihood estimators of the parameters. These are straightforward: $\widehat{\mu}_{k,a}$ is the mean of observed X from class k with $A = a$, $\widehat{\Sigma}$ is the observed covariance matrix (with divisor n) of $X - \mu_{C,A}$, and $g_k(a)$ is estimated by the proportion of examples in class k with $A = a$.

Finite mixture distributions

We can consider larger parametric models, for example mixtures of normals which will allow us to model multi-modal class densities. As this is a way to fit quite general class densities, we defer the most of the details to Chapter 6. However, there is one quite commonly used 'trick' to fit class densities by mixtures, and that is to model sub-populations of the classes. A rather extreme example is that of Oliver *et al.* (1979), who considered 13 cell types in cervical cytology, 5 normal and 8 abnormal.

We have experienced several instances of feature distributions with a clear bimodal structure. Consequently histograms for even well chosen transformations of data are not well described by fitting a normal density. This suggests studying mixtures of two normal distributions as a means of describing class densities. The case of three or more normal components of a mixture is similar, but the number of parameters needed increases quite rapidly. We view the following mixtures of two normals for each class density as still being within the realm of parametric modelling.

Let $X_1, \ldots, X_n$ be a random sample from a density which we intend to describe parametrically by

$$p(x) = p(x; q, \mu_1, \Sigma_1, \mu_2, \Sigma_2) = (1 - q)\, \mathsf{N}_p\{\mu_1, \Sigma_1\}(x) + q\, \mathsf{N}_p\{\mu_2, \Sigma_2\}(x).$$

The important problem of fitting data to this class of densities is a difficult one and is perhaps not yet satisfactorily solved in the literature. The model is not properly defined until a restriction of the parameter set is made to avoid problems of identifiability; we may exchange (μ_1, Σ_1) and (μ_2, Σ_2) and rename q as $1 - q$ to get two representations of the same density. The model is identifiable if one demands $q \leqslant \frac{1}{2}$ or that the first μ_1 component should be to the 'left' of the first μ_2 component, for example. One may check by drawing graphs in the one-dimensional case, however, that curves with rather different sets of parameters may still come close to each other, making estimation of the parameters a more confusing and difficult task than usual. The density is not necessarily bimodal even when μ_1 and μ_2 are different; see Eisenberger (1964).

The maximum likelihood programme does not work as smoothly and automatically as in the earlier examples. First of all it does not exist in the usual sense, since the log-likelihood L_n is unbounded, with many singularities. For example, $L_n \to \infty$ as $\mu_1 = X_1$ and $\Sigma_1 \to 0$, corresponding to the 'explanation' $q = 1 - \frac{1}{n}$, $X_1 \sim \mathsf{N}\{\mu_1, 0\}$, while $X_2, \ldots, X_n$ follow $\mathsf{N}\{\mu_2, \Sigma_2\}$. Clearly this is not the solution we want. The L_n function will usually have several local maxima, and one of these corresponds to the nth element in a sequence of stationary points that converge almost surely to the true parameter values.

For univariate data, Hathaway (1985) establishes consistency for the global minimum under a constraint on the ratio of the variances.

A one-dimensional example of fitting two normals is given in Venables & Ripley (1994, Chapter 9) which illustrates some of the difficulties even in that case. They use direct maximization, with derivatives of the log-likelihood being found by automatic symbolic differentiation.

Updating estimates from unclassified data

We saw at (2.15) that we could include unclassified observations in the likelihood, and this opens the possibility of continuing to estimate the parameters (and the prior probabilities (π_k)) while the classifier is in routine use. This has two important implications: it enables work to start with a minimal training set, and it allows the classifier to adapt to slow changes in the class distributions over time.

The log-likelihood (2.15) involves the marginal density $p(x; \theta)$ which is a mixture of the class-conditional densities. Mixtures are discussed in more detail in Section 6.4, but we only need the application of the EM algorithm (Section A.2). Regard the true classes of the unclassified observations $\mathcal{U}$ as missing data. Then we estimate the posterior distribution of the true class k as $\pi_k(x; \theta) \propto \widehat{\pi}_k p_k(x; \theta)$ and $\widehat{\pi}_k$ as the average of the $\pi_k(x; \theta)$ over all observations. Hence θ is estimated by maximizing a weighted log profile likelihood

$$L(\theta) = \sum_k \sum_j \log p_k(x_{k,j}; \theta) + \sum_j \sum_k \pi_k(x'_j; \theta) \log p_k(x'_j; \theta).$$

The first term comes from the training set $\mathcal{T}$ and the second from the unclassified observations $\mathcal{U}$. This is an iterative process, in that θ and $(\widehat{\pi}_k)$ are updated alternately with $(\pi_k(x'_j; \theta))$.

We can apply this programme to the best linear and quadratic classifiers (Hjort, 1986, §7.2) and to multivariate t distributions. We need only keep the means and covariance matrices of the classified data, but since $\pi_k(x; \theta)$ will change as more examples are collected, all the unclassified data needs to be retained. Approximations for situations where retaining the data is computationally undesirable are discussed by Titterington *et al.* (1985, Chapter 6).

2.3 Logistic discrimination

Let us return to the normal model for classes with a common covariance matrix given by (2.7) with $\mu = \mu_j$ for class j. If we compare class k with class 1 we have

$$2 \log \frac{p(k \mid x)}{p(1 \mid x)}$$

$$= (x - \mu_1)^T \Sigma^{-1} (x - \mu_1) - (x - \mu_k)^T \Sigma^{-1} (x - \mu_k) + 2 \log \frac{\pi_k}{\pi_1}$$

$$= 2(\mu_k - \mu_1)^T \Sigma^{-1} x - (\mu_k + \mu_1)^T \Sigma^{-1} (\mu_k - \mu_1) + 2 \log \frac{\pi_k}{\pi_1}$$

$$= 2\beta_k^T x + 2\alpha_k$$

say, a linear function of x. Thus the posterior probabilities obey a *log-linear model* of the form

$$\log p(k \mid x) = \log p(1 \mid x) + \alpha_k + \beta_k^T x \qquad (2.29)$$

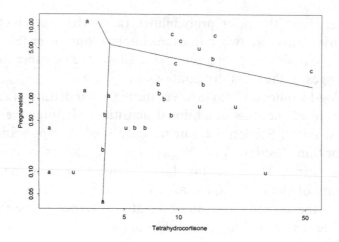

Figure 2.7: The decision regions based on logistic discrimination for the data on Cushing's syndrome.

which is also known as a *multiple logistic* model. The case of two classes is much simplified, as

$$\text{logit } p(2|c) = \alpha + \beta^T x \qquad (2.30)$$

for the *logit* transform $\text{logit}(x) = \log(x/(1-x))$. Thus (2.30) gives the posterior log-odds of class two versus class one, and is known as a *logistic regression*.

Equation (2.29) is illuminating, as it expresses the posterior probabilities, the important quantities in a plug-in rule, directly in terms of the parameters. This suggests that rather than use maximum-likelihood estimation of μ_k, Σ and hence α_k, β_k, we should estimate the latter directly. As (2.29) and (2.30) only concern the dependence of C on X, this is done by conditioning on X. For illustration we consider only the case of two classes here, and defer the general case to Section 3.5. However, comparing Figures 2.2 and 2.7 shows that the two estimates may give quite different classifiers when the common covariance model seems inappropriate.

Conditional on $X = x$, the class C has a Bernoulli distribution with probabilities $p(c|x)$. Thus if we re-express the training set as $\mathscr{T} = \{(C_i, X_i), i = 1, \ldots, n\}$ the conditional log-likelihood for the parameters $\theta = (\alpha, \beta)$ is given by

$$\prod_{i=1}^{n} p(c_i | x_i) = \prod_{i=1}^{n} p(2 | x_i)^{I(c_i=2)} [1 - p(2 | x_i)]^{1-I(c_i=2)}$$

and so if $Y_j = I(C_j = 2)$ the conditional log-likelihood is given by

$$L(\theta; \mathscr{T}) = \sum y_i \log p(2 | x_i) + (1 - y_i) \log[1 - p(2 | x_i)]. \qquad (2.31)$$

Note that maximizing (2.31) will not give the same answer as plugging in the maximum-likelihood estimators to give $\widehat{\beta} = \widehat{\Sigma}^{-1}(\widehat{\mu}_k - \widehat{\mu}_1), \widehat{\alpha} = \log(\pi_2/\pi_1) - \frac{1}{2}(\widehat{\mu}_1 + \widehat{\mu}_2)^T\widehat{\Sigma}^{-1}(\widehat{\mu}_2 - \widehat{\mu}_1)$ as here the likelihood is based on the conditional distribution. As they are based on less information, the direct estimators should be less efficient (that is more variable) although the standard large sample theory applies to show that the estimates are consistent and asymptotically normal. Efron (1975) demonstrates that this is the case, and the loss of efficiency can be appreciable when the class densities overlap, so the classification task is neither easy nor hopeless. If there are two classes with equal prior probabilities, the asymptotic relative efficiencies are a function of the Mahalanobis distance δ between the class means, given by

δ	0.5	1	1.5	2	2.5	3	3.5
ARE	1.00	0.99	0.97	0.90	0.79	0.64	0.49
pmc %	40.1	30.8	22.7	15.9	10.6	6.68	4.01

(The values of pmc are computed from the arguments below (2.10).) On the other hand, the logistic form (2.30) assumes less and is therefore less likely to be biased. (We will see in Chapter 3 that (2.30) can arise from other models of the class densities.)

Logistic discrimination is a very important template for many of the generalizations we will consider, much more so than linear discrimination. We take it up in Chapter 3 as a principle in its own right.

The loss is even higher when there is very little overlap, but then both rules perform well.

The ARE is the ratio in large samples of plug-in error rate minus Bayes risk; this is more relevant than the variability of the parameter estimate itself.

2.4 Predictive classification

Next we discuss the *predictive approach* towards estimation of parametric densities and posterior probabilities. It is Bayesian in inspiration and flavour even though the 'vague prior' versions of the method can be used and motivated outside the Bayesian paradigm. Suppose that we have a parametric family $p(x, c; \theta)$ for the joint distributions of the classes and features; this implies parametric models for $p_k(x; \theta), \pi_k(\theta)$ and $p(k \mid x; \theta)$, although of course not all of these need actually depend on θ. Assume also that we have a prior distribution $p(\theta)$ for θ. Then in principle (and sometimes in practice) we can calculate $\widetilde{p}(k \mid x) = \mathsf{Pr}\{k \mid x; \mathcal{T}\}$. The predictive approach then acts as if $\widetilde{p}(k \mid x)$ were the true posterior probabilities, and uses Proposition 2.1 to calculate the optimal rule, which we will call the predictive classifier. The crucial difference between the plug-in and predictive classifiers is that the former acts as if the estimated θ was the true θ whereas predictive methods average over the uncertainty in $\widehat{\theta}$.

The predictive approach gains very little mention in even comprehensive texts such as Berger (1985) and McLachlan (1992). This may well be because it usually makes little difference within the tightly constrained parametric families we are considering in this chapter, but it will be important when we consider much larger families. The books of Aitchison & Dunsmore (1975) and Geisser (1993) are devoted to the approach. Both contain brief accounts of classification, in Aitchison & Dunsmore's Chapter 11 under the heading of 'diagnosis'.

Within the Bayesian paradigm this needs no further justification: the prescribed way to handle unknown parameters is to integrate them out from the conditional distribution given the data. We may however ask whether the predictive classifier has any optimality properties in terms of risk. In one sense it cannot, for if we knew the true value of θ we must do better. Suppose rather that we extend the framework so θ is an unobserved random variable. Let $\widehat{c}(x, \mathcal{T})$ be a classifier which is allowed to depend on the training set (2.12). Its risk function, the expected loss when using it, is

$$R(\widehat{c}, k, \theta) = \mathsf{E}\left[L(k, \widehat{c}(X, \mathcal{T})) \mid C = k, \theta\right]$$
$$= \mathsf{pmc}(k, \theta) + d\,\mathsf{pd}(k, \theta),$$

where $\mathsf{E}_{k,\theta}$ denotes the expectation for class k and fixed θ. As a function of θ alone the risk function is

$$R(\widehat{c}, \theta) = \mathsf{E}_\theta L(C, \widehat{c}(X, \mathcal{T})) = \mathsf{pmc}(\theta) + d\,\mathsf{pd}(\theta).$$

The *overall risk* in this framework is

$$R(\widehat{c}) = \mathsf{E}L(C, \widehat{c}(X, \mathcal{T}))$$
$$= \int R(\widehat{c}, \theta)\,p(\theta)\,\mathrm{d}\theta = \mathsf{pmc} + d\,\mathsf{pd},$$

where $\mathsf{pmc} = \int \mathsf{pmc}(\theta)\,p(\theta)\,\mathrm{d}\theta$ and $\mathsf{pd} = \int \mathsf{pd}(\theta)\,p(\theta)\,\mathrm{d}\theta$ are unconditional misclassification and reject rates, averaged over the unknown θ. Note that this criterion makes good sense with any reasonable weight function over the parameter space; it does not have to be interpreted as a prior density (although it can be).

Proposition 2.4 *The classifier that minimizes the overall risk under loss (2.2) is*

$$\widetilde{c}(x) = \begin{cases} k & \text{if } \widetilde{p}(k \mid x) = \max_l \widetilde{p}(l \mid x) \text{ and this exceeds } 1-d, \\ \mathcal{D} & \text{if each } \widetilde{p}(k \mid x) \leqslant 1-d, \end{cases} \quad (2.32)$$

where $\widetilde{p}(k \mid x) = \mathsf{Pr}\{C = k \mid X = x; \mathcal{T}\}$. This is also the rule that minimizes the conditional risk given the training data. The extension to other loss functions follows Proposition 2.1.

Proof: We condition on the training set:

$$R(\widehat{c}) = \mathsf{E}L(C, \widehat{c}(X, \mathscr{T})) = \mathsf{E}_{\mathscr{T}}\mathsf{E}\big[L(C, \widehat{c}(X, \mathscr{T})) \mid \mathscr{T}\big]$$

and the conditional expectation is the total risk in the conditional distribution. Now apply Proposition 2.1 to the conditional total risk, to show this is minimized by a classifier of the form (2.32).   □

The full expression for the predictive posterior distribution when the future observation is independent of the training set is

$$\widetilde{p}(k \mid x) \propto \int p(k \mid x; \theta)p(x; \theta)\, p(\theta \mid \mathscr{T})\, \mathrm{d}\theta \tag{2.33}$$

$$p(\theta \mid \mathscr{T}) \propto p(\theta) \prod_{i=1}^{n} p(x_i, c_i; \theta) = p(\theta) \prod_{i=1}^{n} p(c_i \mid x_i; \theta)p(x_i; \theta). \tag{2.34}$$

The posterior density and the integral here readily become intractable. Explicit expressions are given below for some important special cases, and approximations can be provided in other cases.   Note that (2.33) and (2.34) depend on the parametrized marginal density of X; it is at this point that the simplicity of logistic discrimination loses out, unless the assumed form of the marginal density does not depend on θ. Alternative expressions which simplify in that case are

$$\widetilde{p}(k \mid x) = \int p(k \mid x; \theta)\, p(\theta \mid \mathscr{T}, x)\, \mathrm{d}\theta$$

$$p(\theta \mid \mathscr{T}, x) \propto p(\theta)p(x; \theta) \prod_{i=1}^{n} p(c_i \mid x_i; \theta)p(x_i; \theta).$$

and $p(\theta \mid \mathscr{T}, x)$ will not depend on x if $p(x; \theta)$ does not depend on θ.

If we work with parametrized class densities in the sampling paradigm, it is easier to use

$$\widetilde{p}(k \mid x) = \frac{\pi_k \widetilde{p}_k(x)}{\sum_l \pi_l \widetilde{p}_l(x)}, \qquad \widetilde{p}_k(x) = \int p_k(x; \theta)\, p(\theta \mid \mathscr{T})\, \mathrm{d}\theta. \tag{2.35}$$

It is helpful to remember that $\widetilde{p}$ quantities are just conditional densities given $\mathscr{T}$ and so can be manipulated as densities. We consider later what happens if the prior (π_k) is unknown.

Suppose we have to classify $m > 1$ future examples with feature vectors $x'_1, \ldots, x'_m$. The approach so far will not be fully efficient if $p(x; \theta)$ really does depend on θ, since all the x'_j can be used to increase our knowledge of θ (Geisser, 1966). We should use

$$p(\theta \mid \mathscr{T}, x'_1, \ldots, x'_m) \propto p(\theta) \prod_{i=1}^{n} p(c_i \mid x_i; \theta)p(x_i; \theta) \prod_{j=1}^{m} p(x'_i; \theta)$$

in the diagnostic paradigm, and use this omitting the current x'_j in (2.35). (This differs from the proposal of Geisser, 1966, 1993, which is to maximize the joint predictive probability of the m classifications. The latter is optimal under the loss structure that all predictions be correct rather than the number of misclassifications be small. The difference is important in statistical image analysis; Ripley, 1988, p. 114.)

Example: Poisson distributed counts

For a structurally simple example, suppose that a count variable X is modelled by Poisson distribution $p(x; \theta) = \exp(-\theta)\theta^x/x!$. Let θ have a gamma (α, β) prior distribution with density $[\beta^\alpha/\Gamma(\alpha)]\theta^{\alpha-1}\exp(-\beta\theta)$ on $[0, \infty)$, which has mean α/β and variance α/β^2. Assume that independent counts $X_1, \ldots, X_n$ have been observed from $p(x; \theta)$. Then it is not difficult to show that θ given the data has a gamma $(\alpha + n\widehat{\theta}, \beta + n)$ distribution, where $\widehat{\theta} = \overline{X}_n$ is the ML estimate for θ. Hence the predictive density is

$$
\begin{aligned}
\widetilde{p}(x) &= \int_0^\infty \exp(-\theta)\frac{\theta^x}{x!}\frac{(\beta+n)^{\alpha+n\widehat{\theta}}}{\Gamma(\alpha+n\widehat{\theta})}\theta^{\alpha+n\widehat{\theta}-1}\exp[-(\beta+n)\theta]\,d\theta \\
&= \frac{1}{x!}\frac{(\beta+n)^{\alpha+n\widehat{\theta}}}{(\beta+n+1)^{\alpha+n\widehat{\theta}+x}}\frac{\Gamma(\alpha+n\widehat{\theta}+x)}{\Gamma(\alpha+n\widehat{\theta})}.
\end{aligned}
$$

When α and β are sent to zero the expression simplifies to

$$
\widetilde{p}(x) = \frac{1}{x!}\frac{n^{n\widehat{\theta}}}{(n+1)^{n\widehat{\theta}+x}}\frac{\Gamma(n\widehat{\theta}+x)}{\Gamma(n\widehat{\theta})},
$$

and when n is large this is close to $\exp(-\widehat{\theta})\widehat{\theta}^x/x!$, the ML density estimate $p(x; \widehat{\theta})$. The difference is shown in Figure 2.8. $\square$

Example: Normal

Let $p(x; \mu, \Sigma)$ be the $N_p\{\mu, \Sigma\}$ density, (2.7), and assume (at this stage) that Σ is fixed. Then we can consider each class separately. Choose a prior $N_p\{\mu_0, A\}$ for the μ vector. The ML density estimate based on data $X_1, \ldots, X_n$ from $p(x; \mu, \Sigma)$ is $p(x; \widehat{\mu}, \Sigma)$ with $\widehat{\mu} = \overline{X}$. The posterior density is

$$
\mu \mid \mathcal{T} \sim N_p\{\widehat{\mu} - (\Sigma/n)(\Sigma/n + A)^{-1}(\widehat{\mu} - \mu_0), (n\Sigma^{-1} + A^{-1})^{-1}\}.
$$

This follows from the fact that

$$
\begin{pmatrix} \mu \\ \overline{X} \end{pmatrix} \sim N_{2p}\left\{\begin{pmatrix} \mu_0 \\ \mu_0 \end{pmatrix}, \begin{pmatrix} A & A \\ A & A + \Sigma/n \end{pmatrix}\right\},
$$

Figure 2.8: Plug-in (left) and predictive (right) probabilities for four samples with total 20 from a Poisson distribution.

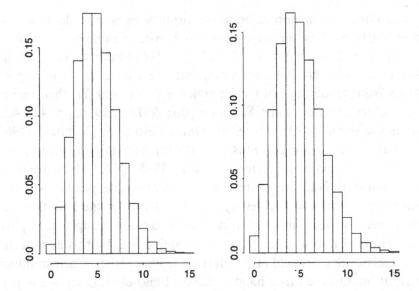

combined with properties of conditional distributions for jointly normal vectors; see for example Mardia *et al.* (1979, §3.2). The predictive density (2.35) can be worked out from this, but gives quite lengthy expressions. It is usual to use instead the simplified version that comes from the uniform (and improper) prior on $\mathbb{R}^p$, and which also corresponds to letting the matrix A tend to infinity (in the sense that all its eigenvalues tend to infinity). With this flat prior $\mu\,|\,\mathcal{T}$ is simply $\mathsf{N}_p\{\widehat{\mu},\Sigma/n\}$. Calculations (Geisser, 1964) give

$$\widetilde{p}(x) = (2\pi)^{-p/2}|\Sigma|^{-1/2}\left(\frac{n}{n+1}\right)^{p/2}\exp\left[-\tfrac{1}{2}\frac{n}{n+1}(x-\widehat{\mu})^T\Sigma^{-1}(x-\widehat{\mu})\right]$$
$$= \mathsf{N}_p\{\widehat{\mu},\tfrac{n+1}{n}\Sigma\}(x).$$

The difference from the plug-in estimate of the density is small when n is moderate or large. The optimal rule is still a linear discriminant and has the same combinations of the variables, but the slightly increased variance will affect the cutpoints if the prior probabilities are unequal.

Unknown covariance matrix/ces

The most important case is when the covariance matrices are also unknown. We start with one class and a non-informative prior of the type $(\mu_k,\Sigma_k) \sim |\Sigma_k|^{-a_0/2}$, where the symmetric covariance matrix is parametrized by its upper triangle, so this is a density on $\mathbb{R}^{p+p(p+1)/2}$. In the quite lengthy calculations that are involved it is more convenient to work with $\Lambda_k = \Sigma_k^{-1}$ instead; the density for (μ_k,Λ_k) is $|\Lambda_k|^{-a/2}$ with $a = 2(p+1) - a_0$.

The choice of a non-informative prior here is not clear-cut, but the majority view is for $a_0 = a = p + 1$ with a minority supporting $a_0 = 2p$, $a = 2$. Even for $p = 1$, Jeffreys' (1961) information principle leads to an answer that he rejects, but the choice $a_0 = a = p + 1$ follows from assuming prior independence of μ_k and Σ_k then seeking a non-informative prior for Σ_k alone (Box & Tiao, 1973, pp. 425–426). This is Geisser's (1993) choice, following Geisser & Cornfield (1963), and that of Aitchison & Dunsmore (1975) who take a limiting case of a Wishart conjugate prior. Berger (1985, §6.6) advocates using right invariant Haar measures as non-informative priors, which he demonstrates for $p = 1$ gives $a_0 = 2$. For larger p the results do not follow directly from his work (since the necessary group isomorphism fails). Hjort (1986) was led to $a_0 = 2p$ and $a = 2$ by this approach, a value Geisser & Cornfield (1963) derived from a Fisher–Cornish fiducial distribution. On the other hand Villegas (1969) obtains $a_0 = a = p + 1$ from a fiducial argument, and this was derived by Fraser (1968) from structural probability.

We can see the difficulties by examining $p = 2$ in more detail. Then Σ can be specified by $\kappa_i = \sigma_i^2, i = 1, 2$ and the correlation ρ; its determinant is $\kappa_1 \kappa_2 (1 - \rho^2)$. Jeffreys (1961, pp. 176, 187) variously advocates the priors $d\sigma_1 d\sigma_2 d\rho / \sigma_1 \sigma_2$ and $d\sigma_1 d\sigma_2 d\rho / \sigma_1 \sigma_2 (1 - \rho^2)^{3/2}$, and the latter corresponds to $|\Sigma|^{-3/2}$.

After extensive manipulation (Geisser & Cornfield, 1963; Hjort, 1986) we find

$$\widetilde{p}_k(x) = \pi^{-p/2}(n_k + 1)^{-p/2} \frac{\Gamma(\frac{1}{2}(n_k + p - a + 1))}{\Gamma(\frac{1}{2}(n_k - a + 1))} |\widehat{\Sigma}_k|^{-1/2}$$

$$\times \left[1 + \frac{1}{n_k + 1}(x - \widehat{\mu}_k)^T \widehat{\Sigma}_k^{-1}(x - \widehat{\mu}_k)\right]^{-\frac{1}{2}(n_k + p - a + 1)}. \quad (2.36)$$

Here $\widehat{\mu}_k$ and $\widehat{\Sigma}_k$ are the usual ML estimators defined at (2.23) on page 36. Figure 2.9 displays three estimates of a one-dimensional normal based on a sample of size 10; the ML plug-in estimate, the estimate which is unbiased on log scale, and the predictive estimate. Note that these three do not estimate comparable quantities, as the predictive estimator takes the uncertainty of the parameters into account in a way that the other two do not, but all are potential estimates to be used in Proposition 2.1.

For $p = 1$ the two approaches to choosing the prior agree. The predictive density is a (scaled) t distribution centred on $\hat{\mu}_k$. The general form is known as a multivariate t distribution on $(n_k + 1 - a)$ degrees

Figure 2.9: Estimates of the density based on a random sample of size 10 from $N\{0,1\}$. The 'plug-in' estimate is shown with a solid line, the predictive estimate (2.36) with a dotted line, and the unbiased estimate on log scale (2.38) with a dashed line.

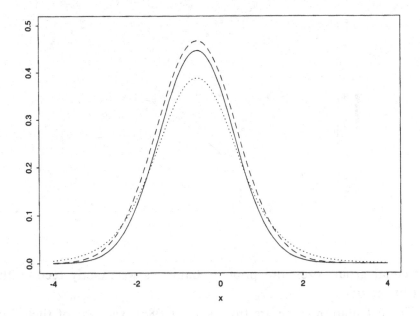

The multivariate t is defined in the glossary (see 't distribution') and on page 39. Aitchison & Dunsmore (1975, p. 255) give a different parametrization which affects the interpretation of their results.

of freedom with location vector $\widehat{\mu}_k$ and scale matrix

$$\frac{n_k + 1}{n_k + 1 - a}\widehat{\Sigma}_k.$$

For the point of view of classification we assume independent non-informative priors for each class. Then the predictive class densities have a larger spread than the normal distribution with variance $\widehat{\Sigma}_k$, but the same shape of contours, and so the optimal classifier is a quadratic rule. The difference from the plug-in quadratic rule is to move the decision boundaries to be more nearly equidistant from the class centres (apart from allowing for unequal prior probabilities of the classes).

Very similar ideas can be applied to the case of a common within-class covariance matrix, leading to

$$\widetilde{p}_k(x) = (N\pi)^{-p/2} \left(\frac{n_k}{n_k + 1}\right)^{p/2} \frac{\Gamma(\frac{1}{2}(N - K + p - a + 2))}{\Gamma(\frac{1}{2}(N - K - a + 2))} |\widehat{\Sigma}|^{-1/2}$$

$$\times \left[1 + \frac{n_k}{N(n_k + 1)}(x - \widehat{\mu}_k)^T \widehat{\Sigma}^{-1}(x - \widehat{\mu}_k)\right]^{-\frac{1}{2}(N - K + p - a + 2)}$$

which is a multivariate t distribution on $N - K - a + 2$ degrees of freedom with location $\widehat{\mu}_k$ and scale matrix

$$\frac{(1 + 1/n_k)N}{N - K - a + 2}\widehat{\Sigma}.$$

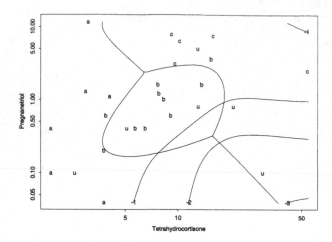

Figure 2.10: The decision regions of the predictive quadratic rule for the data on Cushing's syndrome, together with contours for $\widetilde{p}(x)$ at negative powers of 10 of the average for the training set.

For equal group sizes and prior probabilities we exactly recover the best linear rule.

The calculations here are from Hjort (1986); versions of these formulae are given by Aitchison & Dunsmore (1975) (up to the differences in the meaning of their multivariate t) and Geisser (1993). This approach is originally due to Geisser (1964, 1966).

The differences between the predictive and plug-in approaches will be small or zero for roughly equally prevalent classes. In other cases, for example screening for rare diseases or when very few data are available, the differences can be dramatic as shown by the examples in Aitchison & Dunsmore (1975, §§11.5–11.6). The latter do have groups with n_k only slightly greater than p, for example $p = 8$ and $n_2 = 11$ when fitting a covariance matrix to each class, which would be seen as over-fitting in the plug-in approach. (Indeed, one might choose not to use all the variables, or perhaps to restrict the class of covariance matrices considered.)

Aitchison *et al.* (1977) conducted a small-sample simulation comparison of the plug-in and predictive methods for two multivariate normal populations. They were (correctly) criticized by Moran & Murphy (1979) for using the accuracy of the estimation of the log-odds as the basis of comparison rather than error rates, and for including mainly equal sample sizes of the two classes. Moran & Murphy's results show very little difference in the error rates, and show that for estimation of the log-odds the debiasing methods of Section 2.5 are effective in removing the dramatic optimism of the plug-in method where it occurs.

Vlachonikolos (1990) extended these calculations to some simple cases of the 'location model' for mixed discrete and continuous data discussed on page 41.

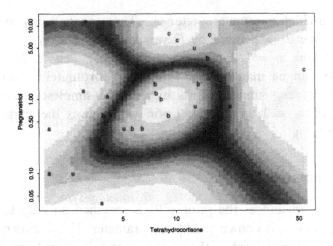

Figure 2.11: The uncertainty of the predictive quadratic rule for the data on Cushing's syndrome. The greyscales represent the maximum posterior probability of a class, with light grey as one and black as zero.

We return to the data on Cushing's syndrome shown on page 11. With the predictive estimates, the classification is less certain, especially of the two apparent outliers which are now both classified as c (with probabilities 61% and 68%). However, both still seem outliers, with values of $\widetilde{p}(x)$ roughly 0.2% and 6% of the average for the training set. These much less dramatic values and the tendency to classify both outliers as c reflect the great uncertainty in the class distribution for that class.

Unknown class prior

We need to consider the effect of unknown prior class proportions (π_k) in (2.35) as we would need to justify plugging-in the 'obvious' estimates. Under random sampling the observed numbers $(n_1, \dots, n_K)$ follow a multinomial$(n, p_1, \dots, p_K)$ distribution. The natural choice for a prior is a Dirichlet(α_i) distribution. The posterior is a Dirichlet$(n_i + \alpha_i)$ distribution so

The Dirichlet distribution is defined in the glossary.

$$p(x, k \mid \mathscr{T}) = \mathsf{E}[\pi_k \, p_k(x; \theta) \mid \mathscr{T}] = \frac{n_k + \alpha_k}{\sum(n_l + \alpha_l)} \, \widetilde{p}_k(x) = \widetilde{\pi}_k \, \widetilde{p}_k(x)$$

say, since the priors and hence posteriors for θ and (π_k) are independent. Then

$$\Pr\{C = k \mid X = x, \mathscr{T}\} = \frac{\widetilde{\pi}_k \, \widetilde{p}_k(x)}{\sum_l \widetilde{\pi}_l \, \widetilde{p}_l(x)}$$

The MLE is $\widehat{\pi}_k = n_k/N$.

so we act as if we had plugged in the estimate $\widetilde{\pi}_k$. The difference between this and plugging in the MLE will be negligible unless the class sizes are very small or the prior extremely strong.

For two classes the Dirichlet distribution reduces to the beta distribution with parameters (α_1, α_2). We can ask how to represent ignorance by the Dirichlet parameter vector α. Three suggestions in the two-class case are $\alpha_i = 1$ (used by Bayes and Laplace), $\alpha_i = 0$ (which is improper) and $\alpha_i = 1/2$ (Berger, 1985, §3.3.4; Geisser, 1984). Good arguments can be made for any of these; fortunately in our setting they will be very similar. This suggests the simplest choice for the Dirichlet of $\alpha_i \equiv 0$ as a vague prior, which gives the simple plug-in rule $\tilde{\pi}_k = n_k/n$.

Hyperparameters

In some circumstances the prior $p(\theta)$ is specified only up to a family $p_\lambda(\theta)$ of priors; λ is known as a hyperparameter. This occurs most often in pattern recognition when the prior is used to express 'smoothness' of the posterior probabilities $p(k \mid x; \theta)$ as a function of x, and λ is then the degree of smoothness. (See, for example, Section 4.3.)

How should λ be chosen? Within the predictive Bayesian framework, the solution is clear; we give a prior to λ, called a *hyperprior*. If this contains parameters, they too are given a prior, and so on. This is sometimes known as *hierarchical Bayes* (Berger, 1985, §3.6, 4.6) and the analysis is in principle obvious, since the effective prior is

$$p(\theta) = \int p_\lambda(\theta)\, p(\lambda)\, \mathrm{d}\lambda.$$

This may be awkward to use if $p_\lambda(\theta)$ has been chosen to simplify computation, and the integration over λ may be postponed to a late stage in the calculation. Thus we may find $\Pr\{k \mid x, \mathcal{T}, \lambda\}$ and then integrate this with respect to the density $p(\lambda \mid x, \mathcal{T})$.

Other approaches have been proposed, and in some cases advocated strongly. *Empirical Bayes* methods use the data to choose λ, which entails a data-dependent prior which purist Bayesians do not allow, but is sometimes seen as an approximation to hierarchical Bayes and sometimes as desirable in its own right. (Maritz & Lwin, 1989, is devoted to empirical Bayes methods; Berger, 1985, §4.5, gives references to many strands.)

Let us consider empirical Bayes as an approximation to $\tilde{p}(k \mid x)$ by $\Pr\{k \mid x, \mathcal{T}, \hat{\lambda}\}$. If $p(\lambda \mid x, \mathcal{T})$ is highly concentrated about one value $\hat{\lambda}(x, \mathcal{T})$, and if we can estimate this value easily and well, empirical Bayes will provide a considerable computational simplification. (Often x will not be at all informative, so $\hat{\lambda}$ can be computed once the training set is given.) How could we find $\hat{\lambda}$? Good (1965, 1983) (the latter a compendium of earlier work) calls one method 'type II maximum likelihood' or 'ML-II'. This is to choose λ to maximize the marginal density of the data; if we ignore x this is

$$m(\mathscr{T}\,|\,\lambda) = \int \ell(\theta;\mathscr{T})\,p_\lambda(\theta)\,\mathrm{d}\theta. \qquad (2.37)$$

Note that ML-II is equivalent to maximizing $p(\lambda\,|\,\mathscr{T}) \propto m(\mathscr{T}\,|\,\lambda)p(\lambda)$ if $p(\lambda)$ is constant, and so its $\widehat{\lambda}$ is likely to be a good estimate of $\lambda_0(x,\mathscr{T})$ if $m(\mathscr{T}\,|\,\lambda)$ has a sharp peak. Deely & Lindley (1981) discuss conditions for such approximations. In the usual empirical Bayes context the assumption is of many problems with different θ but the same λ, so this condition follows from the asymptotic normality of maximum likelihood estimation of λ from independent samples from $m(\mathscr{T}\,|\,\lambda)$. This is not the usual situation in pattern recognition applications, where a single large training set allows increasingly precise inferences about θ but provides just one sample to estimate λ.

The empirical Bayes methods usually ignore the variability in $\widehat{\lambda}$; $p_{\widehat{\lambda}}(\theta)$ will be more concentrated than $p(\theta\,|\,\mathscr{T})$. This may not matter in the centre of the distribution, but may be material in applications such as ours of finding $\widetilde{p}(k\,|\,x)$ since $p(k\,|\,x;\theta)$ is often a highly non-linear function of θ and the empirical Bayes methods often tend to produce fitted probabilities which are too extreme.

2.5   Alternative estimation procedures

The maximum likelihood estimator $\widehat{\theta}_k$ is not always the very best estimator of θ_k (we habitually use a different estimator of the variance of a normal population), and even in cases where it is well-chosen for θ_k the plug-in density estimate $p_k(x;\widehat{\theta}_k)$ is not necessarily the best estimate of $p_k(x;\theta_k)$. In view of (2.5) and (2.4) our interest lies more with the densities themselves than with the parameters that describe them, so here we consider alternative estimation procedures, principally within the sampling paradigm.

Debiasing density estimates

One route is to modify estimators so as to make them unbiased or less biased, if they are not unbiased in the first place. For example, the $\widehat{\Sigma}$ estimator of (2.23) has expected value $\frac{N-K}{N}\Sigma$, and statisticians normally use the modified version with denominator $N-K$ instead of N. The same remark applies to the modification of $\widehat{\Sigma}_k$ of (2.25) which uses denominator $n_k - 1$ instead of n_k.

The densities themselves are more directly involved in the classification problem. It is possible to find an unbiased estimator of a
normal density (Ghurye & Olkin, 1969), but it is more natural to find
an unbiased estimator of the log density, and hence of differences in
log densities. This is the appropriate plug-in estimator if the interest
concentrates on the log-odds or on $\log p(k \mid x)$, as in the experiments of
Aitchison *et al.* (1977) and Moran & Murphy (1979). Suppose $\widetilde{\Sigma}$ is the
unbiased estimator of Σ based on m degrees of freedom (so $m = n_k - 1$
for group k, or $m = N - K$ if a common covariance is assumed). Then

$$\log p^*(x) = -\tfrac{1}{2} p \log(2\pi) - \tfrac{1}{2} \left[\log |\widetilde{\Sigma}| + B_p(m) \right]$$
$$-\tfrac{1}{2} \left[\frac{m - p - 1}{m} (x - \widehat{\mu})^T \widetilde{\Sigma}^{-1} (x - \widehat{\mu}) - \frac{p}{n} \right] \quad (2.38)$$

is the unique unbiased estimator for $\log p(x; \mu, \Sigma)$, where

$$B_p(m) = p \log(\tfrac{1}{2} m) - \sum_{i=0}^{p-1} \psi(\tfrac{1}{2}(m - i)) \quad \text{and} \quad \psi(z) = \Gamma'(z)/\Gamma(z)$$

for the digamma function ψ (Abramowitz & Stegun, 1965, p. 258).
When n goes to infinity there is agreement with the plug-in estimator.
The proof depends on $\mathsf{E}\,\widetilde{\Sigma}^{-1} = m\Sigma^{-1}/(m - p - 1)$ and that $|m\widetilde{\Sigma}|/|\Sigma| = \prod_0^{p-1} Z_i$ where Z_i are independent χ^2_{m-i} random variables (Mardia *et al.*, 1979, pages 85 and 73 respectively).

Moran & Murphy (1979) give explicitly the effect of this bias
correction on the linear and quadratic classifiers. The plug-in version of
the two-class linear discriminant (2.9) is to allocate to class 1 whenever
the estimate of $\text{logit}\,p(1 \mid x)$ is positive, or

$$(\widehat{\mu}_1 - \widehat{\mu}_2)^T \widehat{\Sigma}^{-1}(x - \widehat{\mu}) + \log(\pi_1/\pi_2) > 0. \quad (2.39)$$

Using the unbiased estimator of the log density gives the rule

$$\frac{N - K - p - 1}{N - K} \left[(\widehat{\mu}_1 - \widehat{\mu}_2)^T \widetilde{\Sigma}^{-1}(x - \widehat{\mu}) \right] + \frac{p}{2} \left[\frac{1}{n_1} - \frac{1}{n_2} \right] + \log(\pi_1/\pi_2) > 0.$$
$$(2.40)$$

The effect of the data over the prior class probabilities is reduced, but
the constant term will also be important if the class sizes are very
different in the training set.

For the quadratic discriminant, the effect of the debiasing is to
increase the effective variance by a factor $n_k/(n_k - p - 1)$ over the usual
$n_k/(n_k - 1)$, and to add a constant which depends on n_k.

This debiasing is usually unimportant, but can make a difference if
for some class(es) n_k is only a little larger than $p + 1$, as Figure 2.12
shows for the data on Cushing's syndrome.

Figure 2.12: The decision regions of the debiased quadratic rule for the data on Cushing's syndrome, together with contours for $\hat{p}(x)$ at negative powers of 10 of the average for the training set.

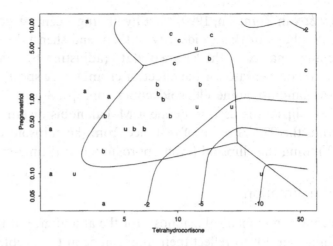

Robust estimation of parametric densities

The normal distribution is a convenient abstraction, but all careful studies show that real distributions do not quite follow a normal distribution but have slightly heavier tails. In addition we should consider the possibility of outliers, that is examples which do not belong to the class under consideration (for example, they might be wrongly labelled in the training set). This has led to discussion of robust estimators of the normal mean and variance for use in 'plug-in' linear and quadratic classifiers (for example, McLachlan, 1992, §5.7)

Our classifiers should be different in the two scenarios. If the distributions are non-normal, then we need to take into consideration that the tails will be longer, and assuming a t distribution will be more appropriate. As we saw on page 39, this leads to robust estimators of the means and variances of the t distribution, but with a common covariance matrix and equal prior probabilities the Bayes rule is still the best linear rule. This suggests that for the linear rule it is reasonable to plug in robust estimators, whereas for the quadratic rule the rate of decay of the densities in the tails is crucial.

On the other hand, if we believe that the true class densities are close to normal but that we have outliers, it will be desirable to plug in robust estimators, since the aim of robust estimation is to characterize the uncontaminated populations, and it is the latter we wish to use in the Bayes rule.

Robust estimation in multivariate problems is trickier than appears at first sight, since simple extensions of univariate methods (based on down-weighting extreme observations as we saw for fitting the multivariate t) can fail completely with a fraction $1/p$ of outliers (Huber, 1981, pp. 227–8). The most recent approaches (for example,

Rousseeuw & van Zomeren, 1990) start by finding a central 'core' of the data, use the shape of this to identify outliers and then take the mean and covariance matrix of the 'cleaned' data (adjusting the scale of the covariance to compensate for the effect of cleaning). A specific method finds the minimum-volume ellipsoid containing $\lfloor (n + p + 1)/2 \rfloor$ data points. This ellipsoid is used to define a Mahalanobis distance, and all points within the 97.5% point of distance from the ellipsoid centre are retained. Finding the ellipsoid (even approximately) is time-consuming.

Weighted estimation

It is quite common in medical diagnosis for the abundance of the classes in the training set not to reflect their importance in the problem. Often when the training data are a random sample from the population, the vast majority of cases are 'normals' yet the cost of mis-classifying a diseased case as normal is ℓ times higher than that of a false positive. In screening problems ℓ can be ten or more.

For clarity, suppose we have just two classes, 'diseased', d, and 'normal', n. The effect of differential costs is to move the decision threshold, so we will declare a positive result when the odds in favour of 'diseased' are not too adverse (better than $1 : \ell$). If we estimate the posterior probabilities $p(k \mid x)$ from the training data by plug-in methods, we would expect to learn $p(n \mid x)$ much more accurately than $p(d \mid x)$. Under some circumstances this can lead to serious bias in the estimators. Consider two normal distributions within the sampling paradigm. For the best quadratic rule, we estimate the class-conditional density $p_k(x)$ from the examples from class k. For the best linear rule, the common covariance matrix Σ is estimated from both populations, and hence is principally determined from the sample of 'normals'. This is fine if the covariance matrix really is the same in each group, but can lead to biased estimates if the covariance matrix in the 'diseased' group is somewhat different from that in the 'normal' group. The effect of this bias on the posterior probabilities is much more pronounced when the classes are unequally represented.

The biases in the diagnostic paradigm are often more serious. The plug-in decision rule is to declare a case 'diseased' if $p(d \mid x; \widehat{\theta}) > c$ for $c = 1/(1+\ell)$ less than 0.5, often much less. We have already noted that plug-in rules tend to produce estimated posterior probabilities which are too extreme, and this will result in a bias when c is small. A further bias results from the disproportion of the two classes in the training set, resulting in underestimation of $p(d \mid x)$ (since there are many more 'normal' cases to be fitted).

There are two ideas to alleviate these biases. A simple idea is to use a biased sample in the training set; this normally means randomly subsampling the 'normal' group. Let n_k denote the numbers of the classes in the training set, and π_k the proportions in the population. (We assume that these are known, for example estimated from the original training set.) The fitted probabilities $p(k \mid x; \widehat{\theta})$ then estimate quantities proportional to $p(k \mid x)n_k/\pi_k$, the posterior probabilities under biased sampling. Thus the decision rule is to declare a case 'diseased' if

$$\frac{p(\mathsf{d} \mid x; \widehat{\theta})\pi_\mathsf{d}/n_\mathsf{d}}{p(\mathsf{n} \mid x; \widehat{\theta})\pi_\mathsf{n}/n_\mathsf{n}} > \frac{1}{\ell}$$

or $p(\mathsf{d} \mid x; \widehat{\theta}) > 1/(1 + \ell\pi_\mathsf{d}n_\mathsf{n}/\pi_\mathsf{n}n_\mathsf{d})$. If we under-sample the 'normal' group by a factor of ℓ, this declares 'diseased' if the odds exceed one. This degree of bias in the training set puts the two groups on an equal footing in the parameter estimation, thereby reducing the estimation biases.

The reduction in the size of the training set can have considerable computational benefits, so sampling might be preferred.

When the biased training set is created by subsampling a larger training set, it seems wasteful to discard data on the 'normal' group. This suggests using weighting rather than sampling. In a weighted procedure, all the 'normal' examples are used, but their contributions to the log-likelihood are weighted by a factor ω and, in the sampling paradigm, the size of the 'normal' sample is regarded as ωn_n. Taking $\omega \approx 1/\ell$ will minimize the estimation biases, since the decision rule declares 'diseased' if $p(\mathsf{d} \mid x; \widehat{\theta}) > 1/(1 + \ell\omega)$.

These palliatives can also be applied when there are several diseased groups. The formulae can be extended quite easily by working with the odds of each disease to 'normal'.

2.6 How complex a model do we need?

Adequacy of a model is not usually an absolute criterion; rather we ask how complex the model needs to be within families of models. For example, we can ask how many input features to use for a linear classifier. In this section we concentrate on the adequacy of a parametric model for the class densities or posterior probabilities. In the next section we look more directly at the effect of model inadequacy on performance, and in the final section of this chapter we consider absolute bounds on performance, averaged over training sets.

We have seen two distinct modelling problems. In Section 2.2 we modelled the class densities $p_k(x; \theta)$; in Section 2.3 we modelled the posterior probabilities $p(k \mid x; \theta)$. It is important to realize that a model may be adequate for the posterior probabilities without being adequate for the class densities (see Figure 2.1 on page 27).

We will consider the traditional statistical approaches to model complexity which are applied to both problems. These fall into two camps.

1 *Iterative selection of a model.* For example, in choosing the number of features to use in a linear discriminant or a logistic discriminant there are many variants of stepwise procedures, modelled on those used for regression problems. *Backward* selection starts with all possible features, and drops them one at a time. *Forward* selection starts with no features, and adds one at a time. *Stepwise* procedures start somewhere (usually with all features) and at each step consider either adding or dropping a feature, and choose the best single step before iterating. There are a large number of families of models considered in later chapters with direct analogues of this, for example selecting centres in radial basis function models, the number of hidden units in a feed-forward neural network and the number of components in a mixture distribution.

   The distinctive feature of this approach is that the selection is made by a series of pairwise comparisons: is the larger model sufficiently superior to the smaller one?

2 *Penalizing the fit* by a measure of the complexity of the model. In this approach the search is in principle over all models within the family. Normally we would expect the largest models to fit best, but the penalty on size will tend to ensure that the smallest adequate model is chosen. In practice we may have to confine the search to only some of the models in the family: this could even be done by a stepwise search as in 1. The penalty is often motivated by predicting the degree of fit on a test set.

We have left open the measure of fit to be used. The most common is the log-likelihood evaluated at the ML estimate. It is often more convenient to work with the *deviance*, minus twice the log-likelihood shifted to be zero for the 'perfect' model. In the classification context the perfect model has $p(k \mid x) = 0$ or 1, with 1 for the class which actually occurs.

This may not be possible as a function of the features actually observed.

In iterative selection we can use likelihood-ratio tests, or equivalently differences in deviances. For a regular problem, the reduction in deviance on adding q further parameters has an asymptotic chi-squared distribution on q degrees of freedom *provided the smaller model is adequate* (Lehmann, 1986, §8.8). Iterative selection normally works by choosing some conventional significance level (often 10%) to decide a comparison.

The penalization methods themselves have three schools. The most common one is based on the idea that the deviance will be smaller on the training set than on a test set of comparable size, since we actually chose the parameters to minimize the deviance on the training set. How large would the difference be on average (over training and test sets)? Akaike's (1973, 1974) AIC criterion is based on the answer $2p$, where p is the number of free parameters. This does assume that the model is correct. The criterion NIC of Murata *et al.* (1991, 1993, 1994) is based on the answer $2p^*$ where

Akaike (1985) reviews the rationale in a more leisurely way than the original papers.

$$p^* = \text{trace}[KJ^{-1}] \tag{2.41}$$

and J and K are defined in Proposition 2.2. As pointed out there, $J = K$ if the model is adequate, so $p^* = \text{trace}(I) = p$. Both criteria follow from (2.20), and are based on asymptotic normality of the parameter estimates. (Moody, 1991, 1992, uses his *effective number of parameters* in the same way.) Note that whereas the deviance plus $2p^*$ may be a poor estimate of the mean test-set deviance because of the variability over training sets, *differences* of this measure between models may be acceptably good estimates of the differences in test-set deviances. Note the *may* in the previous sentence; the fluctuations are normally small enough to distinguish models whose mean test-set deviances differ appreciably, but they can dominate if the mean test-set deviances are nearly the same (see page 34).

Asymptotically in the size of the training set, the use of AIC may choose a model which is at least as large as the correct one with probability one (see Shibata, 1976; Hannan & Quinn, 1979; M. Stone, 1979; this is established for the order of an autoregressive time series, for instance). Thus if there is no correct model in the family, AIC will tend to choose larger and larger models as more training data becomes available. The fact that it tends to overshoot the correct size has led to modifications (BIC) which penalize complex models more severely (Akaike, 1977, 1978; Schwarz, 1978) and to justifications based on allowing the complexity of the true model to depend on n (Shibata, 1980, 1981). An alternative way to estimate the expected deviance on a test set of the same size as the training set is to use cross-validation and the other methods of Section 2.7 on the deviance.

A general programme for measuring and controlling the complexity of fitted models based on minimum description length (MDL) or minimum message length (MML) is given by: Rissanen (1983, 1987,

1989); Wallace & Freeman (1987); Barron (1990, 1994); Barron &
Cover (1991). In this the deviance of a model is penalized via the minimum length of a binary code needed to represent it. Now in most cases
we can only (over-)estimate this length by providing a specific encoding,
and the extension to continuous parameters is via discretization. There
are many variants within this programme.

Cheeseman (1995)
claims 'for most
interesting problems
achieving this goal is
NP-hard', referring to
finding the model with
the minimum message
length.

    Vapnik's (1982) *structural risk minimization* is a similar idea, using
a bound on test-set risk based on the work of Section 2.8 and discussed
further there.

    The third idea is to estimate more directly the performance on a
test set, by cross-validation and allied methods which we discuss in the
next section.

The predictive approach

Bayesian methods provide an interesting view of these measures, as discussed by Smith & Spiegelhalter (1980) for linear models (but asymptotically most methods are locally linear). In the Bayesian formulation,
models are compared via $\Pr\{M \mid \mathcal{T}\}$, the posterior probability assigned
to model M, which requires a prior distribution (p_M) over models
and the ability to integrate out the parameters following the predictive
approach:

$$\Pr\{M \mid \mathcal{T}\} \propto p(\mathcal{T} \mid M)p_M, \qquad p(\mathcal{T} \mid M) = \int p(\mathcal{T} \mid M, \theta)p(\theta)\, d\theta$$

so the ratio in comparing models M_1 and M_2 is proportional to
$p(\mathcal{T} \mid M_2)/p(\mathcal{T} \mid M_1)$, known as the *Bayes factor*. Note that if the
models are nested, the priors will correspond to a prior over the
parameters of the larger model which gives positive probabilities to
zero values of some of the parameters. Then model choice will involve
the controversial testing of 'precise hypotheses', where classical and
Bayesian methods are often in conflict (Berger & Delampady, 1987).

    Note that the predictive approach does not actually select a model,
but averages the predictions of the models, with weights proportional
to the Bayes factors. This seems not to be widely used, possibly for
computational reasons, but can be very effective. It is used for simple
logistic models by Stewart (1987), and in time series prediction by West
& Harrison (1989). Geisser (1993, §4.1) argues that the only possible
loss function which would suggest choosing just one model is one which
embodies an extreme principle of parsimony, that only one model is
acceptable. Even those who argue for restricting the class of models
(such as Madigan & Raftery, 1994) show that averaging is much better

We might want to
restrict attention to one
or a few classifiers for
computational speed.

than any single model. The posterior probability may be spread over many models: Moulton (1991) reports an example in which the top 800 models out of $2^{12} = 4096$ are needed to account for 90% of the posterior probability. In pattern recognition our sole concern is future decision making, but in other applications the use of a single model may be more acceptable or desirable (see, for example, the arguments of Geisser, 1987 and A. F. M. Smith, 1991).

These ideas go back at least to Box & Tiao (1962). Bernardo & Smith (1994) give an overview of the current philosophical discussion, also to be seen in Draper (1995) and its discussion and references.

We will see other ways to choose combinations of models in the next subsection, and examples of using estimated Bayes factors to combine posterior probabilities from different models in Section 5.5. Another idea is to use Markov chain Monte Carlo ideas such as the Gibbs sampler to integrate over both the model space and the parameter space, as considered by George & McCulloch (1993) and Madigan & York (1995).

Suppose we just use the Bayes factor as a guide. The difficulty is in evaluating $p(\mathcal{T} \mid M)$. Asymptotics are not useful for Bayesian methods, as the prior on θ is often very important in providing smoothing, yet asymptotically negligible. We will assume that $p(\theta \mid \mathcal{T})$ is approximately normal with mean $\hat{\theta}$ and covariance matrix V. One approximation is to take $\hat{\theta}$ as the mode of the posterior density and V as the inverse of the Hessian of $-\log p(\hat{\theta} \mid \mathcal{T})$ (since for a normal density this is the covariance matrix); we can hope to find $\hat{\theta}$ and V from the maximization of $\log p(\theta \mid \mathcal{T}) = L(\theta; \mathcal{T}) + \log p(\theta) + \text{const}$. Let $E(\theta) = -L(\theta; \mathcal{T}) - \log p(\theta)$, so this has its minimum at $\hat{\theta}$ and Hessian there of V^{-1}. Then

L is the log-likelihood.

$$p(\mathcal{T} \mid M) = \int p(\mathcal{T} \mid \theta) p(\theta) \, d\theta = \int \exp -E(\theta) \, d\theta$$
$$\approx \exp -E(\hat{\theta}) \int \exp[-\tfrac{1}{2}(\theta - \hat{\theta})^T V^{-1}(\theta - \hat{\theta})] \, d\theta$$
$$= \exp -E(\hat{\theta}) \, (2\pi)^{p/2} |V|^{1/2} \qquad (2.42)$$

Various results are known on the large-sample accuracy of (2.42), but our uses will be as approximations far from asymptotia. Kass & Raftery (1995) suggest that at least $5p$ and preferably $20p$ training samples are required.

from (2.7). (This is sometimes known as a saddle-point approximation or Laplace's method: Lindley, 1980; Tierney & Kadane, 1986. This and other approximate methods are discussed by Evans & Swartz, 1995.) Thus

$$\log p(\mathcal{T} \mid M) \approx L(\hat{\theta}; \mathcal{T}) + \log p(\hat{\theta}) + \tfrac{p}{2} \log 2\pi + \tfrac{1}{2} \log |V|. \qquad (2.43)$$

It may be feasible to use this directly for model choice, as was proposed for nested models by Kass & Vaidyanathan (1992).

If we suppose θ has a prior which we may approximate by $N\{\theta_0, V_0\}$, we have

$$\log p(\mathcal{T} \mid M) \approx L(\widehat{\theta}; \mathcal{T}) - \tfrac{1}{2}(\widehat{\theta} - \theta_0)^T V_0^{-1}(\widehat{\theta} - \theta_0) - \tfrac{1}{2}\log|V_0| + \tfrac{1}{2}\log|V|$$

and V^{-1} is the sum of V_0^{-1} and the Hessian H of the log-likelihood at $\widehat{\theta}$. Thus

$$\log p(\mathcal{T} \mid M) \approx L(\widehat{\theta}; \mathcal{T}) - \tfrac{1}{2}(\widehat{\theta} - \theta_0)^T V_0^{-1}(\widehat{\theta} - \theta_0) - \tfrac{1}{2}\log|H|.$$

If we assume that the prior is very diffuse we can neglect the second term, so the penalty on the log-likelihood is $-\tfrac{1}{2}\log|H|$. For a random sample of size n from the assumed model this might be roughly proportional to $-(\tfrac{1}{2}\log n)\,p$ provided the parameters are identifiable. This is the proposal of Schwarz (1978) derived following the ideas of Smith & Spiegelhalter (1980, §2.1). Others argue for retaining different terms in (2.43); for example Draper (1995, p. 57) retains the second term of (2.43), drops the third and replaces $\log|V|$ by $-\log|H|$. The latter might be damaging (Raftery, 1993) and is often of no computational benefit.

> This is often called BIC. Jeffreys (1961, pp. 248, 272, 277, 343) discussed special cases many years before in earlier editions.

The assumption that the prior can be neglected is a strong one, since we may not obtain much information about parameters which are rarely effective, even in very large samples. For example, suppose we have separate parameters for each class-conditional density $p_k(x; \theta)$. Then we will learn very little about the parameters of very rare classes, and the effective sample size in the expression for BIC for the parameters for class k will be n_k not n. Since we would expect $n_k \propto n$, the leading term in n is still $-(\tfrac{1}{2}\log n)p$, but as $(\tfrac{1}{2}\log n)$ will be quite small for practical n (5.75 for $n = 100\,000$), replacing $\tfrac{1}{2}\log|H|$ by $(\tfrac{1}{2}\log n)p$ can be quite misleading. We should be interested in comparing different models for the same n, and in many problems p will be comparable with n. It seems best to use (2.43) directly.

Kass & Raftery (1995) review many of the approaches to approximating Bayes factors; Gelfand & Dey (1994) provide one of the clearest accounts of the multitude of variations which have been proposed for model choice.

Improper priors over θ (those that are not integrable) lead to difficulties, since $p(\mathcal{T} \mid M)$ will be unknown up to a constant factor, and might be infinite. It may be possible to resolve this by taking limits of results with proper priors (but often improper priors were chosen to make the integrations feasible). Other approaches are discussed by Kass & Raftery (1995, §5.3) including the device of an 'imaginary training sample' used by Spiegelhalter & Smith (1982). It is not clear however that any of these methods is totally satisfactory, especially when the models under consideration have very different numbers of parameters with improper priors. The difficulty shows up in (2.43). With an improper prior we may treat $\log p(\widehat{\theta})$ as constant, but there is no reason to suppose that it is the same constant for different models, although careless workers in the neural networks field often do so.

We can relate these calculations to the derivations of AIC and NIC. The present derivation is less asymptotic and does take account of a prior. It produces a factor of $\log n$ (for some appropriate n) rather than 2 on the penalty for the deviance, which curbs the tendency of AIC to overshoot the true model size. (Remember the Bayes factor is intended for use in averaging not selecting models.) Note that the approach of this subsection necessarily assumes that the model is true while calculating $p(\mathcal{T} \mid M)$. We defer further consideration of NIC to Section 4.3, where we consider other methods of parameter estimation.

Combining models

We have already mentioned that the full predictive approach is to average models by

$$\widetilde{p}(k \mid x) = \sum_m \widetilde{p}_m(k \mid x) \Pr\{m \mid \mathcal{T}\}$$

rather than choosing one model (unless our loss function includes costs on multiplicity of models). We now consider other ideas for combining models, using ideas which are developed in other contexts in Section 2.7.

M. Stone (1974, pp. 126–7) mentioned the idea of using cross-validation not to choose between models but to combine them. In our context this would amount to combining the posterior probabilities (either plug-in or predictive) from a series of M models. The predictive viewpoint motivates Stone's suggestion of

$$\widehat{p}(k \mid x) = \sum_m \alpha_m \widehat{p}_m(k \mid x) \qquad (2.44)$$

for a set of constants (α_m), perhaps confined to a probability distribution and chosen by cross-validation. We could also allow the weights to depend on the class k.

This idea has been developed (independently) and extended by Wolpert (1992), under the name of *stacked generalization*, and applied by Breiman (1992) in the regression context. Breiman's work is precisely within Stone's setting, and shows that in his simple examples it does indeed help to confine attention to non-negative weights (although he seems not to have considered a unit-sum constraint on the weights). The idea of averaging (both simple and weighted) for regression neural networks has been suggested many times. (A selection of references is Baxt, 1992; Benediktsson & Swain, 1992; Bridle & Cox, 1991; Hansen & Salamon, 1990; Lincoln & Skrzypek, 1990; Pearlmutter & Rosenfeld, 1991; Perrone & Cooper, 1993; Srihari, 1992; and Xu *et al.*, 1992.)

LeBlanc & Tibshirani (1993) took up the same thread, but also considered estimation by the bootstrap. Both cross-validation and the bootstrap can be seen as methods to correct the bias of the deviance (or other measure) on the training set as a function of (α_m), and are used exactly as for the apparent error in Section 2.7. The bias-corrected estimate of the deviance is then minimized over (α_m). Their experiments considered non-negativity and unit-sum constraints, but not both together!

Wolpert's ideas were much more general than those picked up by later users. Although (2.44) is suggestive, we might only to want to use it locally in the feature space, and so could allow the weights to vary (slowly) with x. In his general scheme, we use the outputs of all the 'level 0' models (under leave-one-out cross-validation) and the true response as inputs to a 'level 1' procedure which then makes the final decision. At its simplest we could take the predictions from M classifiers and learn how best to combine them to give a single classification. But the outputs of the models can be their posterior probabilities, and we can also pass the inputs through a 'do-nothing' level 0 procedure. So stacked generalization includes any method of combining the outputs of the models, possibly varying with x.

A similar approach is taken by Jacobs *et al.* (1991) in which they train all the classifiers simultaneously and the level-0 classifiers are not required to work well over the whole input space. This approach is discussed in Section 8.5.

2.7 Performance assessment

The title of this section begs the question of what is meant by performance. Since we have identified the Bayes rule on the basis of total risk (expected loss) this seems a suitable basis for performance evaluation. When loss (2.2) is used, it is often helpful to plot the expected rate of misclassification against the expected rate of 'doubt' classification, usually termed the reject rate.

Figure 3.5 on page 114 is an example.

To be explicit, the error rate is the probability of making a definite erroneous classification (including outlier $\mathcal{O}$) for a future randomly chosen sample, previously called pmc, and the reject rate is the probability of declaring doubt, previously denoted pd. Our discussion will concentrate on pmc, but entirely analogous statements follow about pd. Also, statements made about error rates can be replaced verbatim by ones about average cost for other losses. A later subsection on 'confusion matrices' discusses more detailed information on error patterns.

The apparent error rate $\widehat{\text{pmc}}$ is the proportion of errors made when classifying either a training or a test set. If the training set is used, $\widehat{\text{pmc}}$ will (usually) be biased downwards.

Error rate estimation

The easiest way to assess the error rate is to choose a test set independent of the training set (and validation set if used), to classify its examples, count the errors and divide by the size M of the test set. The reject rate is estimated by the proportion of test-set examples which are rejected. These measures are clearly unbiased estimates under all circumstances, but they can be highly variable, and having to use a test set may waste data which could otherwise have been used for training. The idea of a test set is sometimes called the *hold-out* method and goes back at least to Highleyman (1962a).

Simple calculations show that the test set needs to be large for the error rate to be estimated at all accurately. The estimate $\widehat{\text{pmc}} = R/M$ for an error count R has a binomial(M, pmc) distribution. Thus $\widehat{\text{pmc}}$ has variance $\text{pmc}(1 - \text{pmc})/M \leqslant 1/4M$. Suppose that pmc is around 5% and we wish to know it to around 1%. Then we will want

We use use ± 2 standard errors as an approximate 95% confidence interval.

$$2\sqrt{0.05 \times 0.95/M} \approx 0.01$$

or $M \approx 1900$, which is considerable. (Here we use the normal approximation to the binomial, which is justified at such sample sizes.)

Note that the task of comparing the error rates of two classifiers is rather easier as they use the same test set, a point often overlooked in the literature and taken up in a later subsection.

We can also calculate the error rates conditionally for each class, just by counting within each class. If we know the prior probabilities π_k, the estimator

$$\sum_k \pi_k \widehat{\text{pmc}}(k) \tag{2.45}$$

estimates pmc. This form is important when the test set is a deliberately biased sample, which can be a good idea when almost all the errors occur in uncommon classes. In general we would expect it to be less variable, but there is a problem that it will be undefined if $n_k = 0$ for any class, which complicates the theoretical analysis.

Risk averaging

Suppose we knew the posterior probabilities $p(k \mid x)$ and are using the Bayes rule. Consider a random pair (X, C) from the whole population. Then

$$P(\text{correct} \mid X = x) = P\left(C = \arg\max_c p(c \mid x) \mid X = x\right) = \max_c p(c \mid x)$$

and so

$$1 - \text{pmc} = P(\text{correct}) = E\left[\max_c p(c \mid X)\right]. \qquad (2.46)$$

Both $I\left[C = \arg\max_c p(c \mid x)\right]$ and $\max_c p(c \mid x)$ are conditionally unbiased estimators of $P(\text{correct} \mid X = x)$, but the second averages over $P(C \mid X = x)$ and so has a smaller variance. Let $Z = 1 - \max_c p(c \mid X) \leqslant 1 - 1/K$. Then $EZ = \text{pmc}$ and $EZ^2 \leqslant (1 - 1/K)EZ$, so

$$\text{Var}[\max_c p(c \mid X)] = \text{Var}(Z) \leqslant (1 - 1/K)\text{pmc} - \text{pmc}^2$$

$$= \text{Var}\left(I\left[C = \arg\max_c p(c \mid x)\right]\right) - \text{pmc}/K.$$

Note that (2.46) does not depend on knowledge of the true class C, which can be useful if the authenticity of the classifications of the test set is in doubt. This can also be an advantage if examples are cheap but accurate classification is expensive, as in almost any form of automated data collection which needs human classification. This approach is sometimes known as *risk averaging*.

Of course, we only very rarely know the posterior probabilities, but (2.46) can be used if we believe we have accurate estimates of them. It will be much better to use predictive estimates $\tilde{p}(k \mid x)$ rather than plug-in estimates $p(k \mid x; \widehat{\theta})$ as the latter ignore the variability of $\widehat{\theta}$ and so tend to underestimate small probabilities, often quite severely. The estimate from (2.46) can be based on either the training or the test set; if the training set is used there will be some bias, the size of which will depend on the number of parameters. When the probabilities are estimated $\max_c p(c \mid x)$ will be biased, both because we cannot usually find unbiased estimators of $p(c \mid x)$ and because the maximum is a non-linear operation. We would expect the bias to be small for a test set; calculations for one particular classifier are given in Section 6.2.

Compare Figures 2.4 and 2.11.

If this method is to be used, it would seem desirable to check if the estimates $\tilde{p}(k \mid x)$ are reliable, which is itself a check of the adequacy of the model. The methods are part of the more general methodology of verifying if probability forecasters (such as 'the probability of rain tomorrow is 60%') are well calibrated (see, for example, Dawid, 1982, 1986). The basic idea is that if $\widetilde{p}(k\,|\,x) = \eta$ say, amongst the examples for which we predict η, the proportion which occur should be about η. We can apply this either to the posterior probabilities of each class, or to the probability $\max_k \widetilde{p}(k\,|\,x)$ of a correct classification. The test set then provides a set of independent events E_i and predicted probabilities θ_i. We can test the calibration by using a non-linear logistic regression on θ_i (say by the methods of Chapter 4) and test if the identity is adequate. If it is not, we can even use this regression to re-calibrate the probabilities (an idea for a linear logistic regression going back to Cox, 1958). Various methods for fitting such regressions will be described in Chapters 4 and 5, and an example is shown in Figure 3.6 on page 115.

There is also a literature on methods of numerically assessing probability forecasts. The most common measures are (half-)*Brier scores* (Brier, 1950) which is the sum of squared differences between the predicted probabilities and the indicator function that the event occurred (or, equivalently, the sum of $(1 - p)^2$ where p is the forecast probability of the positive or negative event which occurred), and logarithmic scoring (Good, 1983) which sums the negative log probability of the event which occurred. Note that logarithmic scoring computes the conditional log-likelihood as used in logistic regression.

The effect of using risk averaging with inaccurate probability estimates can be severe; on page 228 there is an example in which the error rate is underestimated by a factor of more than two.

Cross-validation

Often the use of a test set is regarded as too wasteful of scarce classified data. Can we avoid this by dividing the training set? If we divide the training set into two halves, we could train with one half and test with the other. As the halves are independent samples, the resulting estimator is unbiased. Furthermore, we can swap the halves and still obtain an unbiased estimator, and so the average of the two estimators remains unbiased.

The drawback of this approach is that the estimate is an unbiased estimate of the performance using just half the data. Can we do better? Yes, at the expense of more computation. Suppose we randomly divide the training set into V pieces. Then we can use one piece to test the performance of the classifier trained on the remaining $(V - 1)$ pieces.

This is again unbiased, and we can average the V such estimates. For moderate V such as 5 or 10, the loss of performance from a smaller training set will usually be small enough, at the expense of V times the computation (although this can be done in parallel if several CPUs are available). (A technical nicety is how to do the average; should we weight by the size of the pieces if they are unequal? Weighting is common practice, as it corresponds to counting the number of errors made by the cross-validated classifiers.)

The extreme version of this strategy is to take V very large. Then each test set will contain zero or one examples. This suggests (but does not quite justify) the *leave-one-out* estimator, in which each observation is tested on the classifier trained on the remaining $(N-1)$ observations, suggested by Mosteller & Wallace (1963), Hills (1966), Lunts & Brailovsky (1967) and Lachenbruch & Mickey (1968) and often rediscovered. The leave-one-out version of cross-validation apparently requires a large amount of computation, but for some classifiers this computation can be reduced to a similar level to classifying a test set of n examples (see pages 100, 184 and 200).

> Leave-one-out cross-validation is also known as *ordinary* cross-validation.

The leave-one-out estimator is a *balanced* version of cross-validation in that the sets are chosen of exactly equal size. This can be applied to the V-fold version as well, and indeed we may choose to balance the subsets on other characteristics such as regions of the feature space or even the numbers taken from each class. Are these variants valid? A test-set error rate is an unbiased estimator of pmc provided that C is sampled from its conditional distribution given X, and that X is sampled with density $p(x)$; no independence is needed to justify the unbiasedness of an average. This justifies all the approaches except choosing fixed numbers from each class, but including the leave-one-out estimator. If we choose fixed numbers from each class, we obtain unbiased estimates of the class-conditional error rates pmc(k), and hence an unbiased estimate if the prior probabilities are known or estimated in the usual unbiased way (from other data).

> This is sometimes called *stratified* cross-validation.

Choosing $V = N$ should give the most accurate assessment, as the true size of the training set is most closely mimicked; it also normally involves the most computation. There is another argument in favour of smaller V. Dropping just one observation assesses the classifier via $O(1/N)$ perturbations from the training set. On the other hand the sampling variations in the parameter estimates are (usually) $O(1/\sqrt{N})$, so for large N we end up extrapolating these from much smaller perturbations. Thus cross-validation estimates of performance for large V might be expected to be (and are often reported to be)

> A good example is a classification tree, where dropping any single example might not alter the chosen topology, only the fitted probabilities at single leaf.

rather variable; taking a smaller V can give a larger bias but smaller variance and mean-square error.

We have concentrated on cross-validation for the error rates, but it is also possible to use cross-validation for the 'smoothed' measures such as $1 - \max_c p(c \mid X)$ discussed in the previous subsection.

Estimation and model choice

Cross-validation is also commonly used for model choice. (Cover, 1969, is the first advocate of which we are aware.) This can be used to estimate either a measure of performance (such as error rate) or a measure of model fit (such as deviance). M. Stone (1974) and Geisser (1975) pointed out that cross-validation could also be used for parameter estimation; just choose the parameter value which minimizes the cross-validated measure of performance or fit. Then if we want to assess the performance of the resulting classifier by cross-validation, we have to do so by a double layer of cross-validation.

M. Stone (1977a, b) considered various asymptotics for model selection by cross-validation. Consider first cross-validating the deviance (under ML estimation). This is a sum of terms D_i over examples in the training set. We follow the usual notation in which (i) refers to a quantity based on the training set with the ith example deleted. From Taylor expansions we have

$$\sum_i D_i(\widehat{\theta}_{(i)}) = D(\widehat{\theta}) + \sum_i [\widehat{\theta}_{(i)} - \widehat{\theta}]^T D_i'(\widetilde{\theta}_i)$$

$$\widehat{\theta}_{(i)} - \widehat{\theta} = D''(\bar{\theta}_i)^{-1} D'(\widehat{\theta}_{(i)})$$

for $\widetilde{\theta}_i, \bar{\theta}_i$ convex combinations of $\widehat{\theta}_{(i)}$ and $\widehat{\theta}$ (and since $\sum_{j \neq i} D_j'(\widehat{\theta}_{(i)}) = 0$ by definition). Under consistency all the estimators converge to θ_0, so

$$\sum D_i(\widehat{\theta}_{(i)}) \sim D(\widehat{\theta}) + \sum D_i'(\theta_0)^T D''(\theta_0)^{-1} D_i'(\theta_0)$$

$$= D(\widehat{\theta}) + \text{trace}\left[D''(\theta_0)^{-1} \sum D_i'(\theta_0) D_i'(\theta_0)^T \right]$$

Stone's argument was also given by Liu (1993, 1995).

and the limit of the second factor on the right-hand side is $2p^*$ by the arguments in Section 2.2. Thus leave-one-out cross-validation of the deviance is asymptotically equivalent to using NIC to correct the deviance. This suggests that model choice by NIC and by leave-one-out cross-validation are asymptotically equivalent. (For finite classes of models this argument will prove so unless two or more models have the same $D(\theta_0)$; all true models have the same value, zero.)

M. Stone (1977a) gave heuristic arguments and examples for asymptotic consistency of cross-validatory assessment (which follows from unbiasedness and a law of large numbers) and asymptotic efficiency of cross-validatory estimation.

Improving on cross-validation

Cross-validation was used to estimate the performance (error-rate, loss, deviance) on a test set by constructing a pseudo test-set from the training set. We now take a different viewpoint, of accepting that the performance measure on the training set is biased, but trying to estimate that bias, and correct it using our estimate. For concreteness we will work with error rates (although the principles apply much more widely). We need to distinguish between the pmc, the true error rate for our classifier trained on this training set, $\mathsf{E}\,\mathrm{pmc}$, its average over training sets, and pmc_0, the true error rate with the 'least false' parameter θ_0 plugged in. We can then aim to correct the bias of $\widehat{\mathrm{pmc}}$ as an estimator of either pmc or pmc_0. The first is most relevant for performance assessment of this classifier, the second if we use pmc_0 as an upper bound for the Bayes risk (which for large parametric families may be close to the Bayes risk) or as a lower bound on the achievable performance within this parametric family. In either case $\widehat{\mathrm{pmc}}$ will be biased. The two biases are $\mathsf{E}[\widehat{\mathrm{pmc}} - \mathrm{pmc}]$ and $\mathsf{E}\,\widehat{\mathrm{pmc}} - \mathrm{pmc}_0$, and in each case we will correct $\widehat{\mathrm{pmc}}$ by subtracting an estimate of the appropriate bias.

How do we estimate the biases? The method of Quenouille (1949), later termed the *jackknife* by Tukey, is sufficiently similar to leave-one-out cross-validation to have caused considerable confusion in the literature. Suppose $\widehat{\theta}_n$ is an estimate of θ based on n observations, and that its mean has the expansion

$$\mathsf{E}\,\widehat{\theta}_n = \theta + \frac{a_1}{n} + \frac{a_2}{n^2} + \cdots.$$

Then each leave-one-out estimator $\widehat{\theta}_{(i)}$ has mean

$$\mathsf{E}\,\widehat{\theta}_{(i)} = \theta + \frac{a_1}{n-1} + \frac{a_2}{(n-1)^2} + \cdots$$

as does their average $\widetilde{\theta}$. Now consider $n\widehat{\theta} - (n-1)\widetilde{\theta}$. This has mean

$$n\mathsf{E}\,\widehat{\theta} - (n-1)\mathsf{E}\,\widetilde{\theta} = \theta - \frac{a_2}{n(n-1)} + O(n^{-3})$$

and so much smaller bias for large n. Thus the jackknife estimator of the bias is $(n-1)[\widetilde{\theta} - \widehat{\theta}]$.

The most obvious application of the jackknife is to reduce the bias of $\widehat{\mathrm{pmc}}$ as an estimate of pmc_0. The expansion needed is valid under very mild regularity assumptions for ML plug-in classifiers, and the biased-reduced estimator of pmc_0 is

$$ n\,\widehat{\text{pmc}} - \frac{n-1}{n} \sum_i \widehat{\text{pmc}}_{(i)} $$

using $\text{E}\,\widehat{\text{pmc}} = \text{pmc}_0 + a_1/n + O(n^{-2})$.

It is less obvious how jackknifing can be used to estimate the bias $\text{E}[\text{pmc} - \widehat{\text{pmc}}]$. Efron (1982, Chapter 7) sketches how to do so. The idea is to compare the error in predicting the omitted sample with that in predicting the $(n-1)$ remaining samples. Let e_i be the indicator of the error of predicting the class of x_i from $\mathcal{T}_{(i)}$, and $\widehat{\text{pmc}}_{(i)}$ the apparent error rate on fitting to $\mathcal{T}_{(i)}$. Then the estimator of the bias is

$$ \frac{(n-1)}{n}\,\text{mean}_i[\widehat{\text{pmc}}_{(i)} - e_i] $$

where the scale factor is a sample-size adjustment. Under mild regularity conditions we would expect $\text{E}\,\text{pmc} = \text{pmc}_0 + b/n + O(n^{-2})$, and so $\text{E}\,e_i = \text{pmc}_0 + b/(n-1) + O(n^{-2})$ and $\text{E}\,\widehat{\text{pmc}} - \text{pmc} = (a_1 - b)/n + O(n^{-2})$. Then our bias estimator has mean

$$ \frac{n-1}{n}\left[\text{pmc}_0 + \frac{a_1}{n-1} + O(n^{-2}) - \text{pmc}_0 - \frac{b}{n-1} - O(n^{-2})\right] $$

$$ = \frac{a_1 - b}{n} + O(n^{-2}) $$

as required. The complete jackknifed estimate of pmc is

$$ \widehat{\text{pmc}} + (1 - 1/n)\,\text{mean}_i\,[e_i - \widehat{\text{pmc}}_{(i)}] $$

which has bias $O(n^{-2})$.

In our current notation the leave-one-out cross-validated estimate of pmc is $\sum e_i/n$, and so it implies $\widehat{\text{pmc}} - \text{mean}_i\,e_i$ as its estimate of the bias. Efron (1982, Chapter 7) gives a suggestive argument why the relative difference between the two bias estimates might be $O_p(1/n)$, and hence there would be little practical difference. Our arguments show that if we drop the sample-size correction, thereby making a relative error of $O(1/n)$, the difference between the two bias corrections is $\widehat{\text{pmc}} - \text{mean}_i\,\widehat{\text{pmc}}_{(i)}$ which has a mean of $O(n^{-2})$, and is often $O_p(n^{-2})$. Note that the computational effort of these two estimators of pmc is almost the same.

The *bootstrap* is loosely related to the jackknife, and conceptually simpler. Suppose we make a new sample of size n by resampling *with replacement* from our sample, and calculate an estimate θ^* from the bootstrap sample (as it is called). Then the variability of $\theta^* - \widehat{\theta}$ should mimic that of $\widehat{\theta} - \theta$, in particular the mean of the first should estimate the bias of $\widehat{\theta}$. This can be used by actually resampling B times and averaging, or sometimes by finding the mean analytically.

In our problem we can bootstrap $\widehat{pmc}$ to estimate pmc_0, or bootstrap $\widehat{pmc} - pmc$ to estimate the bias correction. This bias is the difference between the classifier's apparent error and true error, averaged over training sets. To bootstrap this we replace $\mathcal{T}$ by $\mathcal{T}^*$ and the mean over x by the average over the points in $\mathcal{T}$. Thus the bootstrap estimate of the bias is the average (over bootstrap samples) of the error rate on the training set $\mathcal{T}^*$ minus that on the larger set $\mathcal{T}$. Of course, we will evaluate the error rate on the distinct members of $\mathcal{T}^*$ using weights for multiple members.

Immediately we see a snag, as the example from $\mathcal{T}$ being predicted may be in the bootstrap training set $\mathcal{T}^*$, and if it is we may expect to predict it well, for some classifiers far too well. Efron (1983) proposed the '.632' bootstrap, which considers only the predictions of those members of $\mathcal{T}$ not in $\mathcal{T}^*$; specifically for each point x_i estimate the error by averaging over those bootstrap samples not including x_i, then average over points to get ϵ_0. The final estimate is then

> Note that $\mathcal{T}^*$ contains some of the members of $\mathcal{T}$ more than once, and (usually) some not at all, so as a *set* $\mathcal{T}^*$ is smaller.

$$0.368\,\widehat{pmc} + 0.632\,\epsilon_0.$$

Here 0.632 is shorthand for $(1 - 1/e)$, the limit for large n of the probability that a given observation from $\mathcal{T}$ appears in $\mathcal{T}^*$.

Bootstrap methods may also be used to estimate the precision of the apparent error rate $\widehat{pmc}$, using the variability of $\widehat{pmc}^*$ about $\widehat{pmc}$ to estimate the variability of $\widehat{pmc}$ about pmc, for example to estimate the variance of $\widehat{pmc}$ by the variance of $\widehat{pmc}^*$. But we have to be careful, as we should really be interested in the mean square error of the bias-corrected estimator, not of $\widehat{pmc}$, and the bias correction is itself an estimate. Efron & Gong (1983) suggest the mean square error of the bootstrap samples used to estimate the bias gives a lower bound on the mean square error of the bias-corrected estimator.

Introductions to the bootstrap are given by Efron (1982), Efron & Gong (1983) and Efron & Tibshirani (1993); Efron (1983, 1986) contain comparisons of error rate estimation methods including those based on the bootstrap. Other comparisons within the pattern recognition literature are given by Chernick *et al.* (1985) (for linear classifiers), Crawford (1989) (for classification trees) and Jain *et al.* (1987) and Weiss (1991) (for k-nearest neighbour classifiers). These show some support for the '.632' estimator, but by no means universal improvement over leave-one-out cross-validation. The title of this subsection has been chosen in optimism, since the full power of the bootstrap (for example, used in conjunction with ideas such as (2.46)) seems not to have been fully tested.

Confusion matrices

Thus far we have concentrated on a single measure of performance, the overall error rate. This is natural within our decision-theory framework, but we may want more detail to help understand where a classifier is failing. The next level of detail is the class-conditional error rates previously termed pmc(k), that is the error rate amongst examples of class k. Further, we may want to know which classes are being confused, and so we may wish to know

$$e_{ij} = \Pr\{\text{decision } j \,|\, \text{class } i\}$$

which is sometimes called the *confusion matrix*. Note that the decisions can include 'doubt', $\mathcal{D}$.

The most obvious way to estimate e_{kj} is from the pattern of errors on a test set, and sometimes the term 'confusion matrix' refers to the matrix of counts of the events 'true class i decided as j'.

Almost all the methods we have discussed apply equally to the class-conditional error rates. We just consider only those examples with true class k, and in some cases (such as the jackknife) need to take the sample size as the number of class-k examples. Although Efron (1986) derived the '.632' estimator for the overall error rate for two groups, it extends readily to class-conditional rates (Hjort, 1986). Once the conditioning on class k is accomplished, estimating the confusion matrix merely amounts to accounting for which errors were made.

The one method that has a less obvious extension is the use of the posterior probabilities at (2.46), since this looks at the predicted rather than true class. Extensions were considered by Schwemer & Dunn (1980), Basford & McLachlan (1985) and McLachlan (1992). As at (2.46) we at first assume that the posterior probabilities are known. Then

$$e_{ij} = \Pr\{\text{decision } j \,|\, C = i\} = \Pr\{C = i, \text{decision } j\}/\pi_i$$
$$= \mathsf{E}\{p(i\,|\,X)\,I[c(X) = j]\}/\pi_i = \sum_k \frac{\pi_k}{\pi_i}\mathsf{E}_k\{p(i\,|\,X)\,I[c(X) = j]\}.$$

We can form an unbiased estimator of e_{ij} by replacing the expectations in the final expression by averages over a test set. If the (π_i) are unknown and are estimated as usual by (n_i/n), the estimator simplifies to

$$\widehat{e}_{ij} = \frac{1}{n_i} \sum_{l=1}^{n} p(i \mid X_l) I [c(X_l) = j]$$

which can also be seen to be a ratio of unbiased estimators. This suggested to Basford & McLachlan replacing the conditional probabilities and the Bayes classifier $c(x)$ by estimates. Then it may happen that $\sum_j \widehat{e}_{ij} \neq 1$, so their final estimator

$$\widetilde{e}_{ij} = \sum_{l=1}^{n} \widetilde{p}(i \mid X_l) I [\widehat{c}(X_l) = j] \Big/ \sum_{l=1}^{n} \widetilde{p}(i \mid X_l)$$

is formed by re-normalizing the estimator to sum to one. Note that unlike the estimator of the unconditional error rate, this may be biased even if the posterior probabilities are correct.

Comparing error rates

A study to compare the error rates of difference classifiers is an experiment, and should be designed and analysed as such. There is much known from many years of theory and experiments in the statistics literature; an excellent basic reference is Box *et al.* (1978). Experiments in our field are computer experiments and have much in common with work in the field of simulation; Kleijnen & van Groenendaal (1992) provide a non-technical introduction in that context which is amplified in Kleijnen (1987). Important ideas from that field include importance sampling and stratified sampling, both of which can be used to design 'difficult' test sets and to compensate for the increased difficulty in estimating error rates. For example, we might arrange for rare patterns to be well-represented in the test set, but down-weighted (as in the study of Candela & Chellappa, 1993, Blue *et al.*, 1994).

It should always be possible to give some idea of the variability of a quoted performance estimate. For test-set error counts we can use the binomial distribution or the normal or Poisson approximations to it. For other measures such as the smoothed error counts based on $1 - \max_c p(x \mid x)$ we can use the sample variance, as each example in the test set is assumed to be an independent sample from the population on which we are trying to predict the error rate.

However, a crucial observation is that since the *same* test set is used for each method, the comparisons between methods are usually much more accurate than the standard errors suggest. (In the terminology of the design of experiments we have a paired comparison, or a blocked

methods colour given	4-way yes	sex only yes	sex only no
linear discriminant	8	8	8
linear discriminant on log variables	4	4	4
quadratic discriminant	11	9	8
quadratic discriminant on log variables	9	7	7
. . .		. . .	

experiment if there are more than two methods.) Consider the first two lines of Table 2.1. As frequently happens, the 4 errors made in the second line are also made in the first line. The standard error of the difference between 4/120 and 8/120 assuming separate test sets is $\sqrt{0.0164^2 + 0.0227^2} \approx 3.37/120$ so a naive comparison would conclude that there was no significant difference.

This uses the variance of a binomial distribution.

More appropriate methods are available, such as *McNemar's test* (Fleiss, 1981). Let n_A and n_B be the number of errors made by method A and not method B, and *vice versa*, so in our example $n_A = 4$ and $n_B = 0$. Then McNemar's test (with continuity correction) refers

$$\frac{|n_A - n_B| - 1}{\sqrt{n_A + n_B}}$$

to a $N(0, 1)$ distribution, and an exact test refers n_A to a binomial $(n_A + n_B, 1/2)$ distribution. This suggests that we need $n_A \geqslant 5$ for a significant difference (but this is only sufficient if $n_B = 0$). Thus large test sets are needed to distinguish between classifiers of similar performance; to detect a 1% difference in error rate needs at least 500 examples. So although the difference here is suggestive, the sample size is too small for a definitive conclusion.

An exact test has precisely the distribution claimed under the null hypothesis.

One pitfall to be avoided is to give too much emphasis to statistically significant results. In an experiment in which method A with error rate 29.8% is significantly better than method B with error rate 30.1%, it is clear that the difference is unlikely to be of practical importance, especially if we estimate the Bayes risk as 6% by the methods of Chapter 6.

2.8 Computational learning approaches

One recent strand of theory looks at what Valiant (1984) called 'the theory of the learnable' and has since become known as PAC-learning (for probably almost correct). Suppose we have a training set of n

samples, and use these to fit a classifier g from a class $\mathscr{F}$ of possible classifiers. If the class $\mathscr{F}$ is not too large and includes the true classifier, we would expect that for large enough n the fitted classifier g would be 'close' to the true classifier f. Thus the theory addresses the question of how large the training set needs to be.

How should we measure closeness? The obvious way is to compare the decisions made by f and g for a randomly chosen sample from $\mathscr{X} \times \mathscr{C}$, where $\mathscr{C}$ is the set of classes, and ask that they agree with high probability, that is

$$\Pr\{g(X) \neq f(X)\} < \epsilon, \tag{2.47}$$

say, for some pre-specified ϵ. (This implies that the true error rate of the classifier g exceeds the Bayes risk by less than ϵ.) The left-hand-side of this statement is a random variable, as it depends on the training set. In PAC-learning we ask that (2.47) be true for a high proportion of training sets, say with probability exceeding $1 - \delta$, for a sample size no more than polynomial in $1/\epsilon, 1/\delta$ and that g be fitted in time polynomial in n. Here we are more concerned with the sample size than the computational complexity of finding g.

In the 'noise-free' case when there is a $f \in \mathscr{F}$ which correctly classifies any training set, $\Pr\{g(X) \neq f(X)\}$ is the error rate of g, so (2.47) corresponds to low error rate. In the 'noisy' case we actually study the difference between apparent and true error rates.

The bounds used in studying PAC-learning are often called worst-case bounds since they apply to *any* distribution over $\mathscr{X} \times \mathscr{C}$, provided that both the training set and future samples are drawn (independently) from the same distribution. Much of the theory currently available applies only to two-class problems, and the results are most refined in the noise-free case.

A warning. The results of this section are often misinterpreted. They apply over all possible training sets $\mathscr{T}$, and assert that events occur for most (or few) training sets for a given model. As such they are subject to the usual criticism of frequentist methods, that we cannot know if our particular training set is an exception. But the difficulty here is particularly acute, as the model will have been chosen on the basis of the training set, indeed often on the basis of the sort of performances that these bounds guarantee. A typical claim is

'If our network can be trained to classify correctly a fraction $1-(1-\gamma)\epsilon$ of the n training examples, the probability that its error—a measure of its ability to generalize—is less than ϵ is at least $1 - \delta$.'

However, the probability is in fact guaranteed to be less than $1-\delta$ over all $\mathscr{T}$, including those our model will not fit well, not the conditional probability asserted. To use these results (correctly!) in performance assessment, they have to be applied to the whole procedure including model choice for every problem, with any exceptions included in the probability δ. (This is pointed out by Wolpert, 1994b.) We have never seen this done.

The results are interesting theoretically, and perhaps useful in model choice, but are very conservative. Much better bounds should be possible using the knowledge of the world gained from the examples to fit a class of models. We give examples of the 'dimensions' used in the results in later chapters.

Finite sets of classifiers

The simplest approach is to assume that there is a finite number r of distinct classifiers in $\mathscr{F}$ (Blumer *et al.*, 1987). Then the probability that a g chosen consistent with (that is correctly classifying all of) a training set of size n yet having overall error rate at least ϵ, is at most $r(1-\epsilon)^n$. We can invert this bound to show that

For each g with overall error rate at least ϵ, the probability of getting n correct samples is at most $(1-\epsilon)^n$, and one of at most r classifiers is chosen.

$$\frac{\log r + \log \frac{1}{\delta}}{-\log(1-\epsilon)} \leqslant \frac{\log r + \log \frac{1}{\delta}}{\epsilon} \tag{2.48}$$

training samples are enough to ensure that if we classify the training set correctly, the true error rate is less than ϵ with probability at least $1 - \delta$.

Now suppose that we cannot find a classifier in our class correctly classifying all cases, so the apparent error rate $\widehat{pmc}$ is non-zero. There are several possible bounds. The number of errors is a binomial (n, pmc) random variable so the Bienaymé–Chebychev inequality gives

$$\Pr\{|\widehat{pmc}(g) - pmc(g)| > \epsilon\} \leqslant \frac{pmc(1-pmc)}{n\epsilon^2}$$

which, allowing for the r possible classifiers, gives

$$n \geqslant \frac{r}{4\delta\epsilon^2}$$

which is *much* worse than the noise-free bound for small ϵ or large r.

The bound of Hoeffding (1963) gives for each classifier

$$\Pr\{\widehat{pmc} < pmc - \epsilon\} \leqslant \exp{-2n\epsilon^2}$$

and twice this probability for a two-sided bound, so

$$n \geqslant \frac{\log r + \log \frac{1}{\delta}}{2\epsilon^2}$$

suffices to bound the optimism in the apparent error rate by ϵ for a proportion $(1 - \delta)$ of training sets. This is better, but still has rate $O(\epsilon^{-2})$. To overcome this we have to look at *relative* error with Chernoff's (1952) bound:

$$\Pr\{\widehat{\mathsf{pmc}} < (1 - \gamma)\,\mathsf{pmc}\} \leqslant \exp -\tfrac{1}{2}n\gamma^2 \mathsf{pmc};$$

this gives

$$n \geqslant 2\frac{\log r + \log\frac{1}{\delta}}{\gamma^2\epsilon}$$

if we confine attention to classifiers with $\mathsf{pmc} > \epsilon$. Thus just as for the noise-free case we consider the case in which we are doing badly and are not aware of it from the fit on the training set. Taking $\gamma \to 1$ gives double the previous bound on the sample size for a perfectly-fitted training set.

A combination of absolute and relative error is obtained by using the metric

$$d_v(r, s) = \frac{|r - s|}{v + r + s}$$

used by Pollard (1986) and Haussler (1992). For large v this behaves like absolute error, for small v like relative error. Note that for arguments in $[0, 1]$ (such as error rates) we have

$$\frac{|p - q|}{v + 2} \leqslant d_v(p, q) \leqslant |p - q|,$$

and that $d_\epsilon(p, q) > \tfrac{1}{2}$ implies $|p - q| > \epsilon$, and is equivalent for $p = 0$. We have the bound (Haussler, 1992, Theorem 1)

$$\Pr\{d_v(\widehat{\mathsf{pmc}}, \mathsf{pmc}) > \alpha\} \leqslant 2\exp -nv\alpha^2$$

which translates into the sample size bound

$$n \geqslant \frac{\log r + \log\frac{2}{\delta}}{v\alpha^2}.$$

To compare this bound with the previous ones, note that

$$\Pr\{|\widehat{\mathsf{pmc}} - \mathsf{pmc}| > \epsilon\} \leqslant \Pr\{d_v(\widehat{\mathsf{pmc}}, \mathsf{pmc}) > \epsilon/(v + 2)\}$$
$$\leqslant 2\exp -nv\epsilon^2/(v + 2)^2$$

which is minimized by $v = 2$ as $2\exp -n\epsilon^2/8$. For a Chernoff-like bound we have

$$\Pr\{\widehat{\mathsf{pmc}} < (1 - \gamma)\,\mathsf{pmc}\} \leqslant \Pr\{d_v(\widehat{\mathsf{pmc}}, \mathsf{pmc}) > \frac{\gamma\,\mathsf{pmc}}{v + \mathsf{pmc}}\}$$
$$\leqslant 2\exp -nv\left[\frac{\gamma\,\mathsf{pmc}}{v + \mathsf{pmc}}\right]^2 = 2\exp[-\tfrac{1}{4}n\gamma^2\,\mathsf{pmc}]$$

The precise form of the bound used here is due to Angluin & Valiant (1979). In Bather (1996, p.340) Chernoff says the bound should be named after Herman Rubin.

on taking $v = \text{pmc}$.

Note that none of the bounds in this subsection depend on the number of classes, as they work directly with error events.

Infinite number of two-class classifiers

This concept is discussed very clearly by Pollard (1984).

A key quantity in the PAC-learning results is the Vapnik–Chervonenkis (or VC) dimension d of $\mathscr{F}$. Consider the set of functions $\{0,1\}^n \to \{0,1\}$ induced by evaluating functions in $\mathscr{F}$ at the n points of the training set. (That is, the induced function gives the predicted class for each of the possible class assignments to the training set.) Let the number of distinct functions be $\Delta(n)$. In many cases this number will be 2^n for small n, since all possible functions are induced. Let d be the largest value of n such that $\Delta(n) = 2^n$ for some training set of size n, or infinity if there is no such number. Then it turns out that for $d < \infty$ we have

$$\Delta(n) \leqslant \sum_{i=0}^{d} \binom{n}{i} \leqslant n^d + 1, \quad (en/d)^d \qquad (2.49)$$

the last inequality holding for $n \geqslant d \geqslant 1$ (Blumer *et al.*, 1989, Proposition A2.1). This result is sometimes called *Sauer's lemma*; its history is traced by Assouad (1983). Note that these bounds apply to all training sets; from now on we will use $\Delta(n)$ to denote the maximum over training sets of size n, and to use the results we will replace it by one of the bounds in (2.49).

Anthony & Biggs (1992) give full but opaque versions of the proofs of Blumer *et al.* (1989) for the noise-free case, except for measurability conditions.

We will give the most precise results available for two classes (from Blumer *et al.*, 1989) then sketch how they are derived. (There are benign measurability conditions which we ignore.)

As a simple example of the VC dimension, consider $\mathscr{X} \subset \mathbb{R}^m$ and classifiers of the form

$$\text{sign}\left(\sum_{i=1}^{m} a_i x_i > b\right)$$

which we shall meet under the name of *perceptrons* in Section 3.6. Cover (1965) showed that these rules have VC-dimension $m+1$ (as follows from Proposition 3.1 on page 119). On the other hand, for binary inputs, there are between $2^{m(m-1)/2}$ and 2^{m^2} different functions generated by perceptrons (Muroga, 1971), so the bound given by (2.48) is of the form

$$n \geqslant \frac{m^2 \log 2 + \log \frac{1}{\delta}}{\epsilon}.$$

The following proposition usually gives considerably tighter bounds:

82   *2   Statistical Decision Theory*

Proposition 2.5 (Blumer *et al.*, 1989)

Let d denote the (finite) VC dimension of $\mathcal{F}$.

(i) *Given $0 < \epsilon < 1$, the probability that there is a classifier $g \in \mathcal{F}$ consistent with n training examples and with true error rate greater than ϵ is bounded above by*

$$2\Delta(2n)2^{-n\epsilon/2}.$$

(ii) *If*

$$n \geqslant \max\left(\frac{4}{\epsilon}\log_2\frac{2}{\delta}, \frac{8d}{\epsilon}\log_2\frac{13}{\epsilon}\right) \text{ or } n \geqslant \frac{4}{\epsilon}\left[\log_2\frac{2}{\delta} + d\log_2\frac{12}{\epsilon}\right]$$

the bound in (i) *is less than δ.*

(iii) *For given $0 < \epsilon \leqslant 1/8$, $\delta \leqslant 1/100$, $d \geqslant 2$ and*

$$n < \max\left(\frac{1-\epsilon}{\epsilon}\log_2\frac{1}{\delta}, \frac{d-1}{32\epsilon}\right)$$

fix an algorithm to select a classifier g for each possible training set of size n. Then there is a probability distribution on $\mathcal{X} \times \{0,1\}$ and a function $f \in \mathcal{F}$ which correctly classifies examples with probability one, but the probability exceeds δ that the algorithm gives a classifier with error rate exceeding ϵ.

The lower bound in (iii) was shown by Ehrenfeucht *et al.* (1989) by exhibiting a suitably malicious distribution, which concentrates on d points and gives probability $1 - 8\epsilon$ to one of them. The conclusion from the proposition is that to achieve a high-probability *guarantee* of an error rate of less than ϵ we must take the size of the training set to be at least of order $d/\epsilon \log(d/\epsilon)$. This is, however, very much a worse-case bound, and such empirical evidence as there is (such as Cohn & Tesauro, 1992) suggests that practical performance is closer to the lower bound given by (iii), and can even be well below that bound for any 'normal' distribution of examples.

Similar results are known for the case with no perfect classifier:

Proposition 2.6 *Let d denote the (finite) VC dimension of $\mathcal{F}$.*

(i) (Vapnik & Chervonenkis, 1971) *For any $\epsilon > 0$*

$$\Pr\left\{\sup_{g\in\mathcal{F}} |\widehat{\text{pmc}}(g) - \text{pmc}(g)| > \epsilon\right\} \leqslant 4\Delta(2n)\exp[-n\epsilon^2/8] \qquad (2.50)$$

and the probability is less than δ if

$$n \geqslant \frac{16}{\epsilon^2}\left[\log\frac{4}{\delta} + d\log\frac{32e}{\epsilon^2}\right].$$

The constants in (2.50) can be improved. The factor 1/8 in the exponent can be removed (Parrondo & Van der Broeck, 1993) at the expense of increasing the constant: the claims of Vapnik (1995, pp. 66, 85) are not supported by the belated proofs in Vapnik (1998).

(ii) (Vapnik, 1982, with a slight improvement by Anthony & Shawe-Taylor, 1993) *For any* $\alpha > 0$

$$\Pr\left\{\sup_g \frac{\text{pmc}(g) - \widehat{\text{pmc}}(g)}{\sqrt{\text{pmc}(g)}} > \alpha\right\} \leqslant 4\Delta(2n)\exp -\tfrac{1}{4}n\alpha^2 \qquad (2.51)$$

and the probability is less than δ *if*

$$n \geqslant \frac{8}{\alpha^2}\left[\log\frac{4}{\delta} + d\log\frac{16e}{\alpha^2}\right].$$

(iii) (Blumer *et al.*, 1989) *Given* $0 < \epsilon, \gamma \leqslant 1$, *the probability that there is a classifier* $g \in \mathscr{F}$ *with true error rate* pmc *exceeding* ϵ *and apparent error rate* $\widehat{\text{pmc}} < (1 - \gamma)\text{pmc}$ *is bounded above by*

$$4\Delta(2n)\exp -\tfrac{1}{4}\gamma^2 n\epsilon.$$

This probability is less than δ *if*

$$n \geqslant \max\left(\frac{4}{\gamma^2\epsilon}\log\frac{8}{\delta}, \frac{16d}{\gamma^2\epsilon}\log\frac{16}{\gamma^2\epsilon}\right) \text{ or } n \geqslant \frac{4}{\gamma^2\epsilon}\left[\log\frac{8}{\delta} + d\log\frac{16e}{\gamma^2\epsilon}\right].$$

(iv) (Haussler, 1992, Theorem 3) *For any* $\alpha > 0$

$$\Pr\left\{\sup_{g\in\mathscr{F}} d_v\left(\widehat{\text{pmc}}(g), \text{pmc}(g)\right) > \alpha\right\} \leqslant 4\Delta(2n)\exp[-\tfrac{1}{2}nv\alpha^2]. \qquad (2.52)$$

If we can find a classifier $g \in \mathscr{F}$ which fits the training set exactly we could apply Proposition 2.5(i) or Proposition 2.6(iii) with $\gamma = 1$, but the bound given by Proposition 2.5 is smaller. Part (iv) implies somewhat weaker versions of parts (i) and part (iii) with factors $1/16$ and $1/8$ in the exponent (using $v = 2, \alpha = \epsilon/4$ and $v = \epsilon, \alpha = \gamma/2$ respectively).

Proposition 2.6(i) has been used to give upper bounds for PAC-learning by Pearl (1979) and Abu-Mostafa (1989). This bound can be improved for large $n\epsilon^2$; Devroye (1982) has

$$\Pr\left\{\sup_{g\in\mathscr{F}} |\widehat{\text{pmc}}(g) - \text{pmc}(g)| > \epsilon\right\} \leqslant 4e^{4\epsilon+4\epsilon^2}\Delta(n^2)\exp[-2n\epsilon^2] \qquad (2.53)$$

and Alexander (1984) has

$$\Pr\left\{\sup_{g\in\mathscr{F}} |\widehat{\text{pmc}}(g) - \text{pmc}(g)| > \epsilon\right\} \leqslant 16(\sqrt{n}\epsilon)^{4096(d+1)}\exp[-2n\epsilon^2] \qquad (2\,54)$$

for $n\epsilon^2 \geqslant 64$, which translates to a guarantee for

$$n\epsilon^2 \geqslant \left[\log\frac{16}{\delta} + 1024(d+1)\log\frac{2048(d+1)}{e}\right], 64.$$

So far we have considered these results as bounding the size of the training set. It is possible to change our point of view and seek upper bounds on the true error rate; in fact given the probability framework we will have (conservative) upper confidence limits corresponding to probability $1 - \delta$. From (2.48) we have an upper bound of the form $(\log r - \log \delta)/n$ if we can always fit n training cases exactly. From part (i) of Proposition 2.5, in the same case the upper bound is

$$\epsilon \leqslant \frac{2d}{n} \log_2 \frac{2ne}{d} + \frac{2}{n} \log_2 \frac{2}{\delta}$$

which shows convergence at slightly less than $O(1/n)$. On the other hand, if the Bayes risk is non-zero, we obtain convergence at rates around $O(1/\sqrt{n})$ from Proposition 2.6 parts (i) and (iii). These give

$$|\widehat{\mathrm{pmc}} - \mathrm{pmc}| \leqslant \sqrt{\frac{8}{n} \left[d \log \frac{2ne}{d} + \log \frac{4}{\delta} \right]}$$

$$\mathrm{pmc} - \widehat{\mathrm{pmc}} \leqslant \sqrt{\mathrm{pmc}} \sqrt{\frac{4}{n} \left[d \log \frac{2ne}{d} + \log \frac{4}{\delta} \right]}.$$

Devroye (1988) considered the expected maximal difference (over classifiers) between $\widehat{\mathrm{pmc}}$ and pmc rather than confidence limits for pmc, and also looked at the direct calculation of $\Delta(m)$ for practical families of classifiers.

One way to look at these results is in terms of *empirical risk minimization*. In the noise-free case we select a classifier which makes the minimum number (zero) of errors on the training set. For the noisy case it is convenient to choose a classifier with the same property, as then the upper confidence limits are tightest on the true error rate. This amounts to a parameter estimation strategy, although often it will not lead to a unique parameter estimate. If we then consider families $\mathscr{F}_m$ of models of increasing flexibility, we expect to obtain a lower apparent error rate as m increases, but a confidence limit on the error rate which will decrease and then increase. Vapnik's (1982, 1992) *structural risk minimization* chooses the model class to minimize this bound. (Note that minimizing bounds, especially those as loose as these appear to be, may not be a good idea!)

Outline of proofs

We will only give the main ideas of the proofs, omitting details of measurability. It may be puzzling that $\Delta(2n)$ appears in Propositions 2.5 and 2.6 since we have only n training samples. The reason is an idea that Pollard (1984) calls *symmetrization*. Suppose we consider two independent training sets of size n. Let η be the true error rate and $\widehat{\eta}_i$ be the apparent error rates on the two sets. Then

$$\Pr\{\sup_{g\in\mathscr{F}} |\widehat{\eta}_1 - \eta| > \epsilon\} \leqslant 2\Pr\{\sup_{g\in\mathscr{F}} |\widehat{\eta}_1 - \widehat{\eta}_2| > \tfrac{1}{2}\epsilon\} \qquad (2.55)$$

for $n > 2/\epsilon^2$. This reduces the computation to comparing two independent training sets of size n, thus to events on a training set of size $2n$. This will be used in the proof of (2.50).

Proof of (2.55):

Fix a classifier with true error rate η and consider $n\widehat{\eta}_i$, which has a binomial (n, η) distribution. By the Bienaymé–Chebychev inequality, for $n > 2/\epsilon^2$ we have

$$\Pr\{|\widehat{\eta}_i - \eta| > \tfrac{1}{2}\epsilon\} \leqslant \frac{4}{\epsilon^2}\frac{\eta(1-\eta)}{n} \leqslant \frac{1}{n\epsilon^2} \leqslant \tfrac{1}{2}.$$

Now condition on the first sample and choose a $\widehat{g}$ which maximizes the left-hand side of (2.55). Then conditionally we have

$$\Pr\left\{|\widehat{\eta}_2(\widehat{g}) - \eta(\widehat{g})| \leqslant \tfrac{1}{2}\epsilon \ \Big| \ \widehat{g}\right\} \geqslant \tfrac{1}{2}$$

since the second sample is independent of the first. Thus unconditionally

$$\Pr\{|\widehat{\eta}_1(\widehat{g}) - \eta(\widehat{g})| > \epsilon, |\widehat{\eta}_2(\widehat{g}) - \eta(\widehat{g})| \leqslant \tfrac{1}{2}\epsilon\} \geqslant \tfrac{1}{2}\Pr\{|\widehat{\eta}_1(\widehat{g}) - \eta(\widehat{g})| > \epsilon\}$$

and

$$\begin{aligned}
\Pr\{\sup_{g\in\mathscr{F}} |\widehat{\eta}_1(g) - \widehat{\eta}_2(g)| > \tfrac{1}{2}\epsilon\} &\geqslant \Pr\{|\widehat{\eta}_1(\widehat{g}) - \widehat{\eta}_2(\widehat{g})| > \tfrac{1}{2}\epsilon\} \\
&\geqslant \Pr\{|\widehat{\eta}_1(\widehat{g}) - \eta(\widehat{g})| > \epsilon, |\widehat{\eta}_2(\widehat{g}) - \eta(\widehat{g})| \leqslant \tfrac{1}{2}\epsilon\} \\
&\geqslant \tfrac{1}{2}\Pr\{|\widehat{\eta}_1(\widehat{g}) - \eta(\widehat{g})| > \epsilon\} = \tfrac{1}{2}\Pr\{\sup_{g\in\mathscr{F}} |\widehat{\eta}_1 - \eta| > \epsilon\}
\end{aligned}$$

which suffices. $\qquad\square$

Now consider the class $\mathscr{F}_\epsilon$ of classifiers with true error rate η at least ϵ; we will show

$$\Pr\{\widehat{\eta}_1 = 0 \text{ for some } g \in \mathscr{F}_\epsilon\}$$
$$\leqslant 2\Pr\{\widehat{\eta}_1 = 0, \widehat{\eta}_2 \geqslant \eta/2 \text{ for some } g \in \mathscr{F}_\epsilon\} \qquad (2.56)$$

for $n > 8/\epsilon$. Again condition on the first sample and choose $\widehat{g} \in \mathscr{F}_\epsilon$ which is consistent with it (if possible). Then conditionally

$$\Pr(\widehat{\eta}_2 < \tfrac{1}{2}\eta \ | \ \widehat{g} \in \mathscr{F}_\epsilon) \leqslant \Pr(|\widehat{\eta}_2 - \eta| > \tfrac{1}{2}\eta) \leqslant \frac{4(1-\eta)}{n\eta} \leqslant \frac{4}{n\epsilon} \leqslant \tfrac{1}{2}.$$

Thus as before

$$\Pr\{\widehat{\eta}_1 = 0, \widehat{\eta}_2 \geqslant \tfrac{1}{2}\eta \text{ for some } g \in \mathscr{F}_\epsilon\} \geqslant \Pr\{\widehat{\eta}_1(\widehat{g}) = 0, \widehat{\eta}_2(\widehat{g}) \geqslant \tfrac{1}{2}\eta(\widehat{g})\}$$
$$\geqslant \tfrac{1}{2}\Pr\{\widehat{\eta}_1(\widehat{g}) = 0\}.$$

Consider the right-hand side of (2.56). We condition on the locations of the $2n$ training points, and consider only their order. Of the 2^{2n} possible assignments by classifiers in $\mathscr{F}$ of labels to points, at most $\Delta(2n)$ will occur. Suppose there are ℓ erroneous assignments of labels by an eligible classifier; we must have $\ell > \tfrac{1}{2}n\eta \geqslant \tfrac{1}{2}n\epsilon$. The probability (over permutations) that these all occur in the second sample is $\binom{n}{\ell}/\binom{2n}{\ell} \leqslant 2^{-\ell} \leqslant 2^{-n\epsilon/2}$, so the probability of the event $\{\widehat{\eta}_1 = 0, \widehat{\eta}_2 > \eta/2$ for some $g \in \mathscr{F}_\epsilon\}$ is at most $\Delta(2n)2^{-n\epsilon/2}$, which with (2.56) establishes part (i) of Proposition 2.5 for $n\epsilon > 8$, hence $\Delta(2n) > 8$. The (uninteresting) remaining cases can be proved by showing $\Pr\{\widehat{\eta}_2 < \tfrac{1}{2}\eta\} \leqslant \tfrac{1}{2}$ actually holds for $n > 2/\epsilon$.

Part (ii) of Proposition 2.5 follows from part (i) using the second bound in (2.49). To cover all the cases, consider a bound of the form

$$A\Delta(2n)e^{-Bn} \leqslant A\left(\frac{2en}{d}\right)^d e^{-Bn} \leqslant \delta.$$

Note that $\log x \leqslant x-1$ can be manipulated to give $\log x \leqslant Cx - \log Ce$. Take $C = \alpha B/d$ for $0 < \alpha < 1$. Then

$$\alpha Bn \geqslant d\log n + \log\frac{\alpha Be}{d},$$

so it suffices to choose n satisfying

$$(1 - \alpha)Bn \geqslant \log\frac{A}{\delta} + d\log\frac{2}{\alpha B}.$$

The inequalities come from $\alpha = 1/2$ and $8/\log 2 \leqslant 12$.

We return to (2.55). We condition on the $2n$ examples and only consider their order. Let $e_i = I(\text{error on sample } i)$. Then half the right-hand side of (2.55) is

$$\Pr\{\sup_{g\in\mathscr{F}} |\sum_{i=1}^{n}(e_i - e_{i+n})| > n\epsilon/2\}.$$

Consider random permutations of the total sample. The terms $e_i - e_{i+n}$ are bounded by ± 1 and have a symmetric distribution with mean zero. We need only consider permutations which swap elements between the two sets, so consider $Y_i = \pm(e_i - e_{i+n})$ independently with probability a half for each sign. Hoeffding's (1963) inequality asserts that the probability over these permutations is bounded by

$$\Pr\left\{\left|\sum_{i=1}^{n} Y_i\right| > \tfrac{1}{2}n\epsilon\right\} \leqslant 2\exp -2n(\epsilon/2)^2/4 = 2\exp -n\epsilon^2/8$$

and averaging over the remainder of the permutation distribution does not affect the bound. As before, allowing any $g \in \mathcal{F}$ gives rise to at most $\Delta(2n)$ assignments of errors. With (2.55) this gives (2.50).

Part (iii) of Proposition 2.6 follows immediately from part (ii) on taking $\alpha^2 = \gamma^2\epsilon$. (The reduction of the constant to 4 is from Anthony & Shawe-Taylor, 1993.) We will prove part (ii). Choose $\widehat{g} \in \mathcal{F}$ as before, and assume $\eta = \mathrm{pmc}(\widehat{g}) > \alpha^2$ and $n\alpha^2 > 4$, or the bound is trivial. Since $\widehat{\eta}_2$ has a binomial (n, η) distribution with $n\eta > 4$, $\Pr\{\widehat{\eta}_2 > \eta\} \geqslant 0.32768 > 1/4$. (The worst case occurs with $\eta = 1 - 1/n$ and $n = 5$, hence this value.) Now $\widehat{\eta}_1 < \eta - \alpha\sqrt{\eta}$ and $\widehat{\eta}_2 > \eta$ imply $\widehat{\eta}_2 - \widehat{\eta}_1 > \alpha\sqrt{[\tfrac{1}{2}(\widehat{\eta}_1 + \widehat{\eta}_2)]}$. (Show this by minimizing $[\widehat{\eta}_2 - \widehat{\eta}_1]/\alpha\sqrt{[\tfrac{1}{2}(\widehat{\eta}_1 + \widehat{\eta}_2)]}$ over $\widehat{\eta}_1$, which is clearly achieved by taking this as large as possible, and its bound is least restrictive if $\eta = \widehat{\eta}_2$.) Thus

$$\tfrac{1}{4}\Pr\{\widehat{\eta}_1 < \eta - \alpha\sqrt{\eta}\} \leqslant \Pr\{\widehat{\eta}_1 < \eta - \alpha\sqrt{\eta}, \widehat{\eta}_2 > \eta\}$$
$$\leqslant \Pr\{\widehat{\eta}_2 - \widehat{\eta}_1 > \alpha\sqrt{[\tfrac{1}{2}(\widehat{\eta}_1 + \widehat{\eta}_2)]}\}.$$

As before we consider random swapping permutations. If all Y_i are zero then $\widehat{\eta}_2 = \widehat{\eta}_1$ and the probability is zero, so suppose there is at least one $Y_i \neq 0$. Hoeffding's inequality (1963, Theorem 2) gives

$$\Pr\{\widehat{\eta}_2 - \widehat{\eta}_1 > \alpha\sqrt{[\tfrac{1}{2}(\widehat{\eta}_1 + \widehat{\eta}_2)]}\} = \Pr\{\sum_{i=1}^{n} Y_i > \alpha\sqrt{[\tfrac{1}{2}n\sum_{i=1}^{2n} e_i]}\}$$

$$\leqslant \exp -\frac{2n\alpha^2\tfrac{1}{2}\sum_{i=1}^{2n}e_i}{4\sum Y_i^2} = \exp -\frac{n\alpha^2\sum_{i=1}^{2n}e_i}{4\sum|Y_i|} \leqslant \exp -\tfrac{1}{4}n\alpha^2$$

since $\sum_{i=1}^{n}|Y_i| = \sum_{i=1}^{n}|e_i - e_{i+n}| \leqslant \sum_{i=1}^{2n}e_i$.

More general problems

The methods based on VC dimension are confined to two classes, since the VC dimension itself is. There are several extensions we might need to consider:

1 more than two classes;

2 loss functions other than the error rate;

3   characterizations other than the VC dimension, since this is either unknown or infinite for many of the classifiers we would want to consider in practice.

These are beginning to be addressed, notably by Haussler (1992). An alternative approach for multiple classes is to use the VC dimension of the graph of the classifier, a subset of $\mathscr{X} \times \mathscr{C}$, as in Shawe-Taylor & Anthony (1991).

The loss functions we consider will be bounded by 0 and 1. Since our decision space is finite, this amounts to a possible re-scaling, but excludes working with the deviance, for example, unless probabilities are bounded below. Techniques are available for unbounded loss functions (Pollard, 1990).

The general technique to replace the VC dimension is one of approximating the infinite class of functions by a finite ϵ-cover, that is a set of functions such that any function is within distance ϵ of a member of that set. Let $\mathscr{N}(\epsilon)$ be the smallest number of points in a ϵ-cover, which is closely related to the maximum number $\mathscr{M}(\epsilon)$ of points at least ϵ apart which can be packed in: in fact

Such notions are often referred to as *metric entropy*.

$$\mathscr{M}(2\epsilon) \leqslant \mathscr{N}(\epsilon) \leqslant \mathscr{M}(\epsilon).$$

Then the bounds will be in terms of the expected value of $\mathscr{N}$ applied to the family of functions evaluated at the training set. For example (Haussler, 1992, lemmas 13 and 14)

$$\Pr\left\{\sup_{g \in \mathscr{F}} d_v(\widehat{R}(g), R(g)) > \alpha\right\} \leqslant 2\mathrm{E}\,\mathscr{N}(\alpha v/8) \exp{-\alpha^2 v n/8}$$

where R and $\widehat{R}$ are the risk and estimated risk respectively, and L_1 distance is used between loss functions. Very similar ideas occur in the *method of sieves* (Grenander, 1981; Geman & Hwang, 1982).

The concept of *pseudo-dimension* (Pollard, 1990; Haussler, 1992) generalizes the VC dimension and provides a convenient way to bound covering numbers. Consider orthants of $\mathbb{R}^p$, possibly shifted to a new origin. The pseudo-dimension is the largest n for which there is a training set of n points such that the loss functions evaluated at those points meet all the orthants for some origin. Clearly for $\{0, 1\}$-valued functions the pseudo-dimension *is* the VC dimension. If the set of loss functions for all rules under consideration has pseudo-dimension d then (Pollard, 1984, p. 27; Haussler, 1992, Theorem 6)

An orthant is specified by giving the sign of each coordinate.

$$\mathscr{N}(\epsilon) \leqslant \mathscr{M}(\epsilon) < 2\left[\frac{2e}{\epsilon}\log\frac{2e}{\epsilon}\right]^d$$

for any training set, where ϵ is measured in an L_1 distance. These two results combine to give the bound

$$n \geqslant \frac{8}{\alpha^2 v} \left[\log \frac{8}{\delta} + 2d \log \frac{8e}{\alpha v} \right]$$

to ensure that $\Pr\{d_v(\widehat{R}(g), R(g)) > \alpha\}$ with probability at least $1 - \delta$.

3

Linear Discriminant Analysis

In this chapter we discuss methods which arise in statistics and in pattern recognition based on linear combinations of the feature vectors (so we assume that the feature space $\mathscr{X}$ is contained in $\mathbb{R}^p$). These methods provide the templates for generalization to flexible non-linear methods discussed in the next two chapters as well as being of interest in their own right.

We can identify three distinct ways in which the idea of approximating a function $\mathbf{f}$ from $\mathscr{X}$ to $\mathbb{R}^K$ can be used to produce a classifier, although all have variations on their themes.

1   Take $f_k(\mathbf{x}) = p(k \,|\, \mathbf{x}) = \mathsf{E}[I(Y = k) \,|\, \mathbf{X} = \mathbf{x}]$ and $\mathbf{f}(\mathbf{x}) = (f_k(\mathbf{x}))$. The Bayes rule chooses a maximizer of $f_k(\mathbf{x})$. Define *target* $\mathbf{t}_k$ to be the kth unit vector. Since

$\|\mathbf{y}\|$ denotes the norm of a vector $\mathbf{y}$.

$$\|\mathbf{f}(\mathbf{x}) - \mathbf{t}_k\|^2 = -2f_k(\mathbf{x}) + 1 + \|\mathbf{f}(\mathbf{x})\|^2$$

the Bayes rule amounts to choosing the nearest target to $\mathbf{f}(\mathbf{x})$. This leads to ways to approximate by $\mathbf{f}(\mathbf{x}; \theta)$ based on choosing θ to make the predictions for the training set as close as possible to the targets.

2   Dietterich & Bakiri (1991, 1995) consider coding the class targets $\mathbf{t}_k$ to be widely spaced in $\mathscr{Z} = \{0, 1\}^m$ for $m > k$, and learning a function $\mathbf{f}$ from $\mathscr{X}$ to $[0, 1]^m$. The classifier then chooses the nearest target in $\mathscr{Z}$ to the prediction $\mathbf{f}(\mathbf{x})$ for a new example. The actual coding is done using error-correcting codes, and the distance is L_1. We can view this approach as training a classifier for m pseudo-classes, and then mapping the distribution over pseudo-classes to the K real classes.

The L_1 distance between $\mathbf{x}$ and $\mathbf{y}$ is $\sum_i |x_i - y_i|$.

3   We have seen that the Bayes rule maximizes $\log p(k \,|\, \mathbf{x})$, and the multiple logistic model (2.29) is a linear model for these log posterior

probabilities. Variants which are less principled but commonly used are separate logistic models of each class versus the rest or versus a reference class (see Section 3.5).

We vary slightly our usual notation: there is a training set of n observations (or examples) of a p-variate pattern, and these observations are classified into g groups. Note that the groups need not coincide with the classes, and in the less flexible methods of this chapter it may be desirable to split some of the classes. For example, in symbol recognition we might divide the class for sevens into crossed and uncrossed sevens. If we let the classifier choose the best group and then assign to its class we would be using a cost structure which penalizes the wrong choice of group rather than class. The cost structure based on groups corresponds to adding posterior probabilities over groups to form the posterior probability for the class, then choosing the class with the largest posterior probability.

Let X denote the $n \times p$ matrix of examples, and G the $n \times g$ matrix of indicator functions for the groups (i.e. $g_{ij} = 1$ if and only if observation i belongs to group j). Note that $G^T G = \text{diag}(n_i)$, where n_i is the number of observations for group i. A typical example will be denoted by $\mathbf{x}$ and is a *row* vector where necessary (as it is a row of X), and ^{T} denotes the transpose of a vector or matrix.

Remember $\mathbf{x}$ is a *row* vector.

3.1   Classical linear discrimination

We will normally assume (to ease the notation) in this section and in Section 3.2 that X has been centred; each feature variable has had its average subtracted.

We saw on page 36 that if we assume the probability model in which the observations for group j are normal with mean μ_j and common covariance matrix Σ, the Bayes rule is to allocate a future observation $\mathbf{x}$ to the group for which

$$-2 \log p(j \,|\, \mathbf{x}) = (\mathbf{x} - \mu_j)\Sigma^{-1}(\mathbf{x} - \mu_j)^T - 2 \log \pi_j + \text{const} \qquad (3.1)$$

is smallest. The first term on the right is known as the *Mahalanobis distance* from $\mathbf{x}$ to the group mean. Expanding this out we find

$$-2 \log p(j \,|\, \mathbf{x}) = -2\mathbf{x}\Sigma^{-1}\mu_j^T + \mu_j\Sigma^{-1}\mu_j^T - 2 \log \pi_j + \text{const} + \mathbf{x}\Sigma^{-1}\mathbf{x}^T \quad (3.2)$$

which is a linear term in $\mathbf{x}$ plus a quadratic term which does not depend on the class. Since we wish to maximize $p(j \,|\, \mathbf{x})$ or equivalently to minimize (3.2), we may as well maximize the linear terms

$$LDA_j = 2\mathbf{x}\Sigma^{-1}\mu_j^T - \mu_j\Sigma^{-1}\mu_j^T + 2 \log \pi_j.$$

The space $\mathcal{X} = \mathbb{R}^p$ is partitioned by hyperplanes, another sense in which this is linear discrimination. For the special case of $g = 2$ groups, comparing LDA_2 with LDA_1 amounts to computing the linear function $LDA_2 - LDA_1$ and choosing group 2 if and only if it is positive. More generally, the comparison can be done in a space of dimension at most $g - 1$, and the distances computed in such a space, as we shall show more formally.

In practice the population quantities $\boldsymbol{\mu}_j, \Sigma$ are estimated by $\mathbf{m}_j, W$, where W is the within-group covariance matrix defined below, rather than the maximum likelihood estimates considered in Section 2.2. (Other estimates are considered in Sections 2.4 and 2.5, but very rarely used in applied statistics.)

Fisher's linear discriminant

The classical method of linear discrimination was described by Fisher (1936) for two classes and extended to more by Rao (1948) (but sometimes attributed to Bryan, 1951). It uses a different criterion not based on the decision theory of Chapter 2; it seeks a linear combination $\mathbf{xa}$ of the variables which maximizes the ratio of its *between-group variance* to its *within-group variance*. This is appealing even if multivariate normality is implausible.

These terms come from the analysis of variance and are defined for a variable $\mathbf{y} = (y_i)$ as follows. Let m_j denote the mean of $\mathbf{y}$ in group j, let $\mathbf{m} = (m_j)$ and let $[i]$ denote the group of observation i. The $n \times g$ matrix G indicates which group each observation belongs to, so $g_{ij} = I(j = [i])$. Then the within-group variance is defined to be

$$W_{\mathbf{y}} = \frac{\sum_i (y_i - m_{[i]})^2}{n - g} = \frac{\|\mathbf{y} - G\mathbf{m}\|^2}{n - g}$$

and the between-group variance is

$$B_{\mathbf{y}} = \frac{\sum_i (m_{[i]} - \bar{y})^2}{g - 1} = \frac{\|G\mathbf{m} - \bar{y}\mathbf{1}\|^2}{g - 1}.$$

We can extend these definitions to the multivariate observations X by defining M as the $g \times p$ matrix of group means and

$$W = \frac{(X - GM)^T (X - GM)}{n - g}, \qquad B = \frac{(GM - \mathbf{1}\bar{x})^T (GM - \mathbf{1}\bar{x})}{g - 1}.$$

Then the linear combination $\mathbf{xa}$ has variances $\mathbf{a}^T W \mathbf{a}$ and $\mathbf{a}^T B \mathbf{a}$, and total variance

$$\mathbf{a}^T S_T \mathbf{a} = \mathbf{a}^T \frac{(n - g)W + (g - 1)B}{n - 1} \mathbf{a}.$$

It may be that the training sample is known not to be a random sample from the underlying distribution, but that the numbers n_j of observations from each group were pre-specified. (This is often done to ensure that sufficient information is obtained on rare groups.) Then better estimates are obtained by weighting the observations in group j by $n\pi_j/n_j$ in forming B, W and $\bar{x}$. The interpretation remains unchanged. (Note that the maximum likelihood estimates are not weighted in this circumstance; the weighted versions are thought to be more accurate when the group covariance matrices actually differ.)

The classical computational approach is to seek a rescaling of the variables xS such that their within-group covariance matrix is the identity matrix I and then perform an eigenvalue decomposition of B expressed on these variables. It will suit some of our further computations better to rescale the variables so that the total variance is $nI/(n-1)$. The rescaling is achieved by taking a matrix S such that $S^T X^T X S = nI$. This can be done in a number of ways: a simple one is to use the QR decomposition (Golub & Van Loan, 1989) of X :

$$QX = \begin{bmatrix} R \\ 0 \end{bmatrix}$$

where Q is a $n \times n$ orthogonal matrix and R is a $p \times p$ upper triangular matrix. Take S as the solution to $RS = \sqrt{nI}$. Then on the rescaled variables $X' = XS$ we have $X'^T X' = nI$. There will be difficulties if the covariance matrix does not have full rank, and this can be hard to identify numerically. (For example, a column differing only in the fifth significant digit could be constant up to rounding error or could be an extremely precise measurement of, say, refractance.)

A fairly safe procedure is first to rescale all variables to unit variance (and detect any constant variables) then to use the *singular value decomposition* $X = U\Lambda V^T$. Then small singular values correspond to nearly constant combinations. We would take $S = \sqrt{n}V\Lambda^{-1}$, but small singular values should cause concern, since they correspond to linear combinations which are nearly constant and whose variance is likely to be determined inaccurately from the training set.

See the glossary.

Modified procedures are discussed in Section 3.4.

We now work with these rescaled variables. The matrix $G^T G$ is diagonal containing the numbers n_j of observations on each group. Let $T = \operatorname{diag}(\sqrt{n/n_j})$ so

$$TG^T GT = nI.$$

The group means are given by the $g \times p$ matrix $M = (G^T G)^{-1} G^T X = n^{-1} T^2 G^T X$. Since X has been centred the column sums of M

(weighted by group size) are zero, hence M has rank $r \leqslant \min(p, g-1)$. Consider the singular value decomposition of $T^{-1}M = U\Lambda V^T$. Since we do not wish to assume that either g or p is larger, we will assume that U is $g \times r$, Λ is $r \times r$ and V is $p \times r$.

From the singular value decomposition we find

$$(g-1)B = (GM)^T(GM) = V\Lambda^T U^T(TG^TGT)U\Lambda V^T = nV\Lambda^2 V^T$$
$$(n-g)W = X^T X - (g-1)B = nI - nV\Lambda^2 V^T = nV[I-\Lambda^2]V^T$$

which incidentally shows that the singular values are at most one. (Note that one can occur; it corresponds to a linear combination which is constant within the groups but has different values on two groups. If so it is the desired linear combination.) The original problem reduces to finding a linear combination **xa** of the rescaled variables which maximizes the ratio

$$\frac{\mathbf{a}^T V\Lambda^2 V^T \mathbf{a}}{\mathbf{a}^T V[I-\Lambda^2]V^T \mathbf{a}}.$$

Let $\mathbf{b} = V^T\mathbf{a}$. The ratio is $\sum \lambda_i^2 b_i^2 / \sum(1-\lambda_i^2)b_i^2$, which is maximized by taking only b_1 non-zero. Thus on the original variables **a** is proportional to the first column of SV. The linear combination is unique up to a scale factor unless $\lambda_2 = \lambda_1$.

The linear combination found by this process is called the first *linear discriminant* or the first *canonical variate*. Subsequent columns of SV give further linear discriminants which maximally separate the group means subject to being uncorrelated with previous linear discriminants (since on the variables rescaled by SV both the W and B covariance matrices are diagonal). The first $J \leqslant r$ linear discriminants maximize the ratio of the determinants of the between-group to within-group covariance matrices for J-dimensional linear transformations of the original variables. (This follows immediately since the determinant is the product of the eigenvalues.)

The linear discriminants are usually scaled so that they have within-group variance one (unless constant on groups). We have only defined r of them but a further $p-r$ can be chosen by taking further columns which are orthogonal to the columns of V.

For the linear discriminant variables the group means differ only in the first r variables. The quantity $(n-g)\lambda_i^2/(g-1)(1-\lambda_i^2)$ measures the ratios of the between- to within-group variances on the ith canonical variate. We can show graphically the difference between groups by plotting the data on the first few canonical variates, often the first two. Although the original probability model corresponds to classifying using all r dimensions, it may be better not to use canonical variates which have low discrimination between groups, so we may wish to use only those dimensions for which the ratio of between-group to within-group variance is appreciable.

This is a form of *shrinkage*: see Section 3.4.

Let A be the matrix whose columns define the linear combinations for the canonical variates, specifically

$$A = \text{diag}\left(\sqrt{(n-g)/n(1-\lambda_i^2)}\right)SV.$$

Then the transformed variables XA minus the appropriate group means are uncorrelated with unit variance (since $A^T W A = I$). Thus on the canonical variates, Mahalanobis distance is Euclidean distance. Since the group means differ only on the first r variates, the Mahalanobis distances to the group means can be computed from the distance in the first r dimensions plus a quantity from the remaining dimensions which does not depend on the group. The Bayes rule minimizes the Euclidean distance to the mean in the first r dimensions minus $2 \log \pi_j$. With just two groups this process does find the linear variable $LDA_2 - LDA_1$, for it computes distances only in the first dimension, the only one on which the means differ.

So far we have only considered finding the linear combination(s) required. Fisher's procedure cannot tell us the threshold between the two groups in classification. It seems common practice to classify by choosing the group whose mean is nearest in the space of canonical variates. Since in that space Euclidean distance is the within-group Mahalanobis distance, this corresponds to the Bayes rule if (and only if) the prior probabilities are equal.

The problems of rank-deficiency and various solutions are discussed by Cheng *et al.* (1992); the solutions given here are much simpler and more transparent.

Canonical variates and canonical correlation

The name 'canonical variate' comes from a connection with *canonical correlation analysis*, which seeks linear combinations $\mathbf{xa}$ and $\mathbf{yb}$ of maximal correlation. Since X is centred we have

$$\text{corr}(\mathbf{xa}, \mathbf{yb}) = \frac{\mathbf{b}^T Y^T X \mathbf{a}}{\sqrt{\mathbf{b}^T \text{Var}(Y)\mathbf{b}\, \mathbf{a}^T (X^T X)\mathbf{a}}} = \frac{\mathbf{b}^T Y^T X \mathbf{a}}{\sqrt{\mathbf{b}^T \text{Var}(Y)\mathbf{b}\, n\|\mathbf{a}\|^2}}.$$

Now if we take $Y = G$ we have

$$G^T X = (G^T G)M = (G^T G)TU\Lambda V^T = nT^{-1}U\Lambda V^T. \tag{3.3}$$

Without loss of generality we can centre Y. Let $\mathbf{b}' = U^T T^{-1}\mathbf{b}$. Then $\text{Var}(\mathbf{yb}) = \|G\mathbf{b}\|^2 = n\|T^{-1}\mathbf{b}\|^2 = n\|\mathbf{b}'\|^2$ and $\mathbf{b}^T Y^T X \mathbf{a} = n\mathbf{b}'^T \Lambda V^T \mathbf{a}$,

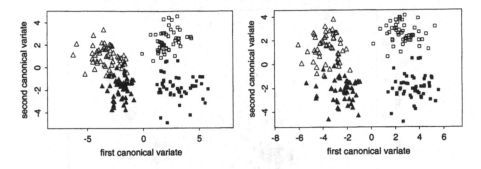

Figure 3.1: Linear discriminant plots for the *Leptograpsus* crab data. The left plot is from the original variables and the right plot from variables on $\log_{10}$ scale. The blue species is shown by triangles, the orange species by squares, and the symbols for females are filled.

so the correlation is maximized by taking **xa** as the first canonical variate and **b**′ proportional to the first coordinate vector $(1, 0, \ldots)^T$. Then **b** is the first column of $\Theta = TU$ and the correlation achieved is λ_1. Subsequent combinations with maximal correlation are given by subsequent discriminant variables and columns of Θ, and achieve correlations λ_i. We refer to the columns of Θ as *scores*. Note for future use that $(G\Theta)^T(G\Theta) = nI$ and so the scores are uncorrelated and have sum of squares n over the training set.

Examples

The crabs data are shown on the first two linear discriminants in Figure 3.1. As the measurements are lengths, we also considered taking logarithms, and as the figure shows this does increase slightly the separation between the groups. As there are four groups, the linear discriminants span three dimensions, and for the variables on log scale the ratios of between- to within-group standard deviations are 25.5, 16.9 and 2.9 on the three discriminants. Thus the first two linear discriminants explain 99.1% of the variance between the groups. Clearly the first linear discriminant expresses the difference between the species, the second that between the sexes. On log scale they are given by

$$L_1 = 72\,\text{FL} + 22\,\text{RW} + 22\,\text{CL} - 151\,\text{CW} + 41\,\text{BD}$$
$$L_2 = -6\,\text{FL} - 56\,\text{RW} + 88\,\text{CL} - 49\,\text{CW} + 13\,\text{BD}$$

which shows that the differences between the sexes are principally in the ratio of length to width, and that the blue form has a wider carapace than the orange form relative to its other measurements.

Next consider the forensic glass dataset. Figure 3.2 shows the first two canonical variates which account for 93.14% of the between-group variation. The variables are not on a common scale, but we can rescale

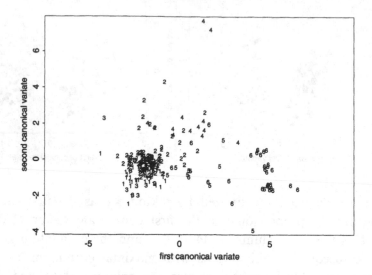

Figure 3.2: Linear discriminant plots for the glass fragments data. The coding is 1 = window float glass, 2 = window non-float glass, 3 = vehicle window glass, 4 = containers, 5 = tableware and 6 = headlamps.

each variable to have unit variance. The weights given by the first two canonical variates are then

```
RI   Na   Mg   Al   Si    K    Ca   Ba    Fe
0.95 1.94 1.07 1.67 1.90 1.02 1.43 1.15 -0.05
0.09 2.58 4.31 0.86 2.33 1.21 3.38 1.71  0.02
```

so the amount of iron appears to be unimportant, and the two most important variables in the second variate are magnesium and calcium. However, as the sum of the compositions is close to 100%, the variables are highly collinear which does not help interpretation. The plot is dominated by groups 4 to 6, with suspicions that groups 4 (containers) and 5 (tableware) are not homogeneous. As the boxplots of Figure 1.5 on page 14 show, this problem has a far from normal distribution, and probably has mixed distributions for some of the composition variables (with a positive probability for zero). Linear discriminant analysis has a cross-validated error rate of 38%. All of the vehicle glass is misclassified as window glass, and there is considerable confusion between float and non-float window glass. The cross-validated confusion matrix is

```
        WinF WinNF Veh Con Tabl Head
 WinF    47    20    3   0    0    0
WinNF    20    49    0   4    2    1
  Veh    11     6    0   0    0    0
  Con     0     6    0   6    0    1
 Tabl     0     3    0   0    5    1
 Head     1     1    0   2    0   25
```

The predictive form of linear discriminant analysis (page 51) gives almost identical results.

Since the within-group Mahalanobis distance is Euclidean distance in the space of canonical variates, it is useful to scale plots so that the axes have equal scales. Then circularly symmetric scatter plots for each group indicate that the assumptions of normality and equal covariances are realistic (or not, as in this example).

The Pima Indians diabetes data have two groups, so plots of the canonical variates are not very useful. The within-group covariance matrices are quite similar, although blood pressure and pedigree are uncorrelated in the non-diabetics group, and strongly negatively correlated in the diabetics group, which generally has slightly higher variability. The correlation is due to just one woman, who has the highest observed pedigree, and the second-lowest blood pressure.

Standard linear discrimination makes 67/332 errors on the test set. Choosing a subset of variables by cross-validation on the training set suggests that no reduction is worthwhile. The predictive version makes one fewer error. Both are making most of their errors (42/109) on the group reported to have diabetes. In this example quadratic discrimination does significantly worse (84/109), making many more errors on the non-diabetics group.

We can see something useful by plotting the first (and only) canonical variate, Figure 3.3, as there is a suspicion of multi-modality in the diabetics group, and skewness in the non-diabetics. The density estimates used were kernel methods (Section 6.1) with automatically chosen bandwidths.

Figure 3.3: Density estimates of the non-diabetic (left) and diabetic group (right) on the canonical variate for the Pima Indians data.

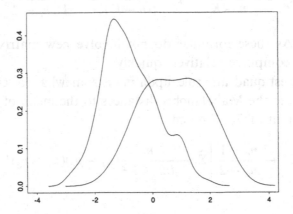

Variable selection

It is sometimes desirable to consider only those variables which make a useful contribution to discriminating between the classes. For classification it may be desirable not to have to measure unimportant variables.

For interpretation it may be easier to concentrate on the important variables. In either case we need a variable-selection procedure. McKay & Campbell (1982a, b) provide a comprehensive introduction to practical issues in variable selection.

Computer programs are widely available to select feature variables to be used in linear discrimination. This is an example of our considerations of Section 2.6, and the simplest stepwise methods are commonly used. There are two main approaches. One is to use significance tests for the value of individual features under the normal model (there are many such tests; McLachlan, 1992, §12.2), the other is to use the error rate (Hermans *et al.*, 1982).

Cross-validation

Linear discrimination is one of those classifiers for which leave-one-out cross-validation can be computed without complete re-fitting (Fukunaga & Kessell, 1971; Hjort, 1986, §12.1). Suppose we wish to re-train the classifier omitting example $\mathbf{x}_j$ which is of class c, and find its predicted class. To do so we need to update our estimates of the Mahalanobis distances Δ_{jk}^2 from $\mathbf{x}_j$ to the group means $\boldsymbol{\mu}_k$. We find

$$\Delta_{jc}^2 \leftarrow \Delta_{jc}^2 \times \frac{n-K-1}{n-K}\left(\frac{n_c}{n_c-1}\right)^2 \Big/ \left[1 - \frac{n_c}{(n_c-1)(n-K)}\Delta_{jc}^2\right]$$

$$\Delta_{jk}^2 \leftarrow \Delta_{jk}^2 \times \frac{n-K-1}{n-K}\left[1 + \frac{\{(\mathbf{x}_j - \boldsymbol{\mu}_c)^T\widehat{\Sigma}^{-1}(\mathbf{x}_j - \boldsymbol{\mu}_k)\}^2}{\{(n-K)(n_c-1)/n_c - \Delta_{jc}^2\}\Delta_{jk}^2}\right]$$

for $k \neq c$. As these formulae do not involve new matrix operations, they can be computed relatively quickly.

For the best quadratic rule, updating is somewhat easier, as we only have to update the Mahalanobis distances to the mean of class c plus the determinant of $\widehat{\Sigma}_c$. We find

$$\widehat{\Sigma}_c \leftarrow \frac{n_c-1}{n_c-2}\left[\widehat{\Sigma}_c - \frac{n_c}{(n_c-1)^2}(\mathbf{x}_j - \boldsymbol{\mu}_c)(\mathbf{x}_j - \boldsymbol{\mu}_c)^T\right]$$

hence

$$|\widehat{\Sigma}_c| \leftarrow |\widehat{\Sigma}_c| \times \left(\frac{n_c-1}{n_c-2}\right)^p \left[1 - \frac{n_c}{(n_c-1)^2}\Delta_{jc}^2\right]$$

$$\Delta_{jc}^2 \leftarrow \Delta_{jc}^2 \times \frac{n_c^2(n_c-2)}{(n_c-1)^3} \Big/ \left[1 - \frac{n_c}{(n_c-1)^2}\Delta_{jc}^2\right]$$

which can be evaluated with modest additional computation.

3.2 Linear discriminants via regression

Consider first the case of two groups, and let Y be the class indicator for class 2. Then the posterior probability for class 2 is

$$p(2 \mid \mathbf{x}) = E(Y \mid \mathbf{X} = \mathbf{x}).$$

Remarkably, although the conditional mean is not linear in $\mathbf{x}$, linear regression of y on $\mathbf{x}$ can be used to find the linear discriminant. With equal prior probabilities and group sizes, future observations can be classified by predicting via the linear regression and selecting group 2 if and only if the prediction exceeds 0.5. This was established by Fisher (1936) by a direct calculation, reproduced by T. W. Anderson (1984, §6.5.4) and Hand (1981). We will take another approach which gives greater insight and provides an extension to more than two groups. Thus for two groups, if the normal model for the population holds, it is more efficient to use linear rather than logistic regression, even though the population regression $E(Y \mid \mathbf{X} = \mathbf{x})$ is logistic not linear.

This follows from results of Efron (1975) discussed in Section 2.3.

We can think of the linear regression as the best linear approximation to the posterior probabilities. As a principle of classifier design this has been used (Duda & Hart, 1973; Devijver & Kittler, 1982; Fukunaga, 1990) under the name of *minimum (mean) square error* classifiers. Unlike the linear discriminant (for more than two groups), that procedure classifies by the nearest target or equivalently the largest component. An alternative would be to find the linear classifier which minimizes the total risk (Highleyman, 1962b). This is much harder, and has only been achieved for two general normal populations (when the Bayes rule is quadratic); see Section 2.2.

The manipulations which follow are of some interest, but would not be used to actually calculate a linear discriminant in preference to the methods of Section 3.1. Their importance lies in the realization of Breiman & Ihaka (1984) that non-linear regressions could be used in place of linear regression, thus providing one way to use non-linear regressions for classification problems.

Once again we assume that X is centred and that in the algebraic formulae we work with the rescaled variables $\mathbf{x}S$, which have covariance matrix $nI/(n-1)$.

The derivation of the canonical variates *via* canonical correlations shows that the 'scores' for classes given by the columns of $\Theta = TU$ have a special place, as these are best predicted by the corresponding canonical variate. However, we can first observe that if we regress the class indicators G on X using the rescaled variables we have regression coefficients

$$\boldsymbol{\beta} = (X^T X)^{-1} X^T G = n^{-1} V \Lambda U^T T^{-1} n = V \Lambda U^T T^{-1} \qquad (3.4)$$

using (3.3). Thus the predicted values are of the form $\mathbf{x} V \Lambda A$ for a full-rank matrix A and so span the same space as the linear discriminants. This implies that the regression will perform the reduction to $r \leqslant \min(p, g - 1)$ dimensions. (Since X is centred, the predicted values cannot predict a constant term. If $g > p+1$ there will be no reduction.) If $g = 2$ then either $r = 1$ or the groups have the same mean, and we see immediately that linear regression will find Fisher's linear discriminant, up to a constant factor that is not needed for classification purposes.

For more than two groups Breiman & Ihaka (1984) showed how to find Θ by minimizing the residual sum of squares over the scores as well as the coefficients, but it seems as easy to apply standard linear discrimination methods to the predicted values. That is, the data are replaced by the fitted values, classification is done by using Mahalanobis distances to the group means based on their within-class covariance matrix, and lower-dimensional plots can be made after a singular value decomposition.

Note that finding the maximum of the linear functions $\mathbf{x}\boldsymbol{\beta}$ does *not* give the linear discriminant classifier, for $\boldsymbol{\beta}_k = n_k/(n-1) S_T^{-1} \mathbf{m}_k$ where S_T is the total variance matrix, whereas LDA_k uses W, the within-group covariance matrix. There are, however, ways to calculate the classifier in the space of predicted values of the regression.

Two groups

In the case of two groups we can achieve a worthwhile simplification. Unless the groups have the same mean, $r = 1$ and we need only consider $\lambda = \lambda_1$. There are only two possible scores with mean zero and sum of squares n over the data; $\theta = \pm(-\sqrt{n_2/n_1}, \sqrt{n_1/n_2})^T$. The coefficients for the regression of the score variable on X are (using (3.4))

$$\boldsymbol{\beta} = (X^T X)^{-1} X^T G \Theta = V \Lambda U^T T^{-1} T U = \lambda V$$

and so the predicted values $\widehat{y}$ are λ times the first column of $(\mathbf{x}S)V$ and have within-group variance $W = n\lambda^2 (1 - \lambda^2)/(n-2)$. The group means are $M = T U \Lambda V^T = \Theta \Lambda V^T$ and these are mapped to $M\boldsymbol{\beta} = \Theta \Lambda \Lambda = \lambda^2 \theta$. The correlation achieved is λ, so the residual sum of squares is $1 - \lambda^2$ times the total sum of squares n, and the residual mean square $s^2 = n(1 - \lambda^2)/(n-2)$.

A regression package will use divisor $n - p$ for s^2.

The linear discriminant chooses class 2 if and only if

$$LDF = \widehat{y}(\mathbf{x}) \frac{m_2 - m_1}{W} - \frac{m_2^2 - m_1^2}{2W} + \log \frac{\pi_2}{\pi_1} > 0.$$

Now

$$
\begin{aligned}
LDF &= \frac{n-2}{n} \left[\widehat{y}(\mathbf{x}) \frac{\theta_2 - \theta_1}{1 - \lambda^2} - \frac{\lambda^2 (\theta_2^2 - \theta_1^2)}{2(1 - \lambda^2)} \right] + \log \frac{\pi_2}{\pi_1} \\
&= \frac{n-2}{n} \left[\left(\widehat{y}(\mathbf{x}) - \frac{\theta_1 + \theta_2}{2} \right) \frac{\theta_2 - \theta_1}{1 - \lambda^2} + \frac{\theta_2^2 - \theta_1^2}{2} \right] + \log \frac{\pi_2}{\pi_1} \\
&= \left(\widehat{y}(\mathbf{x}) - \frac{\theta_1 + \theta_2}{2} \right) \frac{\theta_2 - \theta_1}{s^2} + \frac{(n-2)(\theta_2^2 - \theta_1^2)}{2n} + \log \frac{\pi_2}{\pi_1}. \quad (3.5)
\end{aligned}
$$

Let us change y from the scores to be the indicator of class 2. This is a linear transformation, so we can linearly transform the fitted regression. After some manipulation we find

$$LDF = \frac{\widehat{y}(\mathbf{x}) - 1/2}{s^2} - \frac{(n-2)(n_2 - n_1)}{2 n_1 n_2} + \log \frac{\pi_2}{\pi_1}$$

which is directly computable from the regression. Note that if we use the proportions in the training set to estimate the prior probabilities, the constant term has leading terms in an expansion in powers of $\widehat{\pi}_2 - 1/2$ as

$$\frac{8(\widehat{\pi}_2 - 1/2)}{n} - 32 \left(\frac{1}{3} - \frac{1}{n} \right) (\widehat{\pi}_2 - 1/2)^3$$

so the dividing point for two groups on $\widehat{y}(\mathbf{x})$ will differ negligibly from $1/2$ unless the proportions in the training set are very different. (Remember that s^2 will often be very small since the targets are zero and one and the fit can be very good.)

More than two groups

For more than two groups the simplest procedure is to apply standard linear discriminant techniques in the space of fitted values. However, Breiman & Ihaka (1984) (who appear not to have realized this) extended some of the calculations to more than two groups. We work with the regression of $G\Theta$ on X. Remember that we can replace the data X by the fitted values F of the regression on G on X, and perform a canonical correlation analysis on these to find Θ, since the canonical variates are linear functions of F as well as of X.

The predicted values are proportional to the canonical variates, so W is diagonal expressed in these variables, and the Mahalanobis distance is a sum of r terms similar to that for the two-group case. Let $s_i^2 = n(1-\lambda_i^2)/(n-g)$ be the residual mean square for the ith regression. The linear discriminant between group t and group s becomes (using (3.5))

$$\sum_i \left[\left(\hat{y}_i(\mathbf{x}) - \frac{\theta_{is} + \theta_{it}}{2} \right) \frac{\theta_{it} - \theta_{is}}{s_i^2} + \frac{(n-g)(\theta_{it}^2 - \theta_{is}^2)}{2n} \right] + \log \frac{\pi_t}{\pi_s}$$

$$= \frac{1}{2} \left[\|(\hat{\mathbf{y}}(\mathbf{x}) - \mathbf{t}_s)\operatorname{diag}(1/s_i)\|^2 - \frac{n-g}{n}\|\mathbf{t}_s\|^2 \right]$$

$$\quad - \frac{1}{2} \left[\|(\hat{\mathbf{y}}(\mathbf{x}) - \mathbf{t}_t)\operatorname{diag}(1/s_i)\|^2 - \frac{n-g}{n}\|\mathbf{t}_t\|^2 \right] + \log \frac{\pi_t}{\pi_s}$$

where $\mathbf{t}_s$ is the row of Θ corresponding to group s. Thus we choose the group t to minimize

$$\|(\hat{\mathbf{y}}(\mathbf{x}) - \mathbf{t}_t)\operatorname{diag}(1/s_i)\|^2 - \frac{n-g}{n_t} - 2\log \pi_t \qquad (3.6)$$

since $\|\mathbf{t}_t\|^2 = n/n_t$. (This is a formula of Breiman & Ihaka apart from a difference in divisor for W.) Since

$$\operatorname{diag}(1/s_i) = \sqrt{(n-g)/n}\,[1 - \Lambda^2]^{-1/2}$$

the first term is the Mahalanobis distance between the predicted value and the target in the space of predicted values for the scores Θ (since we saw that $(n-g)W = nV[I - \Lambda^2]V^T$).

Note that (3.6) corresponds to a distance between the predicted values and targets (which are the kth unit vector for class k), with quadratic form $[(n-g)/n]\Theta[1 - \Lambda^2]^{-1}\Theta^T$. (This is not a metric since it ignores variation in the gth coordinate, corresponding to a shift in level of all the predicted values.) Thus the linear discriminant chooses the nearest target in this distance, adjusted by $(n-g)/n_t + 2\log \pi_t$.

Breiman & Ihaka noted that the linear combinations Θ can be found by minimizing the residual sum of squares, since for combination θ this is

$$\|G\theta\|^2 - \|X\beta\theta\|^2 = n\|T^{-1}\theta\|^2 - n\|\Lambda^T U^T T^{-1}\theta\|^2 = n\|\zeta\|^2 - n\|\Lambda^T \zeta\|^2$$

where $\zeta = U^T T^{-1}\theta$, and so is successively minimized subject to $\|\zeta\| = 1$ by taking ζ as the coordinate vectors or θ as the columns of $TU = \Theta$. The condition is that $\|G\theta\| = 1$, so that the scores for the groups have unit variance over the training set. Let F be the matrix of fitted values. Then the problem is to maximize $\|F\theta\|^2$ subject to $\|T^{-1}\theta\| = 1$, or $\|FTU\zeta\|^2$ subject to $\|U\zeta\| = 1$. The solutions are then the columns of TV^* where V^* is the matrix of right singular vectors of FT. But it is simpler to consider the canonical correlations of the fitted values directly.

Hastie *et al.* (1994) have independently re-interpreted the work of Breiman & Ihaka to the same conclusions, via an entirely algebraic route.

3.3 Robustness

When the crabs data were re-entered by a clerk she made an error in row 98 which should have read

```
sp  sex  FL   RW   CL   CW   BD
Bl   F  17.4 16.9 38.2 44.1 16.6
```

but was entered as

```
sp  sex  FL   RW    CL   CW   BD
Bl   F  17.4 16.9 438.2 4.1 16.6
```

We expected this to be disastrous for the discriminant analysis, but in fact it turned out to make rather little difference to the plots on the first two canonical variates. The effect of the errors is to inflate the within-group variance for the variables CL and CW which are then heavily down-weighted in the second canonical variate. It happens that in this example there is enough structure for the remaining variables to show almost the same discrimination.

This example does suggest that it would be wise to have a robust form of linear discrimination. The choice of canonical variates depends on the estimated within-group covariance matrix W and the matrix M of group means. Robust estimators of means and covariance matrices are discussed in Section 2.5, and our experiments used the minimum-volume ellipsoid method discussed there.

The effect of using robust estimators can be seen for the glass fragments data by comparing Figures 3.2 and 3.4. Even more of the variation (97.83%) is explained by the first two canonical variates (using the robust measures of variance) and the central 'core' is more concentrated, with one example of class 2 being shown up as a considerable outlier, perhaps closer to class 4 (containers). But what this example shows is that the groups do not have a common covariance matrix. The cross-validated error is much worse at 46.7%.

We mentioned on page 99 that the covariance matrix for the Pima Indians data was influenced by one unusual observation, so we tried

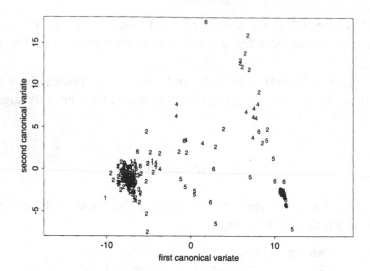

Figure 3.4: Robustified linear discriminant plot for the forensic glass data. Two points of class 4 at $(-4.7, 33.7)$ have been omitted. Compare this to Figure 3.2 on page 98.

robust estimation. This increased the test-set error rate for linear discrimination considerably (but not significantly) to 76/332. For quadratic discrimination there is a large increase to 100/332. Using multivariate t discrimination gives 77/332, better than any other form of quadratic discrimination.

Robustification of discriminant analysis has been considered a number of times in the literature; we first saw it in Campbell (1980b, 1982). Broffit *et al.* (1980) used trimmed estimators of mean and covariance, but these are less satisfactory (as they are not affinely equivariant).

3.4   Shrinkage methods

When discussing the best quadratic rule in Section 2.2 we mentioned that if the training set is not large, we might do better to use the best linear rule than attempt to estimate all the variance matrices Σ_k. There is evidence (for example, Marks & Dunn, 1974) that if the true class covariance matrices Σ_k are similar, a linear discriminant may outperform the quadratic discriminant in small samples. This suggests other compromises. We might take some convex combination of the equal and unequal estimated variance matrices, say

$$\widehat{\Sigma}_k(\alpha) = \frac{(1-\alpha)n_k\widehat{\Sigma}_k + \alpha n\widehat{\Sigma}}{(1-\alpha)n_k + \alpha n} \tag{3.7}$$

with parameter $0 < \alpha < 1$ chosen to maximize performance (using a validation set or cross-validation). We describe this as *shrinking* the covariance matrices towards a common value, and hope thereby to obtain a biased but less variable estimator.

We might wish to shrink our estimator of the common covariance matrix Σ. Recall that our algorithm for linear discriminant analysis was first to rescale each variable to zero mean and unit variance, and then seek a transformation to variables which are uncorrelated (using $\mathbf{x}S$). Suppose we use the singular value decomposition $X = U\Lambda V^T$ to do so; the variables XV are uncorrelated with variances proportional to $\lambda_1^2, \ldots, \lambda_p^2$, and $\text{trace}(X^T X) = \sum \lambda_j^2 = np$. Thus if the original variables were positively correlated, there will be some linear combinations of high variance and some of low variance. The variance of the latter will not be determined at all precisely, so it is conceivable that we will find the first canonical variate taken in the direction of a combination of features that happens by chance to be nearly constant within groups. One way to avoid this is to shrink the λ_i^2 towards a common value, and this is precisely the effect of using the eigendecomposition of $(1 - \gamma)S_T + \gamma I$. In linear discrimination it is more usual to shrink W, that is to use

$$\widehat{\Sigma}(\gamma) = (1 - \gamma)W + \gamma I. \qquad (3.8)$$

Some accounts just consider $W + \lambda I$, but finding the linear discriminants is unchanged by an overall scale change in W.

Of course, this is only appropriate if the variables have been rescaled to unit variance. Many accounts (for example, Campbell, 1980a) replace I by a diagonal matrix, but this is equivalent to rescaling the variables if we use the diagonal matrix of the variances of the variables. Campbell (1980a) suggested that finding the first canonical variate to be in a direction of low within-group variance was common in applications, and that shrinkage was important in interpreting the coefficients of the linear discriminants.

Both aspects of shrinkage were considered by Friedman (1989) under the name of *regularized discriminant analysis*. He took a convex combination of the within-group and pooled covariance matrices to estimate Σ_k, and then shrunk that estimate, to obtain a shrunken quadratic discriminant analysis. Specifically, he used a plug-in quadratic rule with

$$\widehat{\Sigma}_k(\alpha, \gamma) = (1 - \gamma)\widehat{\Sigma}_k(\alpha) + \frac{\gamma}{p}\text{trace}[\Sigma_k(\alpha)]\,I$$

$$\widehat{\Sigma}_k(\alpha) = \frac{(1 - \alpha)n_k\widehat{\Sigma}_k + \alpha n\widehat{\Sigma}}{(1 - \alpha)n_k + \alpha n}.$$

The parameters α and γ are chosen to minimize the cross-validated error rate; Rayens & Greene (1991) point out that the minimum will often occur over a wide range of values of (α, γ). The largest value of (α, γ) is chosen (in lexicographical order, so first the largest α for any γ, then the largest γ for that α).

The idea of shrinkage is better known for regression under the name of *ridge regression* (Hoerl & Kennard, 1970a, b) where the (centred) matrix $X^T X$ is replaced by $X^T X + cD$ where c is an adjustable constant and D is a diagonal matrix which will be the identity if the variables are on a common scale, and otherwise could be the diagonal matrix of the variances of the variables. Given that we have seen that the linear discriminant may be computed by regression, we might wonder if the connection is exact: it is. Other forms of shrinkage are considered in regression (Sen & Srivastava, 1990; Frank & Friedman, 1993), such as dropping small principal components.

This is a special case of the algebra of Hastie *et al.* (1995).

There are other intermediate positions between linear and quadratic discrimination. We can restrict the class of covariance matrices Σ_k, for example to be proportional to each other (Owen, 1984; Flury, 1986; Eriksen, 1987) or to have equal correlation matrices (Manly & Rayner, 1987). More generally, we can parametrize $\Sigma_k = \lambda_k U_k D_k U_k^T$ where U_k are the eigenvectors, λ_k the sum of the eigenvalues and D_k a diagonal matrix of the eigenvalues divided by their sum. The proportionality restriction forces equality of U_k and D_k, and commonality of correlations forces commonality of U_k. Other restrictions which might be imposed are commonality of 'shape' (equal D_k) and/or 'size' (equal λ_k) as discussed by Banfield & Raftery (1993) and Flury *et al.* (1994).

Thus far we have only considered shrinking W. We mentioned on page 96 using $J < r$ canonical variates when allocating future examples, retaining only those dimensions in which the group means are well separated, say those for which the ratio of the between-groups to within-groups variance is around one or more. This is equivalent to setting the difference between the group means to zero in the dropped directions, and so is a form of shrinkage to zero of the between-groups covariance matrix B; we could use a less extreme form of shrinkage in which these directions are down-weighted. However, we can down-weight by either reducing B or increasing W, so this is merely a different perspective.

The most common sources of strong correlations between the variables are when features are measuring essentially the same quantity and when the features are a discretized signal or image. In both cases we may be able to design a better composite feature, and we will certainly be able to design a better form of shrinkage. In the regression context for a discretized signal, we would assume that the quantity of interest was really an integral $\int \beta(t) f(t) \, dt$ which is approximated by a sum at the reported values of t (Hastie & Mallows, 1993), and so would assume that $\beta(t)$ was smooth. There penalized regression could be used (as in Section 4.3). In the discrimination context Hastie *et al.* (1995)

apply this idea by modifying W to $W + \Omega$ for a pre-specified Ω and finding the equivalence to canonical correlation analysis and to ridge regression (for $X^T X + n\Omega$) in the same way as in Section 3.2.

In practice in these high-dimensional situations it may be preferable to re-parametrize the signal, for example by a spline basis, and discard the fine detail before any computation is done. After all, these methods are still assuming normality, and that is likely to be a serious limitation.

We will need to be very careful with shrinkage if some $\widehat{\Sigma}_k$ is singular, since this is likely to make it easy to identify that class, and shrinkage may reduce the performance dramatically.

3.5 Logistic discrimination

Logistic discrimination is important both as a more direct way of estimating posterior probabilities and hence the Bayes rule, and also as a method which is much easier to generalize.

We saw in Section 2.3 that using a normal model for each class (or group) density with a common covariance matrix gave rise to (2.29):

$$\log p(k \mid \mathbf{x}) = \log p(1 \mid \mathbf{x}) + \alpha_k + \mathbf{x}\boldsymbol{\beta}_k.$$

More generally, we could model $\log p(k \mid \mathbf{x}) - \log p(1 \mid \mathbf{x})$ by some parametric family of functions, say $g_k(\mathbf{x}; \theta)$ (with $g_1(\mathbf{x}) \equiv 0$). Then we estimate

$$\widehat{p}(k \mid \mathbf{x}) = \frac{\exp g_k(\mathbf{x}; \widehat{\theta})}{\sum_j \exp g_j(\mathbf{x}; \widehat{\theta})}. \tag{3.9}$$

(For the linear model we take the parameter vector θ to contain all the α_k and $\boldsymbol{\beta}_k$.) In the neural network literature this is known as *softmax* (Bridle, 1990a, b), and in the (earlier) statistical literature as a *multiple logistic* model. Classification is done by using the (estimated) Bayes rule, taking the maximum of $\widehat{p}(k \mid x)$ (under the usual loss function (2.2)). This procedure is known as *logistic discrimination*.

Several remarks are needed. First, (3.9) can arise from other probability models, sometimes with transformed $\mathbf{x}$ variables. Some of its earliest applications were to epidemiological studies where such a direct parametrization was very natural, and other models leading to (3.9) are considered by J. A. Anderson (1972) and Cox & Snell (1989). Another has two classes and independent binary features (Minsky, 1961) with probabilities θ_{ik} under class k, for then

$$\log p(2 \mid \mathbf{x}) - \log p(1 \mid \mathbf{x}) = \sum_i x_i[\log \theta_{i2} - \log \theta_{i1}]$$
$$+ (1 - x_i)[\log(1 - \theta_{i2}) - \log(1 - \theta_{i1})].$$

It is normal to fit logistic regressions and log-linear models by maximum likelihood, but we have to consider carefully the likelihood to be used (as we do below). The standard methods (Cox & Snell, 1989; McCullagh & Nelder, 1989) regard $\mathbf{x}$ as fixed and take the observed class y to have a Bernoulli distribution with probability distribution $(p(1\,|\,\mathbf{x}), p(2\,|\,\mathbf{x}))$ for two classes, and a multinomial distribution for more. Thus the likelihood of the training set is

$$\ell(\theta;\mathscr{T}) = \prod_i p(c_i\,|\,\mathbf{x}_i) = \prod_i \frac{\exp g_{c_i}(\mathbf{x}_i;\theta)}{\sum_j \exp g_j(\mathbf{x}_i;\theta)} \qquad (3.10)$$

with deviance

$$D(\theta) = 2\sum c_i \log p(c_i\,|\,\mathbf{x}_i;\theta). \qquad (3.11)$$

We note that if $g_k(\mathbf{x};\theta) = \alpha_k + \mathbf{x}\beta_k$, the log-likelihood is concave, so any local maximum of the likelihood is a global maximum.

> This is a special case of the computations on page 152 which show that the Hessian is non-positive definite.

In the special case of two groups 0 and 1 the log-likelihood is

$$L(\theta;\mathscr{T}) = \sum c_i\, p(2\,|\,\mathbf{x}_i) + (1 - c_i)\,[1 - p(2\,|\,\mathbf{x}_i)]$$

and hence the deviance is

$$D(\theta) = 2\sum \left[c_i \log \frac{c_i}{p(1\,|\,\mathbf{x}_i)} + (1 - c_i)\log \frac{(1 - c_i)}{(1 - p(1\,|\,\mathbf{x}_i))} \right]. \qquad (3.12)$$

An alternative way to specify the probabilities $p(k\,|\,\mathbf{x})$ is to give $g - 1$ logistic models of the form

$$\log p(k\,|\,\mathbf{x}) = \log p(1\,|\,\mathbf{x}) + g_k(\mathbf{x};\theta) \qquad (3.13)$$

which is the same model as equation (3.9) (as replacing $g_k(\mathbf{x};\theta)$ by $g_k(\mathbf{x};\theta) - g_1(\mathbf{x};\theta)$ leaves (3.9) unchanged). However, the models (3.13) could be fitted separately comparing each group with group 1, and this gives different maximum likelihood parameter estimates. The pairwise comparison uses less information and so will be less efficient, but for linear models Begg & Gray (1984) show the loss is often negligible. An approach which is common in neural networks is to consider a logistic model for each $p(k\,|\,\mathbf{x})$ against the rest, that is

$$\widehat{p}(k\,|\,\mathbf{x}) = \frac{\exp g_k(\mathbf{x};\widehat{\theta})}{1 + \exp g_k(\mathbf{x};\widehat{\theta})}. \qquad (3.14)$$

This has the disadvantage that there is no guarantee that the estimated probabilities will sum to one, and no compensating advantages. This model is however appropriate if the classes are not mutually exclusive, for example if they indicate diseases which could conceivably occur together.

Likelihoods

We derived likelihood (3.10) as a conditional likelihood for a training set of size n which was a random sample from the whole population. It will also be appropriate if the feature vectors are selected rather than sampled. We will lose information if we know $p(\mathbf{x}; \theta)$ as a function of θ, so we are implicitly assuming that $p(\mathbf{x})$ is completely unknown.

The issue is more complicated when the numbers of examples in each group are fixed, which implies that we will not be able to estimate the prior probabilities of the groups. The natural model is for $p_c(\mathbf{x}) = p(\mathbf{x} \mid c)$ not $p(c \mid \mathbf{x})$. However

$$\log p(\mathbf{x} \mid k; \theta) = \log p(k \mid \mathbf{x}; \theta) - \log \pi_k + \log p(\mathbf{x}; \theta)$$
$$= \log p(1 \mid \mathbf{x}) + \alpha_k + \mathbf{x}\boldsymbol{\beta}_k - \log \pi_k + \log p(\mathbf{x}; \theta)$$

and it is clear that we can only estimate $\zeta_k = \alpha_k - \log \pi_k$. The log-likelihood formed by conditioning on all the observed classes is

$$L(\theta; \mathscr{T}) = \sum \log p(\mathbf{x}_i \mid c_i; \theta) = \sum \log p(c_i \mid \mathbf{x}_i; \theta) - \log \pi_{c_i} + \log p(\mathbf{x}_i; \theta).$$

If we assume that $p(\mathbf{x}; \theta)$ is entirely unknown (apart from normalizing to a probability density), we can maximize over this as well as ζ_k and $\boldsymbol{\beta}_k$. The maximum likelihood estimate of $p(\mathbf{x})$ is then the empirical distribution of the observed values (so not a genuine density), leaving the maximization of the profile likelihood for $\boldsymbol{\beta}_k$,

The empirical distribution gives probability $1/n$ to each observed $\mathbf{x}_i$.

$$\sum \log p(\mathbf{x}_i \mid c_i; \widehat{\theta}) = \sum \log p(c_i \mid \mathbf{x}_i; \beta) + \text{const.}$$

This is exactly what we get from the other forms of sampling, and so we will obtain the same maximum likelihood estimates of $\boldsymbol{\beta}_k$. This is a streamlined version of the arguments of J. A. Anderson (1972) and Prentice & Pyke (1979). Note that we have maximized over an infinite-dimensional parameter $p(\mathbf{x})$, so standard likelihood theory does not apply; Anderson avoided this by considering only discrete $\mathbf{x}$, but Prentice & Pyke and Cosslett (1981) prove analogues of the standard asymptotic results.

Scott & Wild (1986) also consider a different approach. Suppose the sample was chosen to have n_k cases of class k, but the prior probability is known to be π_k. A natural idea is to weight the cases of class k by $w_k = N\pi_k/n_k$, that is to use a weighted log-likelihood of the form

$$\sum w_{c_i} \log p(c_i \mid \mathbf{x}_i). \tag{3.15}$$

This is precisely what we would use if combining identical cases in the training set, and is the common practice in survey statistics. Scott &

Wild show that the profile likelihood approach is more efficient if the logistic model is true, but the weighted form may be preferable for estimating the least false parameter if the model is reasonable but not precisely true.

Higher efficiency as usual means smaller variance in large samples.

Ordinal logistic models

The approaches thus far are appropriate if the classifications are purely nominal. Some classifications are naturally ordered, such as grades of foodstuffs and the severity of a disease. For these a model which takes account of the ordering is desirable. McCullagh & Nelder (1989, §5.2.2) advocate models for $\gamma_k = P(Y \leqslant k \mid \mathbf{x})$, on logistic ($\log[\gamma_k/(1 - \gamma_k)]$) or complementary log-log ($\log[-\log(1 - \gamma_k)]$) scales. Suppose there is an unobserved variable Z (say the true quality of the foodstuff) of which Y is a grouped version, so $Y = k$ if and only if $\zeta_{k-1} < Z \leqslant \zeta_k$. Suppose also that Z has a logistic distribution with mean $\eta(\mathbf{x})$; then

$$\gamma_k = P(Y \leqslant k \mid \mathbf{x}) = P(Z \leqslant \zeta_k \mid \mathbf{x}) = \frac{\exp[\zeta_k - \eta(\mathbf{x})]}{1 + \exp[\zeta_k - \eta(\mathbf{x})]}$$

which naturally gives rise to a logistic model

$$\operatorname{logit} P(Y \leqslant k \mid \mathbf{x}) = \zeta_k + \eta(\mathbf{x})$$

for γ_k that differs only in intercept for each category. Giving Z the extreme-value (or Gumbel) distribution rather than the logistic distribution leads to

$$\log\{-\log[1 - P(Y \leqslant k \mid \mathbf{x})]\} = \zeta_k + \eta(\mathbf{x}).$$

In both cases the linear model part is then $\eta(\mathbf{x}) = \mathbf{x}\boldsymbol{\beta}$.

If we regard the intercepts ζ_k as unknown (but necessarily increasing) this analysis can be extended to grouped versions of $\phi(Z)$ for an unknown but monotonic transformation $\phi(\cdot)$, since this will be equivalent to grouping Z.

The likelihood for the parameters (which are (ζ_k) and the parameters in η) then follows from our earlier considerations, since a model for $P(Y \leqslant k \mid \mathbf{x})$ implies one for $p(k \mid \mathbf{x}) = P(k - 1 < Y \leqslant k \mid \mathbf{x})$. In numerically maximizing the likelihood we will have to remember the ordering constraint on (ζ_k).

Anderson & Phillips (1981) illustrate the linear ordinal logistic model for data on severity of back pain. It does lack flexibility, for there must be a linear projection on which the class distributions occur in the assumed order for the fit to be adequate. If we allow a non-linear function η (Mathieson, 1996) this may be much easier to achieve.

Infinite estimates

A difficulty which arises with the conditional likelihood is that the maximum likelihood estimates may be infinite (or, as some prefer to say, fail to exist). Consider just two classes. If there is a direction such that the projection $\boldsymbol{\beta}^T \mathbf{x}_i$ completely separates the two classes, it should be clear that the posterior probabilities can then be made arbitrarily close to one for every example by taking $\boldsymbol{\beta} \to \infty$ with the appropriate sign. Indeed, the methods of Section 3.6 are designed to find such projections where they exist.

Albert & Anderson (1984), Albert & Lesaffre (1986), Santer & Duffy (1986), Silvapulle & Burridge (1986) and Lesaffre & Albert (1989) consider this in detail (apparently unaware of the earlier parallel development given in Section 3.6). With more than two groups, some groups can be completely separable from others on a linear projection, or (quasi-complete separation)

$$p(j \mid \mathbf{x}_i; \theta) \geqslant p(k \mid \mathbf{x}_i; \theta), \qquad k \neq j$$

for all $\mathbf{x}_i$ from class j and some θ. The maximum likelihood estimate has an infinite component. This occurs for the tableware group in the forensic glass dataset.

We feel too much has been made of this. The difficulty is an inappropriate parametrization, and the limits for infinite $\|\boldsymbol{\beta}\|$ of the fitted posterior probabilities remain perfectly suitable fits, albeit sometimes predicting probability zero or one.

Predictive approach

We considered the predictive framework for the diagnostic paradigm at (2.33). Suppose we assume that we know nothing about the marginal density $p(\mathbf{x})$. Then the posterior density is of the form

$$\log p(\theta \mid \mathscr{T}) = L(\theta; \mathscr{T}) + \log p(\theta) + \text{const.} \qquad (3.16)$$

Then for large n the posterior density $p(\theta \mid \mathscr{T})$ will be approximately normal with mean $\widetilde{\theta}$ and covariance matrix V, say, and we can approximate these by the maximizer and the inverse of the Hessian of (3.16). The integration over θ could be performed approximately (Aitken, 1978) or by simulation using importance sampling from the approximate normal distribution (Ripley, 1987, §5.2). Since the approximate normal distribution will have short tails, it is better to choose a longer-tailed distribution, for example by increasing the variance slightly or using a multivariate t distribution.

If the maximum likelihood estimate is (partially) infinite, the local approximation will break down, but with a proper prior the predictive approach will give a sensible fit. Indeed, a proper prior which restrains the size of the parameter vector is a form of ridge regression (see Section 3.4) and has advantages even in non-Bayesian views. For example, it avoids the embarrassment of predicting probability zero for events that happen (as might happen with the plug-in rule if the training set is linearly separable but the populations are not).

Bias correction of the parameters of a linear logistic discrimination has been considered extensively, but Byth & McLachlan (1978) showed that the bias of the plug-in estimates of the posterior probabilities $p(k \mid \mathbf{x}; \widehat{\theta})$ was of smaller order than $1/n$ and so much less important than for the best linear classifier discussed in Section 2.5.

Examples

First we consider the Pima Indians data. The simplest approach is a direct logistic regression on the seven explanatory variables. This made 66 errors on the test set, an error rate of 19.8%. However, some of the features had insignificant coefficients, and a stepwise selection procedure to choose the fit with the smallest value of AIC dropped blood pressure and skin thickness. Its test set performance was also 66 errors (but not the same ones).

The number of pregnancies varies from zero to seventeen, and this seems unlikely to enter linearly. To test this, we allowed separate coefficients for 0, 1, 2, 3, 4, 5 and 6+ pregnancies, but the fit was little better, and the predictions worse (69 errors). If we allow a polynomial in age, AIC chooses a cubic; the number of errors is increased to 69.

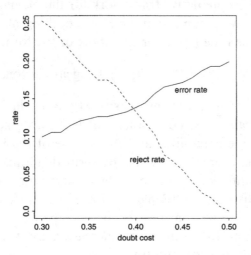

Figure 3.5: Error and rejection rates against doubt cost d for a logistic discrimination model for the Pima Indians data.

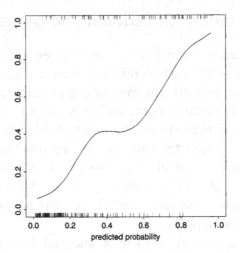

Figure 3.6: Calibration plot for a logistic discrimination model for the Pima Indians data. The 'rug' of ticks shows the events which occurred in the training set against the predicted probabilities. The smooth curve is a kernel regression (see Section 6.1).

We can use this example to illustrate some of the concepts we saw in Chapter 2. If we allow a low enough cost of 'doubt', this will be chosen. Figure 3.5 (computed on the test set) shows a fairly small reduction in error rate with rejection in this example; if we allow 10% of the patients to be rejected, the error rate drops from 19.9% to 15.7%. We can also check if the predicted probabilities look well calibrated. Figure 3.6 was computed on the training set and shows no serious departures.

For the forensic glass data the fitting algorithms converge slowly because of the partial separation of the classes, but produce satisfactory fitted probabilities and an error rate of 26.2% with confusion matrix

	WinF	WinNF	Veh	Con	Tabl	Head
WinF	50	17	3	0	0	0
WinNF	19	55	0	2	0	0
Veh	6	7	4	0	0	0
Con	0	2	0	11	0	0
Tabl	0	0	0	0	9	0
Head	0	0	0	0	0	29

The fit is rather better than linear discriminant analysis, not surprising given the very non-normal nature of these data. The cross-validated estimate of the error rate was 36% with confusion matrix

WinF	46	19	5	0	0	0
WinNF	19	47	2	3	3	2
Veh	7	6	4	0	0	0
Con	0	3	0	9	0	1
Tabl	0	2	0	0	7	0
Head	0	2	0	2	1	24

since the cross-validation runs did not find the 'right' separation vectors.

3.6   Linear separation and perceptrons

Historically, special attention has been given to situations such as Figure 3.1 on page 97 in which the two species are completely separated on the first linear discriminant. We say two groups are *linearly separable* if there is a linear function of the variables, say $\mathbf{x}\mathbf{a} + b$, which is positive on one group and negative on the other. A function which computes a linear combination of the variables and returns the sign of the result is known as a *perceptron* after the work of F. Rosenblatt (1957, 1958, 1962). (There are also publications by Block, 1962; Block *et al.*, 1962.) Their interest now is in their continuing influence on the thinking in the field of neural networks.

Let us add a column of 1's to $\mathbf{x}$ and add b to $\mathbf{a}$. Let $\mathbf{z} = \mathbf{x}/\|\mathbf{x}\|$ on the first group and $\mathbf{z} = -\mathbf{x}/\|\mathbf{x}\|$ on the second group. We then seek a linear combination $\mathbf{a}$ such that $\mathbf{z}\mathbf{a} > 0$ for every example in the training set. Since the training set is finite, we can choose $\delta > 0$ so that $\mathbf{z}\mathbf{a} > \delta$. Indeed, we can achieve this for any $\delta > 0$ by rescaling $\mathbf{a}$.

One approach to the problem would be to choose $\mathbf{a}$ by least squares to make $\mathbf{z}\mathbf{a}$ as near one as possible, or to regress $y = \pm 1$ on $\mathbf{x}$ which as we have seen gives the linear discriminant up to a scale factor. (However, there is no guarantee that the linear discriminant will linearly separate the groups if they are linearly separable, and it is easy to construct examples in which it will not, Figure 3.7.) A more direct formulation is to minimize the number of errors, but as that is a discrete measure, the optimization is difficult. The sum of the degree of error

$$\sum_i [\delta - \mathbf{z}_i\mathbf{a}]_+$$

will be zero if and only if linear separation can be achieved. This is equivalent to solving the linear programming problem $\mathbf{z}_i\mathbf{a} \geqslant \delta$, and linear programming methods can find a solution or show that none

Analytical evidence that optimizing the number of errors made by a perceptron is hard is provided by Höffgen *et al.* (1995) who showed that the problem of determining if there is a solution with at most $k \geqslant 1$ misclassifications is NP-hard. (For $k = 0$ it is reducible to a linear programming problem, so solvable in polynomial time.)

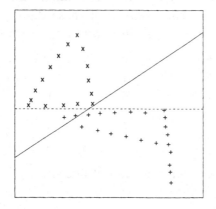

Figure 3.7: A dataset with the least-squares line (solid) and a linear separator (dashed).

exists (Minnick, 1961; Muroga *et al.*, 1961; F. W. Smith, 1968; Grinold, 1969). (We need to introduce δ to avoid the trivial solution $\mathbf{a} = 0$.) We introduce n artificial variables u_i and also split $\mathbf{a} = \mathbf{a}^+ - \mathbf{a}^-$ into positive and negative parts. Then the problem becomes

$$\min_{u_i, \mathbf{a}^+, \mathbf{a}^-} \sum_{i=1}^{n} u_i$$

subject to

$$u_i \geqslant 0 \quad \text{and} \quad u_i - \delta + \mathbf{z}_i(\mathbf{a}^+ - \mathbf{a}^-) \geqslant 0$$

which can be solved in a finite number of steps, and has a zero solution if and only if the classes are linearly separable.

In the late 1950s a number of researchers were interested in simpler but iterative solutions, in which the value of $\mathbf{a}$ was adjusted after each example was presented. The derivative for the least-squares problem $\|\mathbf{y} - X\mathbf{a}\|^2$ is $2X^T(X\mathbf{a} - \mathbf{y})$ and so a steepest descent procedure would be of the form

$$\mathbf{a} \leftarrow \mathbf{a} - \eta \sum_i (\mathbf{x}_i \mathbf{a} - y_i)\mathbf{x}_i^T. \tag{3.17}$$

For small enough η this process converges to the space of least-squares solutions.

Rather than compute the sum on the right-hand side and update $\mathbf{a}$, we could update after each pattern was considered. This gives the rule

$$\mathbf{a} \leftarrow \mathbf{a} - \eta(\mathbf{x}_i \mathbf{a} - y_i)\mathbf{x}_i^T \tag{3.18}$$

known as *Widrow–Hoff learning* (after Widrow & Hoff, 1960) or the *delta rule*. The patterns are presented cyclically until convergence, which will need $\eta \rightarrow 0$.

Rosenblatt's *perceptron learning rule* replaced the term $\mathbf{x}_i\mathbf{a}$ in (3.18) by the output of the perceptron, the sign of $\mathbf{xa}$. Thus $\mathbf{a}$ is changed only if the current pattern is misclassified, and so the rule is of the form

$$\mathbf{a} \leftarrow \mathbf{a} + 2\eta \mathbf{z}_i^T I(\mathbf{z}_i \mathbf{a} \leqslant 0).$$

No generality is lost by taking $\eta = 1/2$, since we can rescale $\mathbf{a}$. Rosenblatt showed that this rule will converge in a finite number of steps to a linearly separating combination *if one exists*. Let $\mathbf{a}^*$ be a suitable combination chosen so that $\mathbf{z}_i\mathbf{a}^* \geqslant 1$ for all members of the training set. If the rule changes $\mathbf{a}$ we have $\mathbf{z}_i\mathbf{a} \leqslant 0$ so

$$\|\mathbf{a} + \Delta\mathbf{a} - \mathbf{a}^*\|^2 = \|\mathbf{a} - \mathbf{a}^*\|^2 + 2\mathbf{z}_i(\mathbf{a} - \mathbf{a}^*) + 1 \leqslant \|\mathbf{a} - \mathbf{a}^*\|^2 - 1.$$

This shows the rule terminates in at most $\|\mathbf{a}_0 - \mathbf{a}^*\|^2$ steps.

This result is known as the *perceptron convergence theorem*. Its limitations were explored by the first edition of Minsky & Papert (1988) published in 1969. They showed that the coefficients needed to achieve linear separation (with fixed δ) could grow very rapidly with the size of the problem and the finite number of steps needed by the perceptron rule could become very large. There is after all another rule which will terminate in a finite number of steps: try all integer-valued **a** in order of increasing length, and no one would advocate that rule.

Minsky & Papert also considered the behaviour of the rule when the two groups were not linearly separable, and stated that $\|\mathbf{a}\|$ would remain bounded. (Their proof was completed by Block & Levin, 1970.) Thus if the **a** belong to a fixed-precision set (as they will do in real computation) the rule will eventually cycle. In particular there is no immediate way to deduce whether the rule will ever terminate, and cycling can be hard to detect, as the cycle length is unknown.

There are a number of variants of the perceptron updating rule. For example, η can be chosen just large enough to correctly classify the current case. Ho & Kashyap (1965) have other algorithms, discussed in detail in Duda & Hart (1973, §§5.6–7). It is also possible to extend the procedure to $K > 2$ categories. In that case the natural classifier would be to choose the largest of K linear discriminants $\mathbf{xa}_k$. Let **a** be the concatenation of the vectors $\mathbf{a}_k$. Then correct classification of pattern **x** in class k is equivalent to $(0\ldots, -\mathbf{x}, 0\ldots 0, \mathbf{x}, \ldots 0)\mathbf{a} > 0$ with the negative element in position j, for each j not equal to k. Thus each example **x** generates $g - 1$ examples in the Kp-dimensional problem. Applying the perceptron updating rule to this problem is equivalent to the updating rule

$$\mathbf{a}_i \leftarrow \mathbf{a}_i + \mathbf{x}, \qquad \mathbf{a}_j \leftarrow \mathbf{a}_j - \mathbf{x}$$

when pattern **x** is from class i, and j is a class with a larger value of $\mathbf{xa}_j$. Since this *is* the perceptron rule in the transformed problem, the convergence proof still holds. F. W. Smith (1969) extends the linear programming approach to more than two classes.

The traditional simplex algorithm for linear programming has exponential worst-case behaviour, but recently a number of provably polynomial algorithms have been devised. Mansfield (1991) has implemented one of these, Khachiyan's algorithm, for perceptron learning. The algorithm is quite simple and related to quasi-Newton methods of optimization (Gill *et al.*, 1981). It is guaranteed to find a solution, if there is one, in $(p + 1)^3 \log(p + 1) + (p + 1)^2 \log[\pi(p + 1)]$ iterations with binary inputs, where an iteration involves finding a misclassified example and updating the perceptron weights.

Prior to Minsky & Papert, Muroga *et al.* (1961) had shown that the weights in a linearly-separating perceptron with n binary inputs could be chosen to be integers less than $(n + 1)^{(n+1)/2} 2^{-n}$, and Muroga (1965) showed that there were problems where integer weights of $\Omega(2^n)$ are required (The Ω notation is defined on page 178.). Hampson & Volper (1986) showed that in some sense an average problem needs integer weights of $\Omega(2^{n/2})$. More recently Håstad found an example which needs integer weights at least as large as $n^{n/2} 2^{-n} e^{O(2^{0.585})}$. These results are reviewed and proved by Parberry (1994).

This bound was reduced to $O(p^2 \log p)$ by Maass & Turán (1994) using a more recent algorithm.

It is possible to extend the ideas of separation by linear planes to other surfaces (for example conic sections) or to piecewise linear surfaces. These have been considered by Mangarasian (1968) and applied to the diagnosis of breast tumours in Mangarasian *et al.* (1990); Wolberg & Mangarasian (1990) and subsequent papers.

Capacity questions

We can ask how many *random* patterns a perceptron with p inputs can learn reliably; that is can be classified without error. There will be a finite limit since the patterns must be linearly separable; this is irrespective of the existence of an algorithm to learn the patterns. Cover (1965) showed that the asymptotic answer is $2p$ patterns. In other words, for large p we expect to be able to store most sets of up to $2p$ patterns without error, but attempts to store more than $2p$ have a very low probability of success.

Proposition 3.1 *The probability that N patterns randomly chosen from any continuous distribution in $\mathbb{R}^p$ and randomly divided into two groups are linearly separable is one for $N \leqslant p+1$ and in general is*

$$2^{1-N} \sum_{i=0}^{\min(N-1,p)} \binom{N-1}{i} \sim \Phi\left(\frac{2p-N}{\sqrt{N}}\right)$$

for large N.

Proof: Let $C(n,p)$ be the number of assignments of classes 1 and 2 to n patterns in p variables which are linearly separable. We will show by induction that this does not depend on the patterns themselves provided they are in general position in $\mathbb{R}^p$ (that is, not collinear). All the assignments (y_i) are linearly separable for $n \leqslant p+1$, since we can solve the system $\mathbf{x}_i \mathbf{a} + b = y_i$ of n equations in $p+1$ unknowns to find a separating hyperplane. (This system will be singular only if the points are not in general position.) Thus $C(n,p) = 2^n$ for $n \leqslant p+1$.

Now consider adding an example to n linearly separable patterns. If the new point lies on the same side of every separating hyperplane, only one label for the new point gives a linearly separable set. In the other case, by continuity, there must exist a separating hyperplane passing through the new point and either label for the new point gives a linearly separable set. Of the $C(n,p)$ sets of linearly separable patterns, precisely $C(n,p-1)$ will pass through the new point (since this reduces the problem to one of dimension $p-1$). Thus

$$C(n+1,p) = 2C(n,p-1) + [C(n,p) - C(n,p-1)] = C(n,p-1) + C(n,p).$$

By induction we find the probability to be

$$C(N,p)/2^N = 2^{1-N} \sum_{i=0}^{\min(N-1,p)} \binom{N-1}{i} = P(X \leqslant \min(N-1,p))$$

where X has a binomial$(N-1, 1/2)$ distribution. The approximation then follows from the normal approximation to the binomial for large $N-1$. $\square$

The approximation on the right-hand side goes rapidly from 1 to 0 as N increases through $2p$, since for large n a binomial$(N-1, 1/2)$ distribution is tightly concentrated about $N/2$.

Support-vector machines

If two classes are linearly separable, there will be a continuum of weight vectors **a** which give rise to separating hyperplanes. Amongst these we can choose a hyperplane with maximal distance to the nearest example, achieved by minimizing $\|\mathbf{a}\|^2$ whilst insisting that $\mathbf{za} \geqslant 1$. Finding this hyperplane is a quadratic programming problem, and the usual Kuhn-Tucker optimality conditions show that there will be a subset of examples $\mathbf{z}_i$ (known as *support vectors*) for which $\mathbf{z}_i\mathbf{a} = 1$ and that the optimal **a** is a linear combination of these $\mathbf{z}_i$.

The advantage is choosing the optimal hyperplane is to reduce the VC-dimension of the space of solutions (which is proportional to a bound on $\|\mathbf{a}\|^2$). If the two classes *are* linearly separable then (Vapnik, 1995, Theorem 5.2) the expected error rate on future examples is bounded by the expected number of support vectors divided by $n-1$. Thus finding a small number of support vectors might indicate good generalization properties.

Of course, linear separation in the original feature space is quite rare, but as for *generalized linear discrimination* (page 121) we can expand the feature space by using polynomials or even radial-basis function networks and sigmoidal functions. These can give rise to very large feature spaces, but generalization may remain acceptable if the number of support vectors remains small, which was the case in the experiments reported by Vapnik (1995, Section 5.7).

By jointly minimizing the sum of the degree of error (page 116) and $\|\mathbf{a}\|^2$ these ideas can be extended to non-separable two-class problems (Cortes & Vapnik, 1995).

4

Flexible Discriminants

Linear combinations of the features will not always suffice to discriminate the groups. It is quite common to include ratios by including features on log scale as we did for the *Leptograpsus* crabs in Chapter 3. We can also allow non-linear functions of the features by including polynomial terms or dividing the range of the feature and including indicator terms for parts of the range. A more sophisticated alternative is to expand a feature on a spline basis such as B-splines (see below). These all amount to linear discrimination in a larger space of features, sometimes called *generalized linear discrimination* (Duda & Hart, 1973).

We will continue to assume n cases from g groups, which may or may not be the classes. We saw at the beginning of Chapter 3 three main ways to use a family of functions $\mathbf{f}: \mathcal{X} \to \mathbb{R}^g$ to approximate the Bayes rule. All these ideas apply equally here, but we will now consider much more general and flexible classes of functions. Another large class is the subject of Chapter 5.

The first approach was to estimate $\mathbf{f}(\mathbf{x})$ from the training set within our parametric family, and choose the class which maximizes $\widehat{f}_k(\mathbf{x})$ or has nearest target $\mathbf{t}_k$. (This includes the Dietterich & Bakiri, 1991, 1995, approach, which specializes the choice of targets.) As $\mathbf{f}(\mathbf{x})$ is a regression, it is natural to fit θ by least squares. Since in the conventional version $\mathbf{f}$ represents $(p(k \mid \mathbf{x}))$, the outputs are sometimes re-normalized to sum to one.

The second approach was to fit $\mathbf{f}(\mathbf{x})$ within the parametric family, but then to use the predicted values as the variables in a linear discriminant analysis. (This appears to have been the motivation of Breiman & Ihaka, 1984.) This amounts to finding the nearest group mean in the Mahalanobis distance given by the within-group covariance matrix of the fitted values (or, equivalently, of the residuals). Note that this differs from the first approach in using a different metric and in minimizing distance to the group means rather than the targets. Formula (3.6) does allow us to consider distances to the targets provided the group sizes are equal, but the metric there is not the Mahalanobis distance. Thus the essential difference is the metric used.

The second procedure appears preferable to the first if the predicted values are approximately normal with a common covariance matrix. However, in practice the fit is often very good except at a few points which therefore dominate the residuals and the estimate of the common covariance matrix. This suggests that it is desirable to use a robust discriminant analysis, and it may be better to accept the safe choice of the Euclidean metric.

The third approach is to use the parametric family within a multiple logistic model of the form (3.9), which is often the most theoretically satisfying but needs the ability to fit by maximum likelihood rather than least squares.

From the predictive viewpoint, these methods are all (in principle) parametric, and so we need to average over the uncertainty in the fitted parameters. Since there will usually be many more parameters than for linear families, it will be more important to average over the greater uncertainty. In practice this can be nigh impossible, as in the high-dimensional parameter space the integration is more difficult and it is unlikely that the asymptotics which suggest a normal approximation will be appropriate unless n is much larger than the number of parameters. We are only aware of such issues having been studied for neural networks, so discuss them in that context in Section 5.5.

4.1   Fitting smooth parametric functions

We discuss some of the possible ways to describe more general functions of the feature variables. We consider first methods using univariate functions $f : \mathscr{X} \to \mathbb{R}$.

Additive models and smoothers

An additive model is of the form

$$f(\mathbf{x}) = \alpha + \sum_{j=1}^{p} g_j(x_j) \tag{4.1}$$

for smooth but unknown functions g_j (Friedman & Silverman, 1989; Hastie & Tibshirani, 1990), which could encompass the effect of transformations (such as square or log or even an arbitrary polynomial) of each feature.

One choice of the smooth functions $g(x)$ of a single feature is to use *splines*. Splines are defined by M *knots* ξ_i which we can consider in increasing order. Then within an interval $[\xi_i, \xi_{i+1}]$ a spline is a polynomial of degree d (often three) and at the knots the first $(d-1)$ derivatives are continuous. This can be written as

The notation $[y]_+$ means $\max(y, 0)$.

$$g(x) = \sum_{i=0}^{d} \alpha_i x^{i-1} + \sum_{i=1}^{M} \beta_i [x - \xi_i]_+^d$$

which shows that there are $M + d + 1$ free parameters. There are other bases which have better numerical properties such as B-splines (de Boor, 1978; Green & Silverman, 1994). In any basis we can write

Sometimes the free parameters are reduced by end conditions; for *natural* cubic splines the second and third derivatives vanish at the boundaries.

$$g(x) = \sum_{i=1}^{M+d+1} \beta_i \phi_i(x). \tag{4.2}$$

It remains to choose the parameters β_i. For a *regression spline* these are chosen by least squares. Cubic *smoothing splines* are the solution to the minimization problem

$$\sum_{i=1}^{M} [y_i - g(\xi_i)]^2 + \lambda \int g''(u)^2 \, du$$

Smoothing splines are a special case of regularization, to be considered in Section 4.3.

and the parameters in (4.2) can be found by solving a sparse system of linear equations. Figure 4.1 shows the effect of the smoothness constraint on a smoothing spline: however many knots are included, over-fitting is prevented by the smoothness term. Thus smoothing splines are normally preferable to regression splines, except that the choice of λ is computationally demanding, and $\lambda = 0$ can be an adequate approximation with a small number of knots.

Other smoothing algorithms are also used, for example the `loess` smoother (Cleveland *et al.*, 1992) which uses robustly-fitted locally-weighted polynomials. Let $h(x; \phi)$ be a polynomial of degree d (usually one or two). Then the fitted value $g(x)$ is found by fitting $h(x; \phi)$ in the neighbourhood of x and reporting the fitted value at x. Specifically, $\widehat{\phi}_x$ is chosen to minimize

$$\sum_i [1 - (|x - x_i|/\tau)^3]_+^3 \, \rho(y_i - h(x_i, \phi))$$

and $\widehat{g}(x) = h(x, \widehat{\phi}_x)$. The loss function $\rho(u)$ could be u^2 but might penalize large departures less severely, in the spirit of robust statistics. Finally, the parameter τ controls the smoothness, and is chosen to

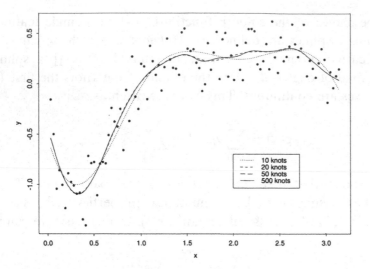

Figure 4.1: The effect of varying the number of knots in a smoothing spline for fixed λ. Based on an example of Wahba & Wold (1975). For 50 or more knots the curves are indistinguishable.

include a neighbourhood of αn points at point x. This approach is easily extended to smoothing d dimensions, for small d.

If the smooth functions in an additive model are written in terms of basis functions, as for polynomials and splines, we have

$$f(\mathbf{x}) = \alpha + \sum_{j=1}^{p} \sum_{k=1}^{df_j} \beta_{jk} \phi_{jk}(x_j). \qquad (4.3)$$

If the smoothing procedure chooses parameters by least squares, we have a linear regression in an extended space of features spanned by the functions $\phi_{jk}(x_j)$. On the other hand, if as in smoothing splines the functions minimize a sum of squares plus a penalty, we will have (approximately or exactly) a penalized linear regression. The method BRUTO used in the examples is described in Hastie & Tibshirani (1990, pp. 262–3). This adaptively chooses the smoothness of the splines and the number of terms in (4.1), including the possibility of linear functions and dropping features completely.

A general procedure to fit additive models is known as *back-fitting* (Hastie & Tibshirani, 1990). This holds all but one of the additive terms constant, removes that term and fits a smooth term to the residuals against the feature. In symbols, the model is

$$f(\mathbf{x}) - \alpha - \sum_{j \neq l} g_j(\mathbf{x}_j) = g_l(\mathbf{x}_l) \qquad (4.4)$$

and any smoothing algorithm (including `loess`) can be applied to the left-hand side. Smoothing is applied a feature at a time until the process converges (which it will under mild conditions).

Thus far we have considered penalized least-squares fitting. Linear models can be used to fit a logistic regression by maximum likelihood via local linearization to give a weighted least-squares problem (McCullagh & Nelder, 1989) and this is often solved iteratively in a handful of iterations. The extension to additive models is immediate (Gu, 1990; Wahba *et al.*, 1995).

An extension of this approach is to allow smooth functions of a small number of the features as terms in the additive model (for example, by C. J. Stone, 1985). Adding pairwise then three-term functions (and so on) is common practice in statistics, and these are known as *interaction* functions. Examples including functions of two features are given by Hastie & Tibshirani (1990, §9.5), Wahba (1995) and Wahba *et al.* (1995).

Pima Indians diabetes

In Section 3.5 we saw that a linear logistic discrimination model worked well for the data on diabetes amongst female Pima Indians, and that a polynomial in age improved the fit. We could consider if smooth non-linear terms in age or other variables such as plasma glucose levels and the body mass index might help the fit.

This was tried with several smoothers. The improvement in fit as measured by AIC or NIC was marginal, but the test-set error rate was unchanged or increased. BRUTO (which fits by penalized least squares) selected only linear terms, dropping blood pressure and skin thickness. Wahba *et al.* (1995) built an additive model including an interaction term in glucose concentration and body-mass index and a categorical term for the number of pregnancies (0, 1 or 2, 3–5, 6 or more). In our experiments this did less well than the linear model.

Projection pursuit regression

Additive models do not allow interactions between the features in $\mathcal{X}$. Perhaps the simplest way to allow interactions is through linear combinations (projections) of features:

$$f(\mathbf{x}) = \alpha + \sum_{j=1}^{r} g_j(\alpha_j + \beta_j^T \mathbf{x}) \tag{4.5}$$

which is *projection pursuit regression* (PPR; Friedman & Stuetzle, 1981). Sometimes the components of (4.5) are called *ridge functions* because a peaked g_j gives a topographic ridge in two dimensions.

This is a surprisingly general class of functions, as it can approximate uniformly arbitrary continuous functions over compacta (Diaconis & Shahshahani, 1984; L. K. Jones, 1987, 1992; Zhao & Atkeson, 1992). (This is sometimes referred to as the 'universal approximation' property.) As PPR encompasses feed-forward neural networks, these results follow from (but are a little easier than) those of Section 5.7 where the functions g_j are restricted to one function, the logistic. However, ridge functions provide better (in the sense of fewer parameters) approximations to some functions than others (Donoho & Johnstone, 1989; Zhao & Atkeson, 1992), which Zhao & Atkeson express as working better for 'angular smooth functions' than for 'Laplacian smooth functions'.

With multivariate regression we have to decide whether to use common non-linear terms for the different independent variables. This is usually done, so that for example for projection pursuit regression we have

$$f_k(\mathbf{x}) = \eta_k + \sum_{j=1}^{r} \gamma_{kj} \phi_j(\alpha_j + \beta_j^T \mathbf{x}). \qquad (4.6)$$

This shows that the fitted values lie in a $(r+1)$-dimensional space. Since the scale of ϕ_j is not otherwise fixed, we can choose $\phi_j(\alpha_j + \beta_j^T \mathbf{x})$ to have zero mean and unit variance over the training set.

Algorithms

The original algorithm for PPR has been superseded by SMART (Friedman, 1984). This constructs the (approximate) least-squares fit iteratively. A maximum value M for r is specified, and terms are added to (4.5) one at a time until M terms are present. Then at each step the least effective term is dropped and the model re-fitted, until r terms are left (and this process can also be used to help select r by looking at the fit). Some of the details can be changed by the user, and we only describe the 'highest' level of optimization.

Backfitting is used to fit the model, and when the jth term is being considered, the direction β_j is optimized by a Gauss–Newton procedure (see Section A.5). This finds a local minimum of the least-squares criterion for a model with r terms. A new term is introduced by an initial direction β_l, and the process continues until convergence. When M terms have been added, the least important (measured by $\sum_k |\gamma_{kj}|$) is dropped, the reduced model re-fitted and the process continued.

The precise algorithm for scatterplot smoothing is not intrinsic to SMART, and we have also use spline smoothers. Friedman used his own 'super-smoother', which uses a local linear fit to $k/2$ data points each to the left and right of the point x at which $g(x)$ is required. (The use of a rectangular window allows fast updating as x scans along.) The value of k is chosen from three possibilities by cross-validation at x, and then this choice is smoothly interpolated between the three smoothers.

See the glossary. Hwang *et al.* (1991, 1992a, b, 1994a, b, 1996) replaced the super-smoother by Hermite polynomials, which tends to produce smoother functions, and Roosen & Hastie (1994) have also tried smoothing splines. Most of the examples were tried both with Friedman's super-smoother and our own implementation of smoothing splines.

Hinging hyperplanes

Breiman's (1993) hinging hyperplanes are the special case $g_j(x) = [x]_+ = \max(x, 0)$ of PPR. These suffice to approximate arbitrary continuous functions, by the results of Section 5.7. It is attractive to use the same projection direction for both positive and negative versions of the function, when we have

$$g(\alpha + \beta^T \mathbf{x}) = a \max(\alpha + \beta^T \mathbf{x}, 0) + b \min(\alpha + \beta^T \mathbf{x}, 0).$$

A little thought shows that this is a linear function plus a multiple of $\max(\pm(\alpha + \beta^T \mathbf{x}))$, which suggests adding an overall linear function to $f(\mathbf{x})$ to avoid a wasteful fit using a projection and two components to recover a linear term.

These have universal approximation properties by the methods of Section 5.7, as using two such functions we can approximate a step ridge function. Further, we will have rate of convergence results for suitable smooth functions of order $O(1/\sqrt{r})$. (Breiman's Theorem 3 is more restrictive than our Proposition 5.3 in assuming a higher degree of smoothness although it may thereby use a smaller r.)

The attraction to Breiman of hinges was a fast way to fit one hinged hyperplane by least squares. This is used within the back-fitting approach of PPR. His reported CPU times seem comparable with the state-of-the-art in fitting neural networks, which have the advantage of using smooth (indeed, infinitely differentiable) functions. The fast algorithm is based on the idea that once we know which side of the hinge the points fall, fitting the hinged hyperplanes is simple (a linear least-squares problem, which can be updated as points change sets). Thus we choose an initial hinge, divide the points, fit, divide again on the fitted hinge and repeat. This gives a local minimum, so may be run from several starting hinges.

Breiman also considers variable selection, that is restricting β_j to pick just some features. This can ease the search (and could also be applied to full PPR) but does assume that the features have some individual meaning.

MARS (multivariate adaptive regression splines)

Friedman's (1991) multivariate adaptive regression splines allow for interactions more explicitly by

$$f(\mathbf{x}) = \alpha + \sum_{m=1}^{M} \beta_m \prod_{k=1}^{K_m} \phi_{km}(x_{v(k,m)}) \qquad (4.7)$$

where

$$\phi_{km} = [x - t_{km}]_+ \quad \text{and} \quad \phi_{k,m+1} = [t_{km} - x]_+$$

$[y]_+$ means $\max(y,0)$.

and t_{km} is an observed value of $x_{v(k,m)}$. The *degree* is the largest K_m; if this is one the model is additive. The functions were chosen to allow fast least-squares fitting algorithms. (Terms ϕ_{km} and $\phi_{k,m+1}$ are added together.) The components are splines corresponding to a penalty on the first rather than second derivative, discussed further in Section 4.3. Once again back-fitting is used, and terms are only considered for a high-order product if they are interactions of terms which occur in the current fit.

However, the splines are fitted by least squares, that is as regression splines rather than smoothing splines.

The precise details of the selection of terms are an 'engineering detail' discussed at length in Friedman's paper and its discussion. His algorithm has a forward phase followed by a backward phase. Terms are chosen to add or delete depending on a lack-of-fit criterion, which is the residual sum of squares divided by $(1 - C/n)^2$. Here n is the number of observations, and C is the number of parameters plus a multiple (2–4) of the number of terms M. The backwards elimination step aims to produce a model with comparable performance but fewer terms. In the additive case, over-fitting is avoided by reducing the number of knots rather than via a smoothness penalty.

This is related to the GCV penalty discussed in Section 4.3.

MARS produces fits which are continuous but not differentiable, which can be visually unappealing. Friedman suggests 'smoothing' out the piecewise linear functions ϕ_{kj}, for example by replacing them by cubics (and re-fitting the coefficients).

PIMPLE

Breiman's (1991) Π-method with program PIMPLE is another way to include interactions (and we will meet a third as interaction splines). As for MARS, Breiman considers a sum of products of splines, this time cubic splines, but the method appears to be geared towards small numbers of products of quite accurate fits, rather than as in MARS large numbers of inaccurate products. Back-fitting is used. The variable-selection strategy starts with a small number of knots in the splines, increases this and then considers deleting individual knots, using generalized cross-validation (Section 4.3) to decide when to stop adding and when to delete.

Breiman suggests that in three or more dimensions the product terms will be identifiable (up to their ordering in the model) and so can be interpreted; his discussants are less confident.

Examples

We first consider the data on *Leptograpsus* crabs. BRUTO selected FL, CL and CW to enter linearly, with a slightly non-linear term for RW but a roughly parabolic term for BD. MARS of degree 1 chose a single break of slope for each feature (and a small departure from linearity) except for BD. If interactions are permitted, several two-term interactions and one three-term interaction are chosen. Although the models are quite non-linear, the discriminant plots are virtually unchanged.

A difficulty (Ripley, 1994c) with our second approach (using LDA on the fitted values of a regression) is that a small number of outliers which are not fitted well can distort the within-group covariance matrix. Figure 4.2 shows the first two canonical variates for a projection-pursuit regression (with $r = 3$) for the crabs data, in which a number of outliers have appeared, together with bunching of points which are predicted particularly well. Sometimes using a robustified discriminant analysis will help, as Figure 4.2 shows.

Figure 4.2: Non-linear discriminants for the crabs data via projection pursuit regression. The left plot uses linear discrimination on the predictions, the right plot a robust version.

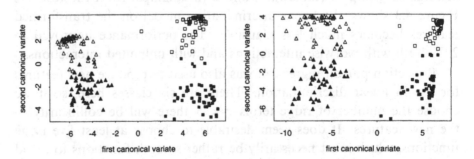

Figure 4.3 shows a series of non-linear discriminant plots for the forensic glass data. The BRUTO fit is very similar to linear discrimination. When we examine the terms of the additive model, we find

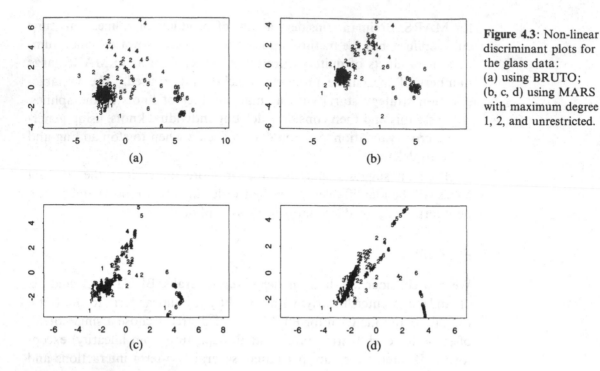

Figure 4.3: Non-linear discriminant plots for the glass data: (a) using BRUTO; (b, c, d) using MARS with maximum degree 1, 2, and unrestricted.

that iron oxide has been omitted, and the term in calcium is slightly non-linear; all the other terms are selected to enter linearly. The cross-validated error rate was 35%, a little better than the full linear discriminant analysis (but not significantly so).

MARS with maximum degree 1 fits an additive model. This introduces non-linear terms in RI, Mg, Si, K, Ca and Ba, and drops the rest. The fitted functions are shown in Figure 4.4. This achieves a cross-validated error rate of 32.2%.

As more interaction terms are introduced in MARS these principally distinguish groups 5 (tableware) and 6 (headlamps) from the rest. As Figure 4.3 shows, the linear discrimination model on the transformed features becomes much less plausible. The performance improved to 29% both with pairwise interactions and with unlimited interactions.

Projection pursuit regression was also used to produce new features for use in linear discrimination. There are six classes here, so if we choose the number of ridge terms $r < 5$ there will be collinearity in the new features. It does seem desirable to choose at least five ridge functions, which must necessarily be rather smooth functions to avoid over-fitting. This was borne out by our experiments, which achieved a cross-validated error rate of 42% if Friedman's super-smoother was used, but 35.5% if smoothing splines were used with a relatively large value of λ. Using just three ridge functions made the performance

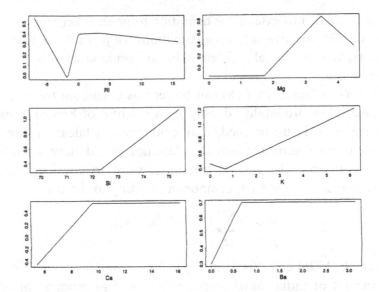

Figure 4.4: Fitted functions for the MARS additive model for the forensic glass data. The y scales are arbitrary.

considerably worse as the two types of window glass were barely separated.

Pima Indians

Fitting MARS and PPR models by least squares to the Pima Indians diabetes data did not improve the fit over linear methods, with a typical test-set error rate of 75/332.

4.2   Radial basis functions

We return to ways of parametrizing f or the log probabilities as a linear combination of basis functions. For a one-dimensional $\mathbf{x}$ splines are a natural choice. For higher dimensions we could use multidimensional splines (Section 4.3), but *radial basis functions* or RBFs have been more widely advocated (Powell, 1987, 1992; Broomhead & Lowe, 1988; Lee & Kil, 1988; Moody & Darken, 1989; Poggio & Girosi, 1990a, b; Musavi *et al.*, 1992).

RBFs are approximations of the form

$$y = \alpha + \sum_j \beta_j G(\|\mathbf{x} - \mathbf{x}_j\|) \tag{4.8}$$

for centres $\mathbf{x}_j$. Examples of G proposed include the Gaussian $G(r) = \exp -r^2/2$, the multiquadric $G(r) = \sqrt{(c^2 + r^2)}$ (Hardy, 1971, 1990; Kansa, 1990) and the thin-plate-spline function $G(r) = r^2 \log r$. (The RCU network of Reilly *et al.*, 1982, has $G(r) = I(r < r_0)$.) It is easy (but not as useful) to extend the definition to general kernels $G(\mathbf{x} - \mathbf{x}_j)$ or $G(\mathbf{x}, \mathbf{x}_j)$. For multivariate approximations we just take α and β_j to be vectors, that is we take different linear combinations of the same basis functions.

When G is Gaussian, (4.8) can be seen as extending the notion of approximating a probability density by a mixture of known densities. The norm $\| \ \|$ is unspecified, and could be Euclidean distance or a Mahalanobis distance, when the densities would have a common covariance matrix. In general we might want to consider different covariance matrices for each component, leading to the form

$$y = \alpha + \sum_j \beta_j G(\|A_j[\mathbf{x} - \mathbf{x}_j]\|). \tag{4.9}$$

Girosi *et al.* (1995) call this form *hyper basis functions*.

A variant of radial basis functions which is sometimes considered (Moody & Darken, 1989; Xu *et al.*, 1994) is the normalized form

$$y = \frac{\sum_j \beta_j G(\|\mathbf{x} - \mathbf{x}_j\|)}{\sum_j G(\|\mathbf{x} - \mathbf{x}_j\|)}. \tag{4.10}$$

Approximation properties

The class of radial basis functions has similarly good approximation properties to those of ridge-function methods (such as projection pursuit regression and feed-forward neural networks). There are many possible cases to consider, depending on whether the centres and G are fixed or adaptive (chosen for each dataset). We will only give a flavour of the results.

Park & Sandberg (1991) studied the subclasses of (4.9),

$$y = \alpha + \sum_j \beta_j G(\|[\mathbf{x} - \mathbf{x}_j]\|/\sigma_j) \tag{4.11}$$

$$y = \alpha + \sum_j \beta_j G(\|[\mathbf{x} - \mathbf{x}_j]\|/\sigma) \tag{4.12}$$

in which G is fixed but the covariance matrices are proportional, including the special case in which the σ_j are identical. (This is still more general than (4.8).) Provided G is continuous, bounded and with a finite and non-zero integral, they show that the class (4.12) is dense in L_p for every $p \in [1, \infty)$, and can uniformly approximate continuous functions on compact sets. (This is sometimes referred to as the 'universal approximation' property.) Thus for any function $f(\mathbf{x})$ there is a set of centres $(\mathbf{x}_j)$ and a $\sigma > 0$ such that (4.12) is close to f

in the appropriate norm (L_p or the maximum difference on a compact set).

The uniform approximation on compact sets of continuous functions for Gaussian RBFs of the form (4.11) can be shown via the Stone–Weierstrass theorem (Girosi & Poggio, 1990; Hartman *et al.*, 1990). Girosi & Poggio (1990, Appendix C) state a more general result for (4.8) and piecewise continuous G arising from (4.14) below, but their proof is of pointwise rather than uniform convergence. (The proof can be completed by consideration of the discretization error in a Riemann integral, using the modulus of (uniform) continuity.) Thus we still have uniform approximation with a fixed basis function G.

Results on the rate of approximation are available for smooth enough target functions f (Girosi & Anzellotti, 1993), using the methods described in Section 5.7. They considered the class (4.8) for $G \in L_2(\mathbb{R}^p)$ and the class of targets f which are in $L_2(\mathbb{R}^p)$ (for Lebesgue measure) and can be expressed in the form

$$f(x) = \int_{\mathbb{R}^p} G(x - y)\, \mathrm{d}\lambda(y)$$

for a signed measure λ of bounded total variation (that is, the difference between two finite measures). Thus functions in the class are at least as smooth as G; for example, for the Gaussian G the sharpness of peaks is firmly controlled. Then the L_2 rate of convergence is bounded by $\|\lambda\|/\sqrt{n}$ for n terms, when G is scaled to $\|G\| = 1$. Further, if G is in $H^{s,2}$ for $s > p/2$, there is uniform convergence at rate $1/\sqrt{n}$.

Here $H^{s,2}$ is the space of L_2 functions all of whose derivatives up to order s are in L_2 and so includes the Gaussian.

As class (4.12) corresponds to the rescaled kernel G, the results of Girosi & Anzellotti also apply to target functions of their class for any rescaling of G (although the constant in the bound will vary with the target function). This class of functions is, however, strictly smaller than L_2. (Consider the function $x^{-1/4} I(|x| < 1)$.)

Fitting

Finding the coefficients α and β_j in (4.8) to (4.12) is usually done by least squares and is easy; for fixed centres x_j and fixed scale parameters σ_j these are linear regression equations, so least-squares fitting reduces to solving linear equations.

This leaves the issues of finding the centres and any scale factors. One possibility is to take every training example as a centre, but this can lead to over-fitting. This suggests taking a representative collection of training examples, for example a random or stratified sample (e.g. Lee, 1991). There are other ways to take a representative collection of points not necessarily within the training set. For example, the k-means algorithm (Section 9.3) of cluster analysis chooses k points in $\mathscr{X}$ to minimize the sum of squares from each training point to the nearest of the k points. These clustering algorithms do not take the classes of the training examples into account.

Musavi *et al.* (1992) designed an agglomerative clustering algorithm (see Section 9.3) which only merges clusters of points with the same class. The cluster means then provide the centres for the basis functions. The algorithm will choose the number of clusters and hence the number of basis functions, but has a 'clustering parameter' α which controls the agglomeration process. Other ideas are the SOM (Section 9.4) and LVQ (Section 6.3) methods of Kohonen. LVQ aims for cluster centres which provide a good set for nearest-neighbour classification.

Moody & Darken (1989) considered finding centres by k-means (using the method of (6.9) on page 202) and this has often been used. Note that this (and most other ways of choosing representatives) depends on the choice of a metric in $\mathscr{X}$, and so is most appropriate for (4.12) or (4.8) rather than for (4.9).

Moody & Darken explored choosing the scale functions σ_j in (4.11) from a heuristic using the P-nearest neighbour distance. Musavi *et al.* (1992) chose the matrices A_i in (4.9) (equivalently the covariance matrix of the Gaussian basis function) by fitting a maximal ellipsoid around the centre which includes no training examples of another class, and taking this as an isodensity surface containing 95% of the probability. Note that this procedure is very sensitive to outliers and faulty training-set classifications.

Leonard *et al.* (1992) extend the Moody–Darken method by adding further outputs designed to signal extrapolation and to give confidence intervals for the predictions.

It is of course possible to minimize the least-squares fit over the parameters σ or σ_j, and even over the centres of the basis functions. As the least-squares fitting is partially linear, this will be computationally quite feasible and seems to be the proposal of Poggio & Girosi (1990b), although considered and rejected as too demanding by Moody & Darken. Wetterschereck & Dietterich (1992) considered optimizing over both parameters σ_j and centres. On their (single) example they found that optimizing over centres was particularly important to achieve a competitive performance by RBF methods, but that choosing the σ_j by a local heuristic was counter-productive.

Yet another idea is to choose cluster centres from a large class of candidates (for example, all training examples) by a stepwise regression procedure. This was the idea of Chen *et al.* (1991). Any sensible selection strategy could be used, and the method extended to choosing from a small number of scale factors at each candidate centre.

Poggio & Girosi (1990b, §IV.D) consider alternatives to least-squares fitting for 'unreliable data'. They replace the square function in the sum of squares or in (4.13) below by

$$V(u) = u^2 - \frac{1}{\beta} \log\left[1 + \exp{-\beta(\epsilon - u^2)}\right]$$

with ϵ positive and β large. This enforces a quadratic penalty only up to about $\pm\sqrt{\epsilon}$, and has the same effect as a re-descending M-estimator in robust regression (Huber, 1981; Hampel *et al.*, 1986; Rousseeuw & Leroy, 1987).

This plethora of methods provides a difficulty in assessing RBF methods; no two workers use the same class of RBFs and method of fitting. There is a range of compromises being made between speed of fitting and accuracy of approximation. RBFs are often claimed to be much faster than ridge-function methods on the basis of their partial linear fitting, yet if the centres are varied (or regularization used; Section 4.3) their computational load seems as large as their competitors. However, there is considerable scope for inspired choices of centres in specific problems (as in Roberts & Tarassenko, 1995).

Potential functions

The concept of *potential functions* has a variety of meanings within the pattern recognition literature. To some users it is synonymous with kernel methods (Section 6.1). Its origins (Bashkirov *et al.*, 1964; Aizerman *et al.*, 1964a, b, 1965; Braverman, 1965; Arkedev & Braverman, 1966) are close to those of radial basis functions. Suppose we consider each observation $\mathbf{x}_i$ as having some 'charge' q_i and measure the 'potential' at another point $\mathbf{x}$. It will be of the form

$$f(\mathbf{x}) = \sum_i q_i K(\mathbf{x}; \mathbf{x}_i)$$

and this provides another way to approximate by a smooth function (except perhaps at the data points). The 'potential' is to be chosen by the user, and so could subsume both kernel methods and radial basis functions.

The potential-function classifier is trained to attempt to correctly classify all the samples of the training set by adjusting the q_i by a

perceptron-like procedure. Indeed, it can be seen as applying perceptron
ideas to generalized discriminant functions.

Potential functions can also be used to approximate probability
density functions (Kashyap & Blaydon, 1968; Tsypkin, 1966).

The method is considered in more detail in books by Meisel (1972)
and Young & Calvert (1974). It seems to have disappeared from view
until revived as the study of radial basis functions.

4.3   Regularization

An alternative to reducing the number of basis functions in fitting
RBFs is to allow one per training example, but to control directly the
smoothness of the fitted function, as is done for smoothing splines in
one dimension. Indeed, since splines are so useful in one dimension,
they might appear to be the obvious method in more. In fact they turn
out to be rather restricted and little used.

This section is dominated by least-squares fitting, since regularization has been most explored in approximation theory. We know of
no exact solutions (such as smoothing splines) for other forms of our
problem such as fitting multiple logistic models by maximum likelihood.

Bishop (1991) considers (4.12) with a centre at every example in the
training set, but adds a penalty term when fitting. Let $f(\mathbf{x})$ denote the
approximating RBF. Then the term to be minimized is

$$\sum_i \| y_i - f(\mathbf{x}_i) \|^2 + \lambda C(f), \qquad C(f) = \sum_i \sum_{j,k} \left(\frac{\partial^2 f_k(\mathbf{x}_i)}{\partial x_j^2} \right)^2$$

where $_i$ refers to the ith training example. This is an example of a
general process termed *regularization* in which other penalties $C(f)$ may $C(f)$ is sometimes
be considered. The parameter λ controls the smoothness and degree   called a *stabilizer*.
of fit. For $\lambda = 0$ the fits will usually be exact (from interpolation
properties of RBFs) and as $\lambda \to \infty$ the fitted function becomes flat.
Bishop chooses λ by trial and error. The sum over examples can be
seen as an approximation to an integral over $\mathcal{X}$.

Adding a penalty $C(f)$ has a Bayesian interpretation. The first term
is proportional to the log-likelihood if we assume that the noise variance
σ_e^2 in a regression is known, so if we take a prior over functions f
which is proportional to $\exp -2\lambda \sigma_e^2 C(f)$, minimizing a penalized sum
of squares is equivalent to maximizing the posterior density over f.
This is a MAP estimator (see Section A.1) and is widely used in image
analysis (following Geman & Geman, 1984). However, the warnings
about MAP estimation given in Section A.1 must be borne in mind.

Suppose we add a penalty of the form $C(f) = \|Pf\|^2$ for a differential operator P. We can then consider the function f minimizing the penalized sum of squares

$$\sum_i \|y_i - f(\mathbf{x}_i)\|^2 + \lambda \|Pf\|^2 \tag{4.13}$$

over all (smooth enough) functions f, not just those represented by a form of RBFs. Exactly as for smoothing splines, the general solution (Poggio & Girosi, 1990b; Wahba, 1990) is of the form

$$\hat{f}(\mathbf{x}) = \sum_i c_i G(\mathbf{x}; \mathbf{x}_i) + n(\mathbf{x}) \tag{4.14}$$

where G is the (symmetric) Green's function of $\widehat{P}P$, $\widehat{P}$ is the adjoint differential operator, and $n(\mathbf{x})$ is a function in the null space of P. If the operator is translation or rotation equivariant, so will G be. Thus equivariance under rigid motions leads to Green's functions of the radial basis function form. The coefficients c_i satisfy the linear equations

$$y_j = \sum_i c_i G(\mathbf{x}_i, \mathbf{x}_j) + \lambda c_j \tag{4.15}$$

and $n(\mathbf{x})$ is chosen by least squares.

The Gaussian RBF arises from the (non-intuitive) penalty functional

$$\sum_{n=0}^{\infty} (-1)^n \frac{\sigma^{2n}}{n!\, 2^n} \sum_{i_1,\dots,i_n} \int \left[\frac{\partial^n f(\mathbf{x})}{\partial x_{i_1} \cdots \partial x_{i_n}} \right]^2 d\mathbf{x}$$

(Poggio & Girosi, 1990b, pp. 95–96), and the null space gives the constant α. The penalty is more obvious when expressed in the Fourier domain, and Girosi *et al.* (1995) consider the class of penalties

$$C(f) = \int \frac{|\tilde{f}(\mathbf{s})|^2}{\tilde{G}(\mathbf{s})} \, d\mathbf{s} \tag{4.16}$$

for some positive symmetric function $\tilde{G}$ that tends to zero as $\|\mathbf{s}\| \to \infty$. Here the tilde denotes Fourier transformation, and it turns out (Dyn, 1987; Madych & Nelson, 1990; Girosi *et al.*, 1995) that $\tilde{G}$ is the Fourier transform of the function $G(\mathbf{x} - \mathbf{x}_i)$ in (4.14). In this formulation the Gaussian RBF arises from $\tilde{G}(\mathbf{s}) = \exp -\beta \|\mathbf{s}\|^2$.

Although regularization is theoretically interesting, it demands the solution of large systems of linear equations (4.15). In the case of smoothing splines in one dimension this is a banded system and can be solved quickly, but in general it will take $O(n^3)$ operations for n

examples and so be prohibitively slow. It remains possible to use a smaller set of basis functions as in the previous subsection *and* to use a penalty functional to control the smoothness. We could use a general-purpose optimizer to minimize the penalized measure of fit over all parameters rather than solve (4.15) within a loop.

Although we have considered only least-squares problems in this subsection, similar considerations apply to other deviance functions, since they can be approximated locally at the optimum by a weighted sum of squares function, and this continues to have a solution of the form (4.14) with appropriate modifications to (4.15).

Another approach to regularization is to add noise during training (see, for example, Sietsma & Dow, 1991). Suppose we add a moderate amount of white noise to the target values y_i. Then (Webb, 1994; Bishop, 1995b) a second-order Taylor expansion shows that the effect is approximately (despite Bishop's title) the same as using the regularizer

$$C(f) = \sum_{i,j} \mathsf{E}\left[\left(\frac{\partial f_i(\mathbf{X})}{\partial X_j}\right)^2 + \tfrac{1}{2}[f_i(\mathbf{X}) - Y_i]\frac{\partial^2 f_i}{\partial X_j^2}\right].$$

Bishop shows that for small added noise *and* a good fit the first term dominates (since $\mathsf{E}[f_i(\mathbf{X}) - Y_i]$ will be small), so the regularization is mainly by the expected squared length of the first derivative. With non-least-squares error functions, local linearization gives a weighted least-squares approximation and hence an expected weighted sum of $\left(\partial f_i(\mathbf{X})/\partial X_j\right)^2$.

Multidimensional splines

Smoothing splines in one dimension arise from the regularization penalty $\int g''(u)^2\, du$ on the sum of squares at the data points. This penalty does not generalize immediately to higher dimensions, and a suitable generalization took some years to emerge. *Thin-plate splines* use the penalty

$$\iint \frac{\partial^2 f(x,y)}{\partial x^2} + 2\frac{\partial^2 f(x,y)}{\partial x \partial y} + \frac{\partial^2 f(x,y)}{\partial y^2}\, dx\, dy$$

on $\mathbb{R}^2$ which is invariant under rigid motions, and

$$\sum_{\alpha_1 + \cdots + \alpha_n} \frac{m!}{\alpha_1! \cdots \alpha_n!} \int \left[\frac{\partial^m f(\mathbf{x})}{\partial x_1^{\alpha_i} \cdots \partial x_n^{\alpha_n}}\right]^2 d\mathbf{x}$$

in $\mathbb{R}^p$. We have to take $2m > p$ to ensure this is a smoothing penalty, since a radially symmetric bump of width h will have penalty proportional to h^{2m-p} which should go to infinity as $h \to 0$.

The solution to the penalized least-squares problem is of the form

$$f_\lambda(\mathbf{x}) = \phi(\mathbf{x}) + \sum_i c_i G(\|\mathbf{x} - \mathbf{x}_i\|)$$

for a polynomial ϕ of total degree at most $m - 1$ and

$$G(r) = \begin{cases} r^{2m-p} \log r & \text{if } 2m - p \text{ is even} \\ r^{2m-p} & \text{otherwise} \end{cases}$$

(Duchon, 1977; Meinguet, 1979; Wahba, 1990, p. 33). Thus univariate smoothing splines and our two-dimensional example correspond to $m = 2$. The higher-dimensional cases lose the computational simplicity of smoothing splines since the matrices are no longer banded. Note that $m = p = 1$ gives a penalty of the integrated square of the first derivative, and the solution is piecewise linear, as used in MARS. We can only use a second-derivative penalty in $p \leqslant 3$ dimensions, and even in two dimensions there are edge effects to consider (Green & Silverman, 1994, Chapter 7).

Additive models, including those with interaction terms such as MARS, can be put within the spline framework by choosing a suitable regularizer (Wahba, 1990, Chapter 10). Summing penalties gives an additive model, and including terms which involve two or more variables gives rise to interaction terms, most simply for $m = 1$ (since for $m > 1$ there are interactions between polynomials and splines to consider). These are known as *additive* and *interaction* splines respectively; the latter are considered by Barry (1986), Gu & Wahba (1991) and Gu *et al.* (1989).

Tensor product splines are similar to MARS in that the interaction terms are products of functions of a single feature. They arise from a penalty of the form (4.16) in which $\widetilde{G}$ is a product of functions of a single feature.

Fitting additive models with interaction terms via splines gives rise to a potentially large number of parameters λ to be considered, one for each term. For $m = 2$ there are 22 λ's to choose even for $p = 4$, which is beyond current methods to select.

Asymptotic theory

Once a regularization term is added to the fitting criterion, for example the deviance, the asymptotic distribution theory of the parameter estimate will be changed, whether or not the model is true, except in the unlikely event that the penalty is completely ineffective. From their intended purpose, we would expect the effect of regularization terms to be to introduce bias, even asymptotically, but to decrease the variance. This raises the possibility of juggling the amount of the penalty so that the bias decreases as $n \to \infty$ but the variance remains under control, indeed decreases to zero at a rate close to $O(1/n)$.

This general programme is often considered 'non-parametric' since arbitrarily complex models will be needed to fit a true model outside the assumed class of functions. (For instance, consider Gaussian RBF functions with centres at data points, and approximating an exactly linear function.) The results of Section 2.8 provide sufficient bounds on the complexity of the model needed for n points which can allow the penalty to be varied with n, and a heuristic outline is given by Geman *et al.* (1992) with a more detailed account by White & Woolridge (1991) and White (1990). As an example of these techniques we can consider a projection pursuit regression with the number r of terms growing at $O(n^{1-\epsilon})$ for $\epsilon > 0$, and control the smoothness of the terms to grow at $O(\log n)$, thereby achieving risk consistency.

This asymptotic programme has little to do with the performance in realistic situations, since there is not usually a large enough training set relative to the complexity of the fitted classifier. For useful results we need an approximation, not a limit theorem. Moody (1991, 1992), Liu (1993, 1995) and Murata *et al.* (1993, 1994) attempt to provide this in a limited set of circumstances. They assume a regularization penalty proportional to n (unrealistic according to the previous paragraph, which suggests it should grow less rapidly, but much better than constant regularization). Then we can use the results of Propositions 2.2 and 2.3 in Section 2.2 with $L_1 = E_1 + \lambda C_1$ replacing $-\log p$, where E_1 is a term in the fit criterion (for example the contribution of an example to the negative log-likelihood) and $C_1 = C/n$. This leads to

$$\mathrm{NIC} = 2[E(\widehat{\theta}) + \lambda C(\widehat{\theta}) + p^*]$$

where $p^* = \mathrm{trace}\left[KJ^{-1}\right]$ and

$$J = \mathsf{E}\frac{\partial^2 L_1(\theta_0; X)}{\partial\theta\,\partial\theta^T} \quad \text{and} \quad K = \mathsf{Var}\frac{\partial L_1(\theta_0; X)}{\partial\theta}.$$

If the penalty $\lambda C(\widehat{\theta})$ measures the smoothness of the fitted function, it is the same for both the training and test sets, so the expected difference in E between test and training sets of size n is also p^* (which is Moody's formulation). Sometimes if the penalty is an integral over the feature space of the form $C(\theta) = \int_{\mathscr{X}} G[f(\mathbf{x};\theta)]\,\mathrm{d}\mathbf{x}$ it is replaced by $\widehat{C}(\theta) = \frac{1}{n}\sum G[f(\mathbf{X}_i;\theta)]$, which has expectation $C(\theta)$ for both training and test sets. Thus a penalty of the form $\widehat{C}(\theta)$ can also be deleted from NIC.

Moody (1991, 1992) calls 'the effective number of parameters', p_{eff}, the estimate of p^* obtained by replacing the expectations in J and K by averages over the training set, and θ_0 by $\hat{\theta}$. There is a potential bias in using the training set to estimate p^* since it shows how many effective parameters are needed to approximate the true distribution over a set of size n, and that could be much smaller than the number needed to approximate the whole distribution. Further, as we suggested in Section 2.2, a divisor $n - p$ for the estimate of K would be more appropriate.

We now return to the approximation of Bayes factors in Section 2.6. We can always regard $p(\theta) \propto -\lambda C(\theta)$ as a prior density for the parameters θ, although it may well be an improper prior (with infinite integral). Then choosing θ to minimize the negative log-likelihod plus $\lambda C(\theta)$ is equivalent to MAP estimation, and we can use (2.43) as an approximation to $\log p(\mathcal{T} \mid M)$.

Choosing λ

To actually use regularization, we have to select all the λ's. If they are derived from Bayesian priors as in Section 5.5, they are already determined, but otherwise they are treated as free parameters (and are also so treated in some empirical Bayes schemes). We can use the methods of Section 2.6, in particular cross-validation. However, this has two disadvantages. The first is computational unless updating schemes are known (as for univariate smoothing splines; Silverman, 1985). The second is that cross-validation is not invariant under orthogonal transformations of the data vector in a regression problem. Generalized cross-validation (GCV) computes the average adjustment of the fit on leaving each point out, rather than producing the adjustment for each point (Craven & Wahba, 1979). Thus in a regression problem we have

$$\text{GCV}(\lambda) = \frac{1}{N} \frac{\sum_{i=1}^{N} \| y_i - f_\lambda(\mathbf{x}_i) \|^2}{[1 - \text{trace}(A(\lambda))/N]^2}$$

where A_λ is the matrix mapping the data vector to the vector of the fitted values. Where several λ's are involved, A and GCV become a function of all of them, and a simultaneous minimization is needed; Wahba (1990) reports using up to 10 λ's for additive interaction splines. Except for univariate splines, the computation is dominated by the computation of $A(\lambda)$. Hastie & Tibshirani (1990, §9.4.3) suggest approximating trace $A(\lambda)$ by one plus the sum of the traces of the matrices for the individual smoothers minus one, which seems *ad hoc* but is much faster. This is used in BRUTO.

For non-least-squares problems there seems no general alternative to V-fold cross-validation. Hastie & Tibshirani (1990, §6.9) propose a version of GCV which weights the deviance by the divisor of the least-squares form, but this seems unsupported theoretically except as a local weighted least-squares approximation (given by Wahba, 1990, p. 113).

5

Feed-forward Neural Networks

A great deal of hyperbole has been devoted to neural networks, both in their first wave around 1960 (Widrow & Hoff, 1960; Rosenblatt, 1962) and in their renaissance from about 1985 (chiefly inspired by Rumelhart & McClelland, 1986), but the ideas of biological relevance seem to us to have detracted from the essence of what is being discussed, and are certainly not relevant to practical applications in pattern recognition. Because 'neural networks' has become a popular subject, it has collected many techniques which are only loosely related and were not originally biologically motivated. In this chapter we will discuss the core area of feed-forward or 'back-propagation' neural networks, which can be seen as extensions of the ideas of the *perceptron* (Section 3.6). From this connection, these networks are also known as *multi-layer perceptrons*.

A formal definition of a feed-forward network is given in the glossary. Informally, they have units which have one-way connections to other units, and the units can be labelled from inputs (low numbers) to outputs (high numbers) so that each unit is only connected to units with higher numbers. The units can always be arranged in layers so that connections go from one layer to a later layer. This is best seen graphically; see Figure 5.1. Each unit sums its inputs and adds a constant (the 'bias') to form a total input x_j and applies a function f_j to x_j to give output y_j. The links have *weights* w_{ij} which multiply the signals travelling along them by that factor. The input units are there just to distribute the inputs, so have $f \equiv 1$. Thus a network such as Figure 5.1 represents the function

What we denote by w_{ij}, the weight on the link from i to j, is more often denoted w_{ji}.

$$y_k = f_k\left(\alpha_k + \sum_{j \to k} w_{jk} f_j\left(\alpha_j + \sum_{i \to j} w_{ij} x_i\right)\right) \qquad (5.1)$$

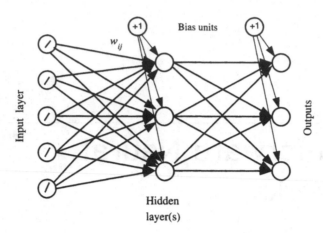

Figure 5.1: A generic feed-forward network with a single hidden layer. To avoid over-crowding bias units are shown for each layer, but they can be the same unit.

from inputs to outputs. The functions f_j are almost invariably taken to be linear, logistic (with $f(x) = \ell(x) = e^x/(1 + e^x)$) or threshold functions (with $f(x) = I(x > 0)$). A variant is to take hyperbolic tangent units with $f(x) = \tanh(x) = (e^x - 1)/(e^x + 1) = 2\ell(x) - 1$, but this only introduces a linear transformation which can be absorbed into the weights (except at the output units). Only threshold units give a genuine multi-layer extension of the perceptron, and such networks were considered in Rosenblatt's work.

The general definition allows more than one hidden layer, and it also allows 'skip-layer' connections from input to output. If all units in a layer have the same function f_h or f_o, we have

$$y_k = f_o\left(\alpha_k + \sum_{i \to k} w_{ik}x_i + \sum_{j \to k} w_{jk}f_h\left(\alpha_j + \sum_{i \to j} w_{ij}x_i\right)\right). \qquad (5.2)$$

The bias terms can be eliminated by introducing a new unit 0 (the *bias unit*) which is permanently at $+1$ and connected to all other units. We set $w_{0j} = \alpha_j$. (This is the same idea as incorporating the constant term in the design matrix of a regression by including a column of 1's.) This is shown in Figure 5.1. The general form is then

$$y_k = f_o\left(\sum_{i \to k} w_{ik}x_i + \sum_{j \to k} w_{jk}f_h\left(\sum_{i \to j} w_{ij}x_i\right)\right). \qquad (5.3)$$

Note that if we have logistic units in the hidden layer, adding skip-layer connections is not really more general, since we can add another unit per output in the hidden layer with input weights w_{ik}/G and output weight G to just unit k. Then for large G we only use the central, linear, part of the range of the logistic function. However, skip-layer connections can be easier both to implement and to interpret.

A neural network with a single logistic output unit can be seen as a non-linear extension of logistic regression. With many logistic output units, it corresponds to linked logistic regressions of each class *vs* the others.

The terminology of neural networks can be very confusing: Figure 5.1 is sometimes referred to as having three layers (which seems visually correct), two layers (as the input layer does nothing) and one hidden layer (as the states of the units in the central layer cannot be inspected from outside the 'black box'). We will refer to the inputs, the outputs and the hidden layer, since we will almost always have only one hidden layer.

We will extend our notation to allow every unit j to have an input x_j and output y_j. The inputs to the whole network are the inputs to the input units, and the outputs from the whole network are those of the output units. The signal paths through the network are determined by the equations

$$y_j = f_j(x_j), \qquad x_j = \sum_{i \to j} w_{ij} y_i. \tag{5.4}$$

We can even drop the condition on the sum by defining w_{ij} to be zero for all non-existent links. When programming it is useful to number the units by layer, so all units in the first layer precede all those in the first hidden layer and so on. Then we know $w_{ij} = 0$ unless $i < j$.

We will briefly consider how such functions came to be suggested, and the theory which shows that they form large and flexible classes of functions. However, in practice the main issues are how the parameters, the weights, should be chosen, and how the *architecture* (the number of layers and the number of units in each, as well as which connections to include) is selected.

5.1 Biological motivation

The original biological motivation for feed-forward networks stems from McCulloch & Pitts (1943) who published a seminal model of a neuron as a binary thresholding device in discrete time, specifically that

$$n_j(t) = I\left(\sum_{i \to j} w_{ij} n_i(t-1) > \theta_i\right)$$

the sum being over neurons i connected to neuron j. Here $n_i(t)$ is the output of neuron i at time t and $0 < w_{ij} < 1$ are attenuation weights. Thus the effect is to threshold a weighted sum of the inputs at value θ_i. Real neurons are now known to be more complicated; they have a graded response rather than the simple thresholding of the McCulloch–Pitts model, work in continuous time, and can perform more general non-linear functions of their inputs, for example logical functions. Readers may worry that this model can only allow non-negative weights. This is so, but neural systems have both excitatory and inhibitory connections, so whereas each connection can effectively have a weight of just one sign, it would be possible to envisage both an excitatory and an inhibitory connection, to simulate the effects of weights of either sign.

There is also a wider motivation based on human abilities in pattern recognition. A driver can recognize that a traffic light has changed to red and take the appropriate action (or decide not to) in well under one second. Neurons are rather slow devices by the standards of electronic computers, with messages travelling at up to 100 m/s, and of low bandwidth—perhaps 100 bits/s (MacKay & McCulloch, 1952). This allows rather few steps in the computation, most famously expressed in Feldman's (1985) concept of a 'one hundred step program', since there is time for at most 100 steps within a human reaction time. Human brains must make up for this lack of speed by massive parallelism, and given the speed of messages this parallel computation must be highly distributed. Brain scientists currently envisage vision as being performed in a series of layers in the brain, which naturally suggests a layered architecture for the distributed computation. (There is some evidence for feedback from later layers to earlier layers in human vision, but this is recent and controversial.)

Further, human beings can *learn* tasks such as driving; it seems extremely implausible that we are pre-programmed to recognize red traffic lights or (except for advocates of Lamarckian evolution) that this is an inherited adaptation. Something in our distributed neural computing system changes through experience. This could involve adding or removing connections or units, but most emphasis has been on exploring the consequences of changes in the *strengths* of connections rather than their topology.

The book of Hebb (1949) has been very influential in thinking about how connection strengths should be changed. His approach is not quantitative, and so is all-embracing. He considers that the connection strength between units should be increased if they are activated together, and this is often taken to suggest reinforcing the connection proportionally to the product of their simultaneous activations, that is

$$w_{ij} \leftarrow w_{ij} + \eta\, y_i y_j.$$

This is known as the *Hebbian learning rule*, widely used in other forms of neural network.

This idea that there could be a simple adaptation rule for connection weights used universally in learning found fruit in the experiments of Rosenblatt (1957, 1958, 1962) and Widrow (Widrow & Hoff, 1960). As discussed in Chapter 3, they used very simple units, in one or few layers, with weights expressed by motor-driven potentiometers. Widrow–Hoff learning (also known as the delta rule) was an iterative algorithm of the reinforcement type to fit a linear regression or discriminant. This seems to have been the inspiration for the rules used for 'learning' in neural networks in their renaissance. Similar rules were proposed for learning in potential function systems by Aizerman *et al.* (1964a, b, 1965).

5.2 Theory

Equations (5.1) and (5.2) are thus far merely suggestive ways to parametrize multidimensional input–output relationships. They are rather general classes of functions, something which took a long time to be appreciated. Cybenko (1989), Funahashi (1989), Hornik *et al.* (1989), Carroll & Dickinson (1989), Stinchcombe & White (1989) and many later authors have shown that neural networks with linear output units and a single hidden layer can approximate any continuous function f uniformly on compacta, by increasing the size of the hidden layer, and this implies many other types of approximation. There are also some results on the rate of approximation (how many units are needed to approximate to a specified accuracy), but as always with such results they are no guide to how many units might be needed in any practical problem. These results are in fact rather easy to prove, very much easier than most published proofs, so we give complete proofs in Section 5.7.

A heuristic reason why feed-forward networks might work well with modest numbers of hidden units is that the first stage allows a projection onto a subspace of $\mathscr{X}$ of much lower dimensionality, within which the approximation can be performed. In this feed-forward neural networks share many of the properties of projection pursuit regression (PPR; Section 4.1). Indeed, for theoretical purposes the two are essentially equivalent. We consider only linear output units, as clearly logistic, softmax or other transforms can be applied to the outputs of either family. Clearly (5.1) is a special case of PPR (4.5), taking the smooth functions to be logistic functions. Conversely, if we have a general PPR

Earlier, Cybenko (1988) had showed that two hidden layers suffice.

Uniform convergence on compacta is defined in the glossary.

$$y_k = \alpha_k + \sum_j g_j(\alpha_j + \sum_i \beta_{ji} x_i)$$

we can approximate each smooth function g_j as a sum of shifted logistic functions (a special case of the results of Section 5.7), so

$$g_j(x) \approx \sum_l \gamma_{jl} \ell(\delta_{jl} + \zeta_{jl} x)$$

$$y_k \approx \alpha_k + \sum_{j,l} \gamma_{jl} \, \ell(\alpha'_{jl} + \sum_i \beta'_{jli} x_i)$$

and on writing j, l as a single index the right-hand side is seen as a special case of (5.1).

5.3   Learning algorithms

Knowing that an approximation exists is useless without some way to find it, and it was this step which held up research in neural networks for many years. The original idea of the Rumelhart–McClelland school was to fit the parametrized function by least squares. Suppose we have examples $(\mathbf{x}^p, \mathbf{t}^p)$, and that the output of the network is $\mathbf{y} = f(\mathbf{x}; \mathbf{w})$. Then the parameter vector $\mathbf{w}$ is chosen to minimize

$$E(\mathbf{w}) = \sum_p \|\mathbf{t}^p - f(\mathbf{x}^p; \mathbf{w})\|^2 \qquad (5.5)$$

as would be done in non-linear regression (Bates & Watts, 1988; Seber & Wild, 1989). (Note that this is a sum of squares over both output units and input examples.) As this is a minimization problem, we can use general algorithms from unconstrained optimization (Section A.5), and we shall see that this seems the most fruitful approach.

Note that $E(\mathbf{w})$ is a differentiable function only for differentiable units, and *from now on we assume differentiable units, thereby excluding threshold units*. Indeed, it was the change from the threshold units of genuine multi-layer perceptrons to logistic units which enabled effective algorithms to train these networks to be found. These algorithms all require the gradient of $E(\mathbf{w})$ with respect to the weights.

The Rumelhart–McClelland group used a form of steepest descent to reduce (5.5), with update rule

$$w_{ij} \leftarrow w_{ij} - \eta \frac{\partial E}{\partial w_{ij}} \qquad (5.6)$$

and since the partial derivative can be written in the form (see the next subsection)

$$\frac{\partial E}{\partial w_{ij}} = \sum_p y_i^p \, \delta_j^p$$

(originally with the sign of δ reversed) this has become known as the *generalized delta rule*. (Here and later the superfix ^{p} refers to calculations involving example p.) Further, as the δ's can be computed from output to input across the network (see the next subsection) both the process of calculating the derivatives and the descent algorithm are known as *back-propagation*.

Alternative discrepancy functions for logistic regressions have been considered in Sections 2.3 and 3.5; most of these have been re-discovered within the neural networks literature. The conditional log-likelihood (2.31) for a two-class problem has been widely suggested (Solla *et al.*, 1988; Bichsel & Seitz, 1989; Hinton, 1989a; Bridle, 1990a, b; Holt & Semnani, 1990; Spackman, 1992; van Ooyen & Nienhuis, 1992). This is often summed over multiple logistic output units to give

$$E = \sum_p \sum_k \left[t_k^p \log \frac{t_k^p}{y_k^p} + (1 - t_k^p) \log \frac{1 - t_k^p}{1 - y_k^p} \right] \quad (5.7)$$

the terms in $t \log t$ being chosen so that $E \geqslant 0$ with equality only for a perfect fit.

The log-linear approach to classification gives rise to what Bridle (1990a, b) termed *softmax*. This is no different from the multiple logistic models considered several times in earlier chapters, but for convenience we will recap the notation here. We have

$$p(k \mid \mathbf{x}) = \frac{\exp g_k(\mathbf{x}; \widehat{\theta})}{\sum_j \exp g_j(\mathbf{x}; \widehat{\theta})} \quad (5.8)$$

and the classifier chooses the class maximizing $p(k \mid \mathbf{x})$ and hence $g_k(\mathbf{x}; \widehat{\theta})$. To use this with a neural network, we take the y_k to the outputs of a network with linear output units, and compute probabilities by

$$p_k = \frac{\exp y_k}{\sum_j \exp y_j}. \quad (5.9)$$

Minus the log-likelihood for the multinomial distribution is then

$$E = \sum_k t_k \log \frac{t_k}{p_k}$$

summed as usual over examples. The targets t_k will usually be one for the correct class, and zero for the others, and the probabilities are given by (5.9) computed from (y_k). From now on we will assume that exactly one of the targets is one, all the others are zero, so $\sum_j t_j = 1$. The outputs (y_k) can all be changed by the same additive constant without changing the probabilities or the fit, so there is a degree of redundancy. It is often convenient to take $g_k(x; \theta) \equiv 0$ for some preferred class k.

The error criteria of robust statistics (Huber, 1981; Hampel *et al.*, 1986) may be used to replace least squares, as in Chen & Jain (1994). Liu (1994) uses a t distribution for regression errors, and hence a different robust error criterion.

Back-propagation

Recall that each unit has input $x_j = \sum_{i \to j} w_{ij} y_i$ and output $y_j = f_j(x_j)$. Each form of the fit criterion E is a sum over examples, so we calculate the derivatives of E^p, which can then be summed over examples. For the rest of this subsection we consider one example and drop the superfix p.

In these calculations we will take partial derivatives of E with respect to weights w_{ij} and with respect to inputs x_i and outputs y_i of units. We have to make clear precisely what is kept fixed. (The literature with very few exceptions does not. Werbos, 1994, has a concept of *ordered* derivatives for this.) When we take partial derivatives with respect to weights, we regard E as a function of all the weights, so changes in a weight w_{ij} affect the input and output of unit j and all units connected to j, including some output unit(s). When we take partial derivatives with respect to an input or output, we allow all other signals in the network which depend on the input or output to follow their usual dependence. Thus all weights and all inputs (and hence outputs) of other units in the same and earlier layers are kept fixed. We evaluate $\partial E / \partial x_j$ by noting that x_j only affects the outputs through y_j, and this only acts through connections to output units.

For the first derivatives we have

$$\frac{\partial E}{\partial w_{ij}} = \frac{\partial E}{\partial x_j} \frac{\partial x_j}{\partial w_{ij}} = y_i \frac{\partial E}{\partial x_j} = y_i f_j'(x_j) \frac{\partial E}{\partial y_j} = y_i \delta_j \qquad (5.10)$$

if we define $\delta_j = \partial E / \partial x_j$. The first equality comes from the dependence of E on the weights only through the outputs, the second from $x_j = \sum w_{ij} y_i$ (5.4). We note that

$$\delta_j = \frac{\partial E}{\partial x_j} = \frac{\partial E}{\partial y_j} \frac{\partial y_j}{\partial x_j} = f_j'(x_j) \frac{\partial E}{\partial y_j}.$$

For output units $\partial E / \partial y_j$ can be calculated directly from the form of E. It is customary to express $f_j'(x_j)$ in terms of y_j; for logistic units

Sometimes δ_j is given a sign change, to set

$$w_{ij} \leftarrow w_{ij} + \eta \sum_p y_i^p \delta_j^p$$

from (5.6).

we have $f'(x) = y(1-y)$. For an output unit o we have the expressions

$$\delta_o = 2y_o(1-y_o)(y_o - t_o), \qquad \text{logistic output unit, least squares}$$

$$\delta_o = (y_o - t_o), \qquad \text{logistic output unit, entropy fit}$$

$$\delta_o = 2(y_o - t_o), \qquad \text{linear output unit, least squares}$$

$$\delta_o = \left(\sum_j t_j\right) p_o - t_o, \qquad \text{softmax}$$

For units in earlier layers we have

$$\delta_j = f'_j(x_j)\frac{\partial E}{\partial y_j} = f'_j(x_j) \sum_{k:j\to k} w_{jk}\frac{\partial E}{\partial x_k} = f'_j(x_j) \sum_{k:j\to k} \frac{\partial E}{\partial x_k}\frac{\partial x_k}{\partial y_j}$$

$$= f'_j(x_j) \sum_{k:j\to k} w_{jk}\delta_k, \qquad (5.11)$$

the sum being over units k fed by unit j. (The first and last equalities follow from definitions; the second traces the effect of the output of an internal unit via the units to which it is connected.) Since the formula (5.11) for δ_i only contains terms in later layers, it is clear that it can be calculated from output to input on the network. This simple idea has been re-discovered many times, and much credit has been given for it. Werbos (1974) had the idea of organizing such computations as a recursive computation, but its modern use stems from Rumelhart *et al.* (1986) and Rumelhart & McClelland (1986, Chapter 8). It is often discussed as a *forward pass* to calculate the outputs from the inputs, followed by a *backward pass* to calculate (δ_i) and hence $\partial E/\partial w_{ij}$. In control theory (Bryson & Ho, 1969, §2.2) the idea occurs in a more general form if the weights are considered as control inputs to the layers.

Second derivatives

The results here are only needed by the writers of programs, so others may wish skip to the next subsection. Our development follows Ripley (1994b) and extends that of Bishop (1992).

We can find the Hessian of the fit criterion E with respect to the weights by extending the derivation of (5.10). We use the symmetry of the Hessian matrix to require that j is never in a later layer than l in these expressions. (This ensures that x_j is changed by w_{ij} but not by w_{kl}, although x_l could be changed by both.) Then we have

$$H(\mathbf{w})_{ij,kl} = \frac{\partial^2 E}{\partial w_{ij}\partial w_{kl}} = y_i \frac{\partial}{\partial x_j}\frac{\partial E}{\partial w_{kl}} = y_i \frac{\partial(y_k\delta_l)}{\partial x_j}$$

$$= y_i \left[\delta_l \frac{\partial y_k}{\partial x_j} + y_k \frac{\partial \delta_l}{\partial x_j}\right] = y_i \delta_l \frac{\partial y_k}{\partial x_j} + y_i y_k h_{jl} \qquad (5.12)$$

where $h_{jl} = \partial\delta_l/\partial x_j = \partial^2 E/\partial x_j\partial x_l = \partial\delta_j/\partial x_l$. The first term is zero unless $j=k$ or there is a path from j to k.

For a general network with a single hidden layer (allowing connections from input to output) the first term must be zero unless $j = k$ is a unit in the hidden layer. Thus we have:

1   If both j and l are output units the first term is zero, and so we have

$$\frac{\partial^2 E}{\partial w_{ij} \partial w_{kl}} = y_i \, y_k \, h_{jl}. \tag{5.13}$$

2   If j is in the hidden layer and l in the output layer

$$\frac{\partial^2 E}{\partial w_{ij} \partial w_{kl}} = y_i \left[\delta_l I(j = k) f_j'(x_j) + y_k \, h_{jl} \right]$$

$$= y_i \, f_j'(x_j) \left[\delta_l I(j = k) + y_k \sum_{j \to m} w_{jm} h_{ml} \right] \tag{5.14}$$

on differentiating (5.11) with respect to x_l.

3   If j and l are both in the hidden layer

$$\frac{\partial^2 E}{\partial w_{ij} \partial w_{kl}} = y_i \, y_k \, \frac{\partial \delta_l}{\partial x_j} = y_i \, y_k \, \frac{\partial}{\partial x_j} \left[f_l'(x_l) \sum_{l \to m} w_{lm} \delta_m \right]$$

$$= y_i \, y_k \left[I(j = l) \, f_j''(x_j) \sum_{j \to m} w_{jm} \delta_m \right.$$

$$\left. + f_j'(x_j) f_l'(x_l) \sum_{j \to m} \sum_{l \to n} w_{jm} w_{ln} h_{mn} \right]. \tag{5.15}$$

These expressions only involve h_{jl} for units in the output layer. If we differentiate the expressions given for δ_o we find

$$h_{oo} = 2y_o(1 - y_o)(2y_o - t_o + 2t_o y_o - 3y_o^2),$$
$$\text{logistic output unit, least squares}$$

$$h_{oo} = y_o(1 - y_o), \qquad \text{logistic output unit, entropy fit}$$

$$h_{oo} = 2, \qquad \text{linear output unit, least squares}$$

$$h_{jl} = p_j I(j = l) - p_j p_l, \qquad \text{softmax}$$

In the first three cases the off-diagonal terms of (h_{jl}) are zero.

The double sum in the third case simplifies for softmax, for if we define $H_j = \sum_{j \to m} w_{jm} p_m$ we have

$$\frac{\partial^2 E}{\partial w_{ij} \partial w_{kl}} = y_i \, y_k \left[I(j = l) \, f_j''(x_j) \sum_{j \to m} w_{jm} \delta_m \right.$$

$$\left. + f_j'(x_j) f_l'(x_l) \left\{ \sum_{j \to m} w_{jm} w_{lm} p_m - H_j \, H_l \right\} \right].$$

Buntine & Weigend (1994) give rules for finding second derivatives in more general networks, in particular those with more than one hidden layer, but the results here suffice for the networks used in practice.

Several algorithms use the Hessian $H(\mathbf{w})$ only to compute $H(\mathbf{w})\mathbf{v}$ for a few directions $\mathbf{v}$ (for example as part of a line search along direction $\mathbf{v}$, although that only needs $\mathbf{v}^T H(\mathbf{w})\mathbf{v}$). Pearlmutter (1994) shows how to compute $H\mathbf{v}$ without computing the whole matrix H, by a technique he calls $\mathcal{R}$-backpropagation (and which also appears in Werbos, 1988). Define the operator

$$\mathcal{R}[f(\mathbf{w})] = \frac{\partial}{\partial r} f(\mathbf{w} + r\mathbf{v}) \tag{5.16}$$

as the directional derivative in direction $\mathbf{v}$.

We want to compute $H\mathbf{v} = (\mathcal{R}[\partial E/\partial w_{ij}])$. Let $\Delta_i = \partial E/\partial y_i$. We compute $\partial E/\partial w_{ij}$ by

$$x_j = \sum_{i \to j} w_{ij} y_i, \qquad y_i = f_i(x_i)$$

$$\frac{\partial E}{\partial w_{ij}} = y_i \delta_j, \qquad \delta_j = f'_j(x_j)\Delta_j, \qquad \Delta_j = \sum_{j \to k} w_{jk}\delta_k$$

after computing δ_o from the formulae above (5.11). Applying the $\mathcal{R}$ operator using the normal rules for a derivative operator, we find

$$(H\mathbf{v})_{ij} = y_i \mathcal{R}[\delta_j] + \mathcal{R}[y_i]\delta_j$$
$$\mathcal{R}[\delta_j] = f'_j(x_j)\mathcal{R}[\Delta_j] + f''_j(x_j)\mathcal{R}[x_j]\Delta_j$$
$$\mathcal{R}[\Delta_j] = \sum_{j \to k} v_{ij}\delta_k + w_{jk}\mathcal{R}[\delta_k]$$
$$\mathcal{R}[y_i] = f'_i(x_i)\mathcal{R}[x_i]$$
$$\mathcal{R}[x_j] = \sum_{i \to j} v_{ij}y_i + w_{ij}\mathcal{R}[y_i]$$

The last two equations then form a forward pass, the first three a backward pass (started by $\mathcal{R}[\delta_o]$ for the output units o). Whether these are easier to use than (5.13–5.15) will depend on the application, and in particular on how special $\mathbf{v}$ is.

The classic algorithm

The basic back-propagation algorithm (5.6) has been modified in many ways. In the original Rumelhart & McClelland experiments (1986,

p. 330) 'momentum' was added, that is exponential smoothing was applied to the correction term, so we have

$$w_{ij} \leftarrow w_{ij} - \eta \left[(1 - \alpha) \frac{\partial E}{\partial w_{ij}} + \alpha (\Delta w_{ij})_{\text{previous}} \right]. \qquad (5.17)$$

They also considered the 'on-line' version of (5.17), that is

$$w_{ij} \leftarrow w_{ij} - \eta' y_i^p \delta_j^p + \alpha' (\Delta w_{ij})_{\text{previous}} \qquad (5.18)$$

Exponential smoothing is a method of smoothing time series, taking an exponentially decaying weighted average over past values. We can compute $Y_t = (1 - \alpha) \sum_0^\infty \alpha^i X_{t-i}$ by $Y_t = (1 - \alpha) X_t + \alpha Y_{t-1}$.

and updating the weights after every example. This only makes sense if the examples are presented in a random or unstructured order, in which case the momentum creates an approximation to (5.6). In contrast, (5.6) and (5.17) are sometimes known as 'batch' algorithms.

This algorithm can be implemented by a form of distributed computing on a (two-way) network, with outputs being passed forward and then δ's being passed back.

There seem to be three motivations for the 'on-line' algorithm. One is the biological motivation of learning from every experience. Another is that it can converge faster than the batch version. Suppose that the training set contains large numbers of exact or near duplicate examples. Then the average over a small proportion of examples will provide a good approximation to E or its derivatives, and we would expect the on-line methods with a small momentum term to do well. However, in that circumstance the alternative is to use a small sample of examples in the batch algorithm, at least in the early stages of training. The third is a belief that by introducing 'noise' into the algorithm (the random choice of which example to present) local minima in the optimization are more likely to be avoided. (We return to this on page 156.)

Iterative algorithms need both a starting point and a stopping rule. The starting point is usually taken to be a random set of weights. Some care is needed that they are not taken to be too large, for if all the combinations $\sum_j w_{ij} x_i^p$ are initially large, the hidden units start in a 'saturated' state (with outputs very near zero or one).

The stopping rule does need a form of central control. The earliest idea was to stop when (if) E became small. This is often fine in logical problems, where no example is ever mis-classified, but can result in poor generalization. Very many *ad hoc* stopping rules have been proposed. One which seems popular is to have a validation set, and stop training when the error measure on the validation set starts to rise. This is dangerous, as we have often encountered examples in which after an initial drop the error on the validation set rises slowly for a

Generalization is defined in the glossary; it refers to the test-set performance.

large number of iterations, then falls dramatically to a small fraction of its previous minimum. Thus one can never know if the minimum error on the validation set has yet been attained. It is also not uncommon to use the test set rather than a validation set, as the use of a validation set is thought wasteful. (One example in a textbook is in Thornton, 1992, p. 199.)

The issue of when to stop is important, and we know of no satisfactory rule for this algorithm. Folklore suggests that disasters have been saved because its convergence is so notoriously slow that users cannot afford the computer time to overfit the training set. Much has been made of the idea of stopping before convergence (for example by Finnoff *et al.*, 1993). Note that the fitted weights will depend on the starting point if *early stopping* is used. This complicates the analysis of early stopping procedures. Wang *et al.* (1994) attempt to study early stopping of a linear regression problem, which is a neural network with no hidden layer and a single linear output unit. The algorithm studied is (5.6), batch learning without momentum. (They also allow fixed transformations of the inputs, which adds nothing to the analysis.) They assume that the data were generated by random samples from a linear regression. Then there is an optimal stopping point before convergence, but this has a delicate dependence on the size n of the training set and the starting point; their actual results are useful only if the starting point is taken to converge to the true value as n increases. That the starting point must be critical can be seen by considering what happens if the initial weights happen to be the true weights, when the expected test-set performance will normally steadily decline during training.

The batch version of the classic algorithm can converge for fixed η, but the on-line version will continue to wander unless η is reduced to zero. During training we want η to be large to approach the (local) minimum rapidly, but small to avoid large excursions about the local minimum. Both Amari (1967) and Heskes & Kappen (1991) studied the effect of the choice of η, specifically finding the covariance matrix of $\hat{\mathbf{w}}$ at time t to be proportional to η, for large t and small η. From this various rules have been used to adapt η; Amari's original suggestion was to increase η if successive steps had an angle of less than 90°, and to decrease η otherwise. White (1989b) uses results from stochastic approximation to give conditions under which convergence is bound to occur to a local minimizer ($\sum \eta_n$ diverges, $1/\eta_n - 1/\eta_{n-1}$ be bounded and $\sum \eta_n^d < \infty$ for some $d > 1$) which are satisfied by $\eta_n \propto n^{-\kappa}$ for $0 < \kappa \leqslant 1$.

Variants of the classic algorithm

A number of ideas have been proposed to speed up the convergence of
(5.6). Many are reviewed by Jacobs (1988); for example, the constants
η and α in (5.17) can be chosen adaptively for each weight w_{ij}. Some
further references are Schmidhuber (1989), Silva & Almeida (1990),
Tollenaere (1990), Darken & Moody (1991), Salomon (1991) and Eaton
& Oliver (1992). One scheme that is popular is *Quickprop* (Fahlman,
1989) which uses a crude line-search over η for each parameter. It
retains the immediate past value of the weight update, and fits a
quadratic using the past and current derivatives. If this has its minimum
at a sensible value the latter is used as the new weight, otherwise a
number of heuristics are used. The details are given at the end of this
subsection.

Analogies have been drawn between the on-line algorithm and
stochastic approximation (for example by White, 1989a, b, 1992), which
can be seen as an algorithm of the form

$$w_{ij} \leftarrow w_{ij} - \eta_n\, y_i^n\, \delta_j^n$$

with $\eta_n \to 0$ for a sequence of randomly chosen examples, and which
then converges to a local minimum of the least-squares criterion. In-
jecting further noise (Kushner, 1987; White, 1989a; Styblinski & Tang,
1990; Gelfand & Mitter, 1991), for example

$$w_{ij} \leftarrow w_{ij} - \eta_n \left(y_i^n\, \delta_j^n + \epsilon_n \right)$$

for independent Gaussian ϵ_n and $\eta_n \propto 1/\log(n+1)$, can lead to
a global minimum, analogously to simulated annealing. However,
stochastic approximation is not a very effective way to solve a least-
squares problem, not least because the magnitude of the current $f(\mathbf{x}^p; \mathbf{w})$
is ignored.

See the glossary for
simulated annealing.

One of the difficulties encountered by the classic algorithms is that
logistic units may become 'stuck' at the wrong extreme, in which case it
takes many steps to change them to the opposite extreme, since in (5.10)
we have $f'(x) = y(1 - y)$ which is small if y is very near zero or one.
Fahlman (1989) has suggested an offset, using say $f'(x) = 0.1 + y(1-y)$.
van Ooyen & Nienhuis (1992) argue for the entropy fit because for the
output units δ_o does not go to zero for y_o tending to the extremes
unless the fit is correct, and demonstrate that this leads to faster
convergence in their examples. (However, saturation can still occur
at the internal units.) This effect has led to a number of claims of
algorithms reaching local minima when they are merely in very nearly

flat regions. Genuine local minima do occur (Section 5.4), and this can be checked by considering the Hessian at a supposed minimum.

One way to avoid saturation is to discourage large weights and hence large inputs to units. *Weight decay* (Hinton, 1986) modifies the classic algorithm to

$$w_{ij} \leftarrow w_{ij} - \eta \sum_p y_i^p \delta_j^p - 2\eta \lambda w_{ij} \qquad (5.19)$$

For a regression neural network with no hidden units this will be equivalent to ridge regression.

which tries to reduces the magnitude of the weights at each step. We can see that (5.19) is steepest descent applied to

$$E + \lambda \sum_{ij} w_{ij}^2 = E + \lambda C \qquad (5.20)$$

say, a form of regularization. This will only make sense if the inputs and outputs have been (roughly) rescaled to the range $[0, 1]$ to be comparable to the outputs of the hidden units. In other problems it may make sense to used a weighted sum of weights, and/or to use different weight decay parameters for groups of weights. Other functions C have been used, for example (5.22) on page 170.

With linear output units and the least-squares error criterion, the selection of the *output* weights is a linear least-squares problem, and as in a regression or an RBF network this can be solved without iteration; this also applies to skip-layer weights. (In the parlance of non-linear regression we have a 'partially linear' problem.) This has been incorporated into a number of variant algorithms, for example by Shepanski (1987) and Hrycej (1992, Chapter 9).

Another approach has been to turn the discrete-time update into the system of continuous non-linear differential equations

$$\frac{dw_{ij}}{dt} = -\eta_n \frac{\partial E}{\partial w_{ij}} \qquad (5.21)$$

(Owens & Filkin, 1989; Weiss & Kulikowski, 1991). Then the classic algorithm is seen as a very simple fixed-step Euler integrator for this system, and much more sophisticated integration schemes can be used, especially those which are designed for *stiff* systems, those in which the Hessian has eigenvalues of very different magnitudes. Effectively these schemes allow long steps for some linear combinations of the weights, and short steps for others.

A similar approach for the discrete-time update has been to use versions of the Kalman filter; see for example Singhal & Wu (1989), Ruck *et al.* (1992) and Chandran (1994).

There have been numerous published comparisons of these variants, but these must be treated with circumspection, as the effectiveness of *ad hoc* devices can depend on the problem to be treated and how free parameters are chosen, as well as on the starting point and the quality of the implementation.

Details of Quickprop

Fahlman (1989) modifies the gradient $\partial E/\partial w_{ij}$ both by including his offset and a small (10^{-4}) weight decay; let the modified gradient be denoted $g(k)$ at the kth iteration for weight w_{ij}. We take a quadratic approximation along the line between the gradients $g(k-1)$ and $g(k)$ and look for its minimum. This amount to finding the zero for a linear approximation to the gradient, which occurs at $w + \alpha(k)\partial E/\partial w(k-1)$, where

$$\alpha(k) = \frac{g(k)}{g(k-1) - g(k)}.$$

This is replaced by 1.75 if it exceeds 1.75 or is uphill along $g(k)$ (when the quadratic would give a maximum). A learning rate is needed to start, to re-start for $\alpha \approx 0$ and is also used if the gradient and the last update have the same sign. Thus the update rule for w_{ij} becomes

$$w_{ij} \leftarrow w_{ij} - 0.55\, I\,[g_{ij}(k)\Delta w_{ij}(k-1) > 0]g_{ij}(k) + \alpha_{ij}(k)\Delta w_{ij}(k-1).$$

The values of η and the weight decay are taken from Fahlman's publicly distributed code.

This can be seen as a combination of a line-search strategy through the dependence on the last step, and gradient descent to re-start when the the line search is close to a minimum. Beware that both the value of $\eta = 0.55$ and the weight decay are not scale-free in E, and so will need to be adjusted for regression networks.

Other algorithms

Many other algorithms have been proposed, but by far the most effective in our experience are those which treat the minimization of the fit criterion E or $E + \lambda C$ as a general optimization problem. Steepest descent is generally regarded as a poor strategy for optimization, and the most widely used methods for optimizing differentiable functions are based on approximations by quadratic functions.

Introductions to these methods are provided in Section A.5 and by Fletcher (1987), Gill *et al.* (1981), Nash (1990) and Press *et al.* (1992). For realistic numbers of weights (up to thousands) quasi-Newton methods work well. For larger problems the storage of the approximate Hessian can be too demanding, and conjugate gradient methods or the limited-memory BFGS quasi-Newton method (see Section A.5) can be used. Our experience is that these all work well; a quasi-Newton algorithm was used for all the examples in this book. There are specialized algorithms for non-linear least squares, but these are designed for exactly-fitting functions and in general use are less effective.

There are at least two reasons for the superiority of these algorithms. One is that they try to solve the original minimization problem, not the system of equations setting the derivatives to zero, and so are able to use the information provided by the size of the objective which is unavailable to equation-solving algorithms. Another is that a Taylor expansion will show that locally the objective is well approximated by a quadratic function, and this enables the algorithms to have super-linear convergence (see Section A.5). In practice this means that once they get close to a local minimum, they reach it to machine accuracy in a few iterations, this being especially effective with quasi-Newton methods.

The field of numerical optimization is as large as that of feed-forward neural networks and the practical experience very substantial. Two important issues are when conjugate gradient algorithms are re-started, and how the line searches are done.

There is much collective wisdom in implementing these methods well, and the reader is advised to use a well-tested implementation from an expert. They are part of almost every package of numerical analysis procedures, and both Nash (1990) and Press *et al.* (1992) publish code. Unfortunately many of the published comparisons in the neural networks field have used their own implementations without fully understanding the issues nor documenting the precise procedures used.

Fefferman & Markel (1994) consider more hidden layers.

With algorithms that can actually solve the optimization problem to machine accuracy in a modest time, we can explore whether we have found a local minimum, by looking at the eigenvalues of the Hessian at the solution (which should all be positive, at least up to computing tolerances), and also if more than one local minimum exists. We find it to be the norm that choosing enough different starting values will lead to more than one local minimum being found. Now there will always be a number of local minima of the same value, since the hidden units are not identifiable, and can be permuted without changing the functional form. Further, the signs of all input and output weights to a single hidden unit can be reversed, and with suitable changes to the biases the fitted function is unchanged. Sussmann (1992) and Albertini *et al.* (1993) consider precisely when different sets of weights can give the same fitted function for a single hidden layer. However, we expect to find local minima with different values of the objective function. (Some simple examples are given by Gori & Tesi, 1992, and we will see examples in the next section.)

Weight decay helps the optimization in several ways. When weight decay terms are included, it is normal to find fewer local minima, and as the objective function is more nearly quadratic, the quasi-Newton and conjugate gradient methods exhibit super-linear convergence from much farther from the local minimum and so converge in many fewer iterations. There seems no reason ever to exclude a regularizer such as weight decay. If no regularizer is used, the Hessian is usually almost singular at a local maximum, and this slows the convergence of second-order optimization methods (Saarinen *et al.*, 1993).

The idea of using general-purpose optimization algorithms is a very obvious one, much re-discovered for neural network fitting. Some early references are Watrous (1987), Battiti & Massuli (1990) and Beigi & Li (1990, 1993) for quasi-Newton methods and Kramer & Sangiovanni-Vincentelli (1989), Makram-Ebeid *et al.* (1989) and Johansson *et al.* (1991) for conjugate gradient methods; Battiti (1992) gives a review.

It is very easy to give an algorithm guaranteed to reach a global minimum (Baba *et al.*, 1994), for example by taking a random step in the weight space from a distribution with positive density (for example, any Gaussian) and accepting the step if the new weights are better than the old. Such algorithms will not be practical ones in the weight space of a non-trivial network. (The proof that this algorithm works is simple. Fix $\epsilon > 0$. We assume that there is a minimizing $\mathbf{w}$, say $\mathbf{w}^0$, and $E(\mathbf{w})$ is continuous. Then there is a ball around $\mathbf{w}^0$ with $E(\mathbf{w})$ within ϵ of the minimum. The random step will hit that ball with positive probability, and in an infinite sequence of steps will hit with probability one. The move will be accepted unless the current solution has $E(\mathbf{w}) - E(\mathbf{w}^0) < \epsilon$.)

5.4 Examples

We start with the data on Cushing's syndrome. If we add just two hidden units to direct input–output connections and use a softmax output layer, we find many solutions that fit the data exactly (that is, predict probability one for the observed class for each example). Figure 5.2 (a) and (b) show two such solutions; the lines are rough because the posterior probabilities vary so fast with $\mathbf{x}$ that the contouring routine's interpolation is inadequate. Adding even a minimal amount of weight decay produces a much smoother solution (see Figure 5.2(c)). With a weight decay of $\lambda = 0.01$ the solutions are quite smooth, but seventeen local minima with distinct values of $E + \lambda C$ were found. Two commonly found solutions are shown in the figure, with values of $E + \lambda C \approx 4.18$ and 5.93. The largest local minimum found had $E + \lambda C \approx 7.65$. The Hessian at the minimum showed that in each of the seventeen cases

There are 21 weights.

Figure 5.2: Neural network fits to the data on Cushing's syndrome. The network used had two hidden units and connections from the input to output layer. Figures (a) and (b) are two solutions without weight decay which fit perfectly. Figures (c) and (d) show two local minima each, for $\lambda = 0.001$ and $\lambda = 0.01$ respectively. The dashed lines correspond to the local minimum that fits less well.

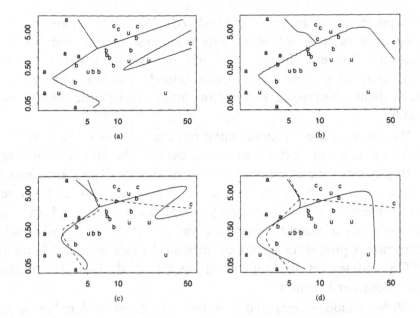

these are well-defined local minima, since it was positive-definite with eigenvalues which were well away from zero.

Adding more hidden units while keeping the weight decay constant makes only a little difference to the solution, as Figure 5.3 shows.

Figure 5.3: Neural network fits to the data on Cushing's syndrome with $\lambda = 0.01$. Part (a) had five hidden units, part (b) twenty.

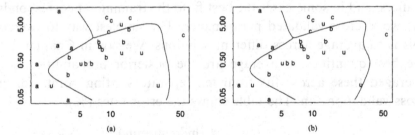

For the Pima Indians diabetes data we tried fitting a neural network with a single logistic output unit, using the conditional likelihood (2.31). Omitting the hidden layer is equivalent to fitting a logistic discriminant. Adding weight decay (up to $\lambda = 0.01$) changed the performance marginally. Adding even one hidden unit increased the error rate to the mid 70s/332, and no non-linear neural network fit approached the (linear) logistic discrimination.

Forensic glass

We have been assessing the performance of classifiers on the forensic glass data by 10-fold cross-validation. Since we will expect multiple local minima to occur, we will have to be careful to define precisely what procedure is to be cross-validated. We could choose to start fitting at a random set of weights (the same random set for each cross-validation experiment, or a different one for each) or to start from the fitted weights to the whole dataset. The latter will bias the results slightly, but may mean that the fitting can be done slightly more quickly.

We chose to use the same initial random weights for all the cross-validation runs, in part so we could explore the effects of different starting points. We know that tableware and headlamp glass can be linearly separated from the remaining classes, and this leads to very slow convergence if weight decay is not used. To avoid this, the smallest value of weight decay used was $\lambda = 10^{-4}$. The quasi-Newton optimization procedure used took around 4 times as many iterations (BFGS updates) as the number of weights, and about 1.2 function evaluations per iteration.

With no hidden layer and $\lambda = 0.001$ the cross-validated error rate was 37.8%, slightly worse than that of multiple logistic regression (the same procedure without the weight decay). Adding two hidden units reduced this to about 30.4 to 34.1%, and four and eight hidden units gave solutions in the range 24.8 to 29.9% depending on the starting point. Clearly the different local minima have quite different performances, and in some cases the best fit to the training set corresponds to the worst cross-validated performance. The simplest way to overcome this is to average across different solutions. We saw in Chapter 2 that the best quantities to average are the posterior probabilities, so we averaged these across fits from ten separate starting points for each cross-validation run. The table shows the error rates (%).

This averaging is quite time-consuming since 100 fits are required; it took about an hour per run on a Sparc 20 workstation.

	# (hidden units)		
λ	2	4	8
0.0001	30.8	23.8	27.1
0.001	30.4	26.2	26.2
0.01	31.8	29.9	29.9

This hows that only a small amount of regularization by weight decay is needed in this dataset; this is related to the near-linear separation of some of the classes. These results are the best for fitting a flexible discriminant model (and by far the most time-consuming). The sharp variation in performance with λ and the number of hidden units in this example is unusual, and can be traced to the success in separating the rare classes.

This is the best performance found in this example for a flexible discriminant method, but it is comparable with the much simpler nearest neighbour rule of page 201, which took around a second. The cross-validated confusion matrix is quite different:

	WinF	WinNF	Veh	Con	Tabl	Head
WinF	56	11	3	0	0	0
WinNF	10	59	3	3	0	1
Veh	8	1	8	0	0	0
Con	0	4	0	8	0	1
Tabl	1	0	0	1	7	0
Head	1	1	0	1	1	25

and different decisions were made in 47 examples. This suggests that combining the two classifiers might well improve the overall performance, but it did not do so appreciably.

5.5 Bayesian perspectives

The Bayesian view of decision theory can add considerable insight to the fitting of neural networks, although this insight has sometimes been clouded in the literature by the confusion of poor approximations with exact calculations.

Setting the weight decay

An important question when using weight decay is how to set the parameter(s) λ. A Bayesian perspective (Buntine & Weigend, 1991; Ripley, 1994b) helps. Suppose E is the negative log-likelihood, up to a constant, or half the deviance. Then if we take a prior distribution over weights with density $p(\mathbf{w}) \propto \exp{-\lambda C(\mathbf{w})}$, the minimizer of (5.20) will maximize the posterior density for the weights. For the weight decay of (5.20) this prior corresponds to independent Gaussian weights with mean zero and variance $1/2\lambda$. As the logistic function saturates for inputs beyond around ± 3, the standard deviation of the total input might be expected to be around 2. (Remember that one motivation for weight decay is to avoid unnecessary saturation of logistic units.) If there is a small number of inputs scaled to the range $[0, 1]$, this suggests that the standard deviations of the weights should be around 5, which corresponds to $\lambda = 1/50$. This argument is rather conservative, as we do want some weights to saturate, but suggests the range $\lambda \approx 0.001$–0.1 as a basis for exploration. Experience shows that λ is not critical to within a factor of 5.

For the sum-of-squares error criterion, E is not half the deviance, and must be rescaled. In that case the deviance is of the form $P \log(E/P)$ for P examples. Suppose σ_e^2 expresses the value of E/P that we expect to achieve. Then

$$P \log(E/P) = P \log(\sigma_e^2) + P \log(E/P\sigma_e^2) \approx P \log(\sigma_e^2) + P[E/P\sigma_e^2 - 1]$$

so an appropriate scaling is $E/2\sigma_e^2$. This suggests choosing λ in the range $(0.002\text{--}0.2)\sigma_e^2$. When least-squares fitting is used with outputs in the range $[0, 1]$, this suggests values of $\sigma_e^2 \approx 10^{-4}\text{--}10^{-2}$ might be appropriate.

With the softmax criterion (5.9) we will lose the symmetry of the classes if we set the output for one class to zero, so it usual to include all classes in the network. The weight decay resolves the redundancy over shifting all outputs, and gives a local minimum of $E + \lambda C$ (rather than a saddle point).

One criticism of weight decay emerges from this interpretation. Because it implies independence and the number of weights is often large, the opinion expressed about $S(\mathbf{w}) = \sum w_{ij}^2$ is strong, being a rescaled chi-squared distribution with degrees of freedom the number of weights. It is therefore potentially dangerous if λ is set incorrectly.

The predictive approach

The Bayesian perspective goes much deeper, and has been the subject of partial implementations and considerable controversy (Buntine & Weigend, 1991; MacKay, 1992a–e; Wolpert, 1993). We will confine attention here to classification problems. Our aim is to model $p(k \mid \mathbf{x})$ by a K-output neural network. For $K = 2$ we will use one logistic output unit to model $p(2 \mid \mathbf{x})$, and for $K > 2$ unordered classes we will use a multiple logistic model, also known as softmax. The weights are parameters, and are given a prior. It is usual to use the weight-decay prior. This itself has parameters, one or more λ, often called *hyperparameters*.

The predictive Bayes approach approximates $p(k \mid \mathbf{x})$ by averaging $p(k \mid \mathbf{x}; \mathbf{w})$ over the posterior distribution for the weights $\mathbf{w}$. This is of the general form (2.34), that is

$$p(\mathbf{w} \mid \mathcal{T}) \propto \prod_{p=1}^{n} p(y^p \mid \mathbf{x}^p; \mathbf{w}) p(\mathbf{x}^p; \mathbf{w}) \, p(\mathbf{w}), \quad p(\mathbf{w}) = \int p(\mathbf{w}; \lambda) \, p(\lambda) \, d\lambda.$$

Note that we are assuming no model for the marginal distribution $p(\mathbf{x}^p; \mathbf{w})$, and so assume that $\mathbf{x}$ carries no information about $\mathbf{w}$. (If this is false or unreasonable it can lead to fallacies.)

Such formulae can be misleadingly simple. For the moment consider fixed λ. Fitting a neural network by minimizing $E + \lambda C$ is equivalent to maximizing the posterior density over $\mathbf{w}$. Since we normally find several quite sharp local minima for $E + \lambda C$, the posterior density will normally be sharply peaked at more than one point. Averaging the non-linear function $p(k \mid \mathbf{x}; \mathbf{w})$ over such a density is computationally difficult. As we saw in Chapter 2, the 'plug-in' approach ignores this difficulty, and uses $p(k \mid \mathbf{x}; \widehat{\mathbf{w}})$ for one fitted set of weights. A more general way to approximate the posterior is to take a multivariate Gaussian distribution about each local maximum of $p(\mathbf{w} \mid \mathscr{T})$ (Buntine & Weigend, 1991; Ripley, 1994c). At each local minimum of $E + \lambda C$ we can find the Hessian $H(\widehat{\mathbf{w}})$, and take a local Gaussian approximation to the likelihood surface, of the form $N(\widehat{\mathbf{w}}, H(\widehat{\mathbf{w}})^{-1})$. The total mass of that Gaussian is then proportional to

$$|H(\widehat{\mathbf{w}})|^{-1/2} \exp -E(\widehat{\mathbf{w}}) \, (2\pi)^{n_w/2}$$

where n_w is the number of freely-varying weights. Normalizing over all the local minima found gives a mixture of Gaussian distributions as an approximation to the posterior distribution of $\mathbf{w}$. (This is closely related to the single Gaussian approximations discussed in Section 2.6.)

We can average $p(j \mid \mathbf{x}; \mathbf{w})$ for future $\mathbf{x}$ by simulating $\mathbf{w}$ from this posterior density. (Often the effect of the spread about the peaks is so small that it is sufficient to average over the peaks. The weights given to the peaks can be radically different from their relative heights.) More exactly our mixture of Gaussians can be used as a density for importance sampling (Ripley, 1987, §5.2) in integrating $p(j \mid \mathbf{x}; \mathbf{w})$, since $p(\mathbf{w} \mid \mathscr{T})$ can be calculated (up to a constant) from the fit of the network at $\mathbf{w}$. Finally, a general approach is Monte Carlo integration which can be very inefficient unless the sampling is chosen to 'fill' the space of $\mathbf{w}$ effectively (as in Neal, 1993, 1996), and only very small examples have been demonstrated at a very large computational cost.

Example

We can compare the seventeen local minima found for the Cushing's syndrome data with two hidden units and $\lambda = 0.01$. Six of them carry 5% or more of the total mass, with fitted values and weights

$E + \lambda C$	4.544	5.928	6.198	4.502	4.269	4.180
%	24.4	21.1	17.5	6.9	6.1	5.7

The second and sixth are shown on Figure 5.2, and the (approximate) predictive classifier is shown in Figure 5.4.

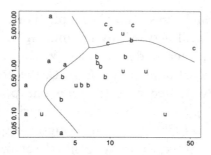

Figure 5.4: Predictive neural network fit to the data on Cushing's syndrome with two hidden units and $\lambda = 0.01$.

The choice of priors

In the last subsection a weight decay prior was used with a λ which was assumed known. If we have a hyperprior for λ, the extension is quite easy: we can sample from this hyperprior, apply the procedure for each sample based on $p(\mathbf{w}; \lambda)$, and average over samples. (It may be better computationally to use a weighted average over a very coarse grid of values of λ.) Two other approaches are to use a vague prior for λ (Buntine & Weigend, 1991) and to estimate λ by ML-II empirical Bayes (Berger, 1985, p. 99) as advocated by MacKay (1992a–e). The usual vague prior for a (squared) scale parameter such as λ has a uniform density on log scale and so has improper (un-normalizable) density $1/\lambda$ on $(0, \infty)$. This can be integrated out, to give the prior density

$$p(\mathbf{w}) \propto \left[\sum w_{ij}^2 \right]^{-n_w/2}$$

which is the vague prior on $S(\mathbf{w}) = \sum w_{ij}^2$ and is again improper. Effectively there is no regularization assumed, since this density for $S(\mathbf{w})$ has all its mass at infinity. This makes intuitive sense, since $1/2\lambda$ is the prior variance of the parameters, and if we express no opinion about it, it will be allowed to be as large as needed. This does not accord with our prior beliefs. Whereas fixed λ is a strong opinion, perhaps too strong, a vague hyperprior is too weak an opinion. A more sensible choice of hyperprior would be a gamma distribution for λ. Note that a gamma hyperprior for $1/\lambda$ would be a scaled chi-squared distribution for the variance of the weights, and so the prior distribution of the weights would be a multivariate t distribution centred at zero.

This hyperprior was also proposed by Neal (1996).

Although MacKay sometimes claims to use a uniform prior for $\log \lambda$, he uses ML-II empirical Bayes (which omits $p(\lambda)$) to choose $\widehat{\lambda}$ and so is effectively using a uniform hyperprior on $(0, \infty)$ for λ. The usual asymptotic justification for ML-II empirical Bayes is missing here, but the assumed prior independence of w_{ij} may allow the posterior for λ to be quite concentrated, since we have n_w pieces of information

n_w is the number of weights.

about λ. However, prior independence is a dubious assumption, and if n_w is large and comparable with n (which it often is) there is no reason to suppose that $p(\lambda \mid \mathcal{T})$ will be highly concentrated about one point. Our experiments showed this often not to be so.

The use of the weight decay prior is convenient, but our prior beliefs are really on the functions represented by the network and not on the parameters (weights) *per se*. This is the approach of the regularization penalties, which can be alternatively expressed as priors over the family of functions realized by the network. These have been most explored for regression networks, but Buntine & Weigend (1991) do give a small example for a classification network. Bishop (1993) uses the regularizer discussed in Section 4.3 for RBFs,

$$\sum_i \|y_i - f(\mathbf{x}_i)\|^2 + \lambda C(f), \qquad C(f) = \sum_i \sum_{j,k} \left(\frac{\partial^2 f_k(\mathbf{x}_i)}{\partial x_j^2} \right)^2.$$

This is intended for regression networks, with linear output units. (The implied prior is $\exp -\lambda C(f)$.) For this choice of regularizer Bishop shows that $\partial C / \partial w_{ij}$ can be calculated quite easily via the chain rule. For classification problems such a regularizer might be appropriate when applied to the total inputs to the logistic or softmax output stage.

Nowlan & Hinton (1992a, b) motivate a prior for the weights which is a mixture of Gaussians by its encouragement for similar values of the weights. This had proved useful in networks with a hand-tuned architecture (but not for arbitrary groupings of weights). They choose a fixed number of mixture components (not necessarily centred at zero) and optimize the penalized log likelihood over the parameters in the prior as well as the weights.

MAP estimation

MAP is a common abbreviation for *maximum a posteriori*, summarizing a predictive distribution by its mode. Most of the controversy in this area comes from inappropriate use of MAP estimates. The predictive approach is quite clear; we should average over the hyperprior (if any) and prior on the weights. This is done by mapping the posterior distribution over weights to one over $f_k(\mathbf{x}) = p(k \mid \mathbf{x}; \mathbf{w})$, and averaging over the weights $\mathbf{w}$ and hyperparameters. Finally we maximize the expected cost over decisions. As we have seen, this programme is computationally daunting, and early work (such as Buntine & Weigend, 1991) implicitly or explicitly approximated the posterior density $p(\mathbf{w} \mid \mathcal{T})$ by a mode.

This approximation is often not appropriate and can be misleading. It needs to be stressed that plugging in the MAP estimator of the weights does *not* give the MAP estimator of the function f from inputs to outputs. If it is really necessary to use MAP as an approximation, it would be better to take the MAP estimate of $f = (f_k)$ rather than that of $\mathbf{w}$. Wolpert (1994a) gives a computational programme for finding a correction term to the posterior density of $\mathbf{w}$ to skew the mode towards the MAP estimate of f, but the problem of multiple maxima remains, and in practice it appears to be easier to average, as well as being theoretically correct.

One difficulty with approximating distributions by their modes is that the mode depends on both the underlying measure for the density and the parametrization. For example, in the weight-decay regularization $\log \lambda$ and $1/2\lambda$ (the implied variance) are equally plausible parameters, yet their modes will not map to the mode of λ. Indeed, approximating by the mode seems safe only when the parameter itself has a physical meaning and so a natural scale, or when the posterior distribution is so concentrated that the parameter is effectively constant.

Optimization over both the weights and hyperparameters has been advocated by MacKay (1992b) and Nowlan & Hinton (1992b). This is further from the full Bayesian procedure, and so seems even more likely to mislead.

5.6   Network complexity

One of the simplest and most commonly asked questions is

> 'How many hidden units should I use? Are there any rules of thumb?'

There *are* rules of thumb, but these are as unreliable as those for the complexity of a multiple or polynomial regression. The answer depends on the unknown underlying function f which the neural network $f(\mathbf{x}; \mathbf{w})$ is approximating. There are three ways to control the complexity of the functions represented by a network: to cut out links, to change the number of hidden units, and to change the regularization (such as weight decay) parameter. We consider each in turn.

The statistical background in Sections 2.2 and 2.6 is needed to appreciate these methods. Many of those results can be extended to methods of parameter estimation which minimize some criterion E (including any penalty λC) and so in particular to penalized likelihood methods and the Bayes MAP estimate. However, the extensions are not very useful, as asymptotically the criterion E which grows proportionally to n (the size of the training set $\mathcal{T}$) will swamp any penalty and so not alter the results. Since the point of a penalty is to produce better behaviour for moderate n, a different asymptotic regime is needed. It follows from what we have said about the prevalence of local minima that the asymptotic theory is not normally relevant when fitting neural networks; n is often not much greater than the number of parameters.

Neural networks are 'black box' models used for prediction. For prediction performance, it is almost always better to make smooth changes (shrinkage or regularization) than to select parts of the model: it is often computationally preferable too. It is important to remember the merits of combining models discussed in Section 2.6.

Pruning networks

The usual selection procedures can be used, including stepwise selection and AIC. However, it makes little sense to set individual weights to zero, and whole internal units and the weights on their connections are added or deleted. Often some form of cross-validation (Section 2.6) is used to decide how many hidden units to use.

The neural network community has developed some fanciful names for these ideas. *Optimal Brain Damage* (Le Cun *et al.*, 1990b) and *Optimal Brain Surgeon* (Hassibi & Stork, 1993; Hassibi *et al.*, 1994; Buntine & Weigend, 1994) are both methods to 'prune' weights, that is to choose to set some of the weights to zero. Both are approximate versions of the Wald test, which considers the ratio of a parameter to its standard error. (The large-sample theory computes the standard error as the appropriate diagonal element of the Fisher information matrix K.) The methods differ in how crudely the standard errors are estimated. OBD uses just the inverse of the diagonal of the observed information matrix. OBS replaces the Fisher information matrix by the covariance matrix of the *scores* $\partial L(\widehat{\theta}; X_i)/\partial\theta$. The empirical covariance matrix can then be inverted incrementally by the Sherman–Morrison–Woodbury formula. It seems simpler to compute and invert the Hessian matrix, and Proposition 2.2 gives us more accurate asymptotic formulae which do not assume the model to be true.

Reed (1993) gives a partial survey of pruning algorithms in the neural network literature.

There are well-known difficulties in setting parameters to zero. Near-collinearities can mean that if one weight is set to zero, the standard errors for others are drastically reduced, so it can be unsafe to set more than one weight to zero at a time. For classification problems the Wald

Wald tests are large-sample equivalents of likelihood ratio tests: see Cox & Hinkley (1974) or Lehmann (1986).

See the glossary.

test is known to have little power when the true weight is large (Hauck & Donner, 1977), and so can be most misleading.

Another approach is to encourage small weights during training, so these can subsequently be set to zero, and perhaps hidden units and all their connections removed thereby. This is done by an extension of weight decay, for example with penalty

$$C = \sum \frac{w_{ij}^2/W^2}{1 + w_{ij}^2/W^2} \tag{5.22}$$

(Weigend *et al.*, 1990, 1991, 1992) (which corresponds to a very improper prior with free parameter W) or to base the penalty on the total squared weights reaching a hidden unit (Chauvin, 1989; Hanson & Pratt, 1989).

Levin *et al.* (1994) extend the idea of principal components regression, in which the inputs are first linearly transformed to their principal components and then only some of the principal components are used in the final regression. Like all pruning methods, this reduces the variance of predictions at the expense of bias. This idea is applied to the input layer, reducing the inputs to the principal components, and then to subsequent layers in the neural network. Thus no weights nor units are actually removed, but the weights are restricted to lie in a lower-dimensional space.

Selecting the number of hidden units

Selecting the number of hidden units in a neural network is *in principle* no different from selecting regressors in a linear regression or the order of a polynomial regression. The main ideas that have been developed in that field are pruning by small steps (backward selection) discussed in the last subsection, incremental construction (forward selection; the next subsection), and minimizing some measure of performance over the class of possible models.

All the candidate measures aim to predict the performance on a test set, and so to select the model with the best performance on the test set. The most general idea is to use cross-validation (Section 2.7). There is a particular difficulty in cross-validation for neural nets. How do we move from the whole set to a subset for training? Do we start at the fitted weights for the whole data? This could bias the procedure. If we start at another random starting point, we could end up at a very different solution even with the whole data, let alone with a subset. This shows that the learning procedure for a neural network is not well defined, as there are often multiple local minima of rather different performance.

The motivation of Moody & Utans (1995) seems to have been to save time by re-training from a local minimum on all the data. For the forensic glass example with 10-fold cross-validation we found re-training took at least half as long as training from scratch.

Moody & Utans (1992, 1995) suggest viewing each local minimum as a separate model. This is appealing, but flawed. Consider a real example we encountered of around 100 examples and a binary classification. All but two of the examples could be fitted very well by a simple logistic regression model with no hidden units, but the other two examples had apparently been attributed to the wrong class. Fitting a net with two hidden units did much better overall, but had at least ten non-equivalent local minima. Now dropping either of the badly-fitted examples changed the nature of the fitted function completely. If we cannot track the local minimum over dropping just one example, a local minimum cannot be sufficiently well defined to be used within a cross-validation procedure. The approach taken in Section 5.4 of averaging across the local minima is much stabler, and only a little more time-consuming.

It is important to realize that penalty terms such as weight decay and regularization change the problem completely, as they often impose a limit on the complexity of the fitted functions irrespective of the number of hidden units. The use of splines for smoothing in one dimension (Section 4.1) provides graphic evidence of this. The smoothness of the fit can be controlled either by restricting the number of knots or the degree of regularization; when regularization is used (that is, smoothing splines) the fitted function will remain smooth however many knots and data points are taken. Figure 5.3 shows a similar phenomenon for fitting a neural network with weight decay.

The NIC penalty of Murata *et al.* (1991, 1993, 1994) and Moody's (1991, 1992) p_{eff} discussed in Section 4.3 were developed for this application, but do assume a strong single local minimum. For the fits of Section 5.4 for the data on Cushing's syndrome this condition is not met. The fits shown with weight-decay constant $\lambda = 0.1$ have p_{eff} in the range 2–5 depending more on the local minimum than on the number of hidden units. There is too much uncertainty in p_{eff} for it to be useful in such a small example. Ripley (1995) found it quite effective in a regression problem with one input and a training set of 100 examples.

Incremental network construction

There have been a number of ideas to grow networks incrementally, by adding hidden units one at a time in the same or extra layers. Many of the ideas were first developed for perceptron (threshold) units and are surveyed in Gallant (1993, Chapter 10). The 'pyramid' algorithm of Gallant (1990) adds units one at a time in a new layer, each unit being connected to all previous units. Frean's (1990) 'upstart' algorithm is for binary outputs. It starts with a perceptron, then adds two hidden layer units to attempt to correct (separately) the positive and negative mistakes. As these algorithms are of very limited scope (binary outputs, threshold units), we refer the reader to the references for detailed descriptions. Other early construction algorithms are described by Ash (1989) and Moody (1989).

Moody & Utans (1995) call 'SNC' the heuristic construction algorithm which adds units to the hidden layer in groups of C units, and first trains the new units before re-training all the units. They implemented this for $C = 1$, when it is an example of back-fitting (described in Section 4.1).

Cascade correlation (Fahlman & Lebiere, 1990) is a particular way to grow a 'pyramid' network, and seems the only iterative construction algorithm that is at all widely used. Initially just input–output connections are used. At each subsequent stage a new unit is added which has as inputs the original inputs and the outputs of all the previous units, so effectively a further hidden layer containing just one unit is added, with a prescribed set of skip-layer connections. However, only the additional weights (the input and output weights of that unit) are fitted at that step; weights to hidden units once fitted are never altered. Further units are added until some pre-specified measure of fit is achieved. (Their examples are of noiseless classification, and units are added until the whole training set is correctly classified.) This is called by its authors a *cascade* architecture.

The 'correlation' in the name of the algorithm comes from the way that the new unit's weights are selected, although this involves a covariance not a correlation. If there is just one output unit, the weights are selected to maximize the absolute value of the covariance (over training-set examples) between the output value of the unit and the prediction error before that unit is added. If there are multiple outputs, the sum of this measure over outputs is used. In what can be seen as a means to avoid bad local minima and 'flat spots', several attempts are made to maximize this objective from random starting points, and the best taken. Then all the weights to the output units (including that from the new unit) are re-trained.

It is illuminating to contrast this algorithm with the SMART algorithm for projection pursuit regression (page 126). Cascade correlation uses a layered architecture; SMART uses only the original inputs but also fits to the residuals. In cascade correlation no global optimization is done, whereas in SMART all the weights are periodically re-fitted, and a pruning stage removes units one at a time. This re-fitting of weights will take more CPU time, but is likely to produce a smaller network with better generalization properties. This is borne out by the empirical comparisons of Hwang *et al.* (1994b) on the example used by Fahlman & Lebiere, which shows that projection pursuit regression produces a smoother fit and hence a better generalization.

Lee *et al.* (1990) take an approach they term *structure-level adaptation*. This adds units to the network during training where they appear to be most useful, and also removes units which appear not to be effective. The rules to do so are *ad hoc*; for example a new unit is added in parallel to one whose input weights appear to fluctuate continually during on-line training based on the heuristic idea that that unit is trying to follow two (or more) different features.

Bayesian model choice

Ripley (1995) found approximate Bayes factors by importance sampling using the multiple Gaussian approximation to the posterior density $p(\mathbf{w} \mid \mathcal{T})$ discussed on page 165. This was for a curve-fitting problem (one input, one output) selecting both the number of hidden units and a single weight decay parameter λ. The results are similar to those using NIC.

5.7   Approximation results

We first prove the 'universal approximation' result. We are concerned with functions $f : \mathbb{R}^n \to \mathbb{R}^p$ for n inputs and p outputs. We wish to approximate a given function f by g from some specified class of functions, such as (5.1). Uniform approximation on compacta means that given a compact set $K \subset \mathbb{R}^n$ and $\epsilon > 0$ we can find a function g within our class such that $\|f(x) - g(x)\| < \epsilon$ for all $x \in K$. We do not write out all the fine details, on the understanding that readers who are interested will be familiar with simple arguments in mathematical analysis. By 'ramp units' we mean

$$g(x) = \max(0, \min(x, 1))$$

which can be made up of two of Breiman's (1993) hinge functions.

If we can approximate the posterior probability function $f_k(x) = p(k \mid x)$ uniformly on compacta, we certainly will have enough to solve any practical classification problem. The restriction to compacta obviates the need to extrapolate correctly. It is reasonable to assume that this f is continuous, but it need not be (consider the task of classifying real numbers as rational or irrational).

Proposition 5.1 *Any continuous function* $f : \mathbb{R}^n \rightarrow \mathbb{R}^p$ *can approximated uniformly on compacta by functions of the form* (5.1) *with linear output units and logistic units in the hidden layer, and also by networks with threshold units or ramp units in the hidden layer.*

Proof: Our proof proceeds by building up the class of functions we can approximate (uniformly on compacta).

The idea of this proof is based on that of Diaconis & Shahshahani (1984), who used it to show the universal approximation property for projection pursuit regression.

(a) Our first step is to take $n = p = 1$. A compact set is contained in a bounded interval, say $[a, b]$, and any continuous function f can be uniformly approximated on $[a, b]$ by a step function with steps of size less than $\epsilon/2$. (For each $x \in [a, b]$ define the interval $I(x) = (\ell(x), u(x))$ where $\ell(x) = \max\{y < x : |f(y) - f(x)| \geqslant \epsilon/4\}$ and $u(x) = \min\{y > x : |f(y) - f(x)| \geqslant \epsilon/4\}$. The open sets $I(x)$ cover $[a, b]$, hence so does a finite collection $I(x_i)$. Sort the x_i into increasing order, and let g take the value $f(x_i)$ on $[\ell_i, u_i) = [u(x_{i-1}), u(x_i))$. Then $|\Delta g(\ell_i)| = |f(x_{i-1} - f(x_i)| \leqslant |f(x_{i-1} - f(\ell_i)| + |f(x_i) - f(\ell_i)| \leqslant 2\epsilon/4$.) A step function is in class (5.1) for threshold units, and sums of logistic or ramp functions can approximate a step function arbitrarily closely except at the steps, and certainly to within one half of the largest step size.

(b) We then extend the result to trigonometric functions of the form $\prod_{i=1}^{n} \cos(\omega_i x + \psi_i)$ for any n. By repeated use of the $\cos(A+B)$ formula, this can be written as a sum of the form $\sum_j a_j \cos(\omega'_j x + \psi'_j)$. Each term in this sum is continuous, and so can be approximated by step (a); hence the whole sum can, as well as linear combinations of these functions, including arbitrary trigonometric polynomials on $\mathbb{R}^n$.

(c) Any continuous function $f : \mathbb{R}^n \rightarrow \mathbb{R}$ can be approximated by a trigonometric polynomial. This is a well-known result in Fourier theory, but we give an elementary proof as Proposition 5.2.

(d) Fix the compact set K and $\epsilon > 0$. Each component function f_j of f is continuous, so we can find functions g_j within our class such that

$$\sup_{x \in K} |f_j(x) - g_j(x)| < \epsilon/\sqrt{p}$$

and so

$$\sup_{x \in K} \|f_j(x) - g_j(x)\| < \epsilon.$$

The function $(g_1(x), \dots, g_p(x))$ is within the class (5.1), since we can take separate groups of hidden units for each coordinate function.

□

We can easily extend the proof to other types of unit in the hidden layer; all we need is the ability to approximate a step function in the sense used in the proof. It is clear that, for example, any cumulative distribution function could be used. The literature sometimes refers to such functions as *sigmoidal*, particularly if they are continuous, although that term is also used for just the logistic and hyperbolic tangent functions.

We now give a probabilistic proof of the only slightly tricky step. This is an easy consequence of the (advanced) Stone–Weierstrass theorem (Simmons, 1963, p. 160).

Proposition 5.2 *Any continuous function $f:\mathbb{R}^n \to \mathbb{R}$ can be uniformly approximated on compacta by trigonometric polynomials.*

Proof: Fix the compact set K. By an affine transformation of the coordinate system we may ensure that $K \subset [0,1]^n$. Now re-parametrize each coordinate invertibly by $p_j = \sin(x_j/\max\{2,n\})$, and use the result of the second half of the proof to approximate $h(\mathbf{p}) = f(\mathbf{x})$ by a polynomial $h_N(\mathbf{p})$ on the simplex

$$\mathscr{P} = \{p \mid p_i \geqslant 0, p_1 + \cdots + p_n \leqslant 1\}.$$

Then $f_N(\mathbf{x}) = h_N(\mathbf{p})$ is a trigonometric polynomial which approximates f to the required accuracy.

Let $\mathbf{Y} = (Y_1,\ldots,Y_n)$ be a sample of size N from a multinomial distribution with parameters $\mathbf{p} = (p_1,\ldots,p_n) \in \mathscr{P}$. (As the probabilities need not sum to one, we complete the definition by taking a category $n+1$ which can occur with probability $1 - \sum p_i$.) Define

$$h_N(\mathbf{p}) = \mathsf{E}_\mathbf{p}\, h\left(\frac{Y_1}{N},\ldots,\frac{Y_n}{N}\right)$$
$$= \sum_{\substack{i_1,\ldots,i_r \\ \text{sum}\leqslant N}} h\left(\frac{i_1}{N},\ldots,\frac{i_n}{N}\right) \prod p_j^{i_j}(1-\textstyle\sum p_i)^{N-i_1-\cdots-i_n}.$$

We will show that h_N converges uniformly to h on $\mathscr{P}$. Let $M = \max_{\mathbf{p}\in\mathscr{P}} |h(\mathbf{p})| < \infty$ and choose $\delta > 0$ so that

$$\max_{\|\mathbf{p}_2-\mathbf{p}_1\|<\delta} |h(\mathbf{p}_2) - h(\mathbf{p}_1)| < \frac{\epsilon}{2},$$

which we can by continuity and compactness. Then

$$h_N(\mathbf{p}) - h(\mathbf{p}) = \mathsf{E}_\mathbf{p}\left[\{h(\mathbf{Y}/N) - h(\mathbf{p})\}I(\|\mathbf{Y}/N - \mathbf{p}\| < \delta)\right]$$
$$+ \mathsf{E}_\mathbf{p}\left[\{h(\mathbf{Y}/N) - h(\mathbf{p})\}I(\|\mathbf{Y}/N - \mathbf{p}\| \geqslant \delta)\right]$$

gives

$$|h_N(\mathbf{p}) - h(\mathbf{p})| \leqslant \frac{\epsilon}{2} + 2M \Pr_{\mathbf{p}}\left(\sum_{i=1}^{n}\left|\frac{Y_i}{N} - p_i\right|^2 \geqslant \delta^2\right)$$

$$\leqslant \frac{\epsilon}{2} + 2M \Pr_{\mathbf{p}}\left(\text{any } \left|\frac{Y_i}{N} - p_i\right| \geqslant \frac{\delta}{\sqrt{n}}\right)$$

$$\leqslant \frac{\epsilon}{2} + 2M \sum_{i=1}^{n} \Pr_{\mathbf{p}}\left(\left|\frac{Y_i}{N} - p_i\right| \geqslant \frac{\delta}{\sqrt{n}}\right)$$

$$\leqslant \frac{\epsilon}{2} + 2M \sum_{i=1}^{n} \frac{n}{\delta^2 N} p_i(1 - p_i)$$

$$\leqslant \frac{\epsilon}{2} + 2M \frac{n}{\delta^2 N} < \epsilon$$

for large enough N. (The fourth step uses the Bienaymé–Chebychev inequality.) $\qquad\square$

Uniform convergence on compacta is a strong form of convergence, and implies many others. In particular, it implies $L_2(\mu)$ convergence for any probability measure μ on $\mathbb{R}^n$. For the purposes of classification this is also sufficient, since it implies that decisions made by the classifiers derived from f and from g agree with high probability (except in artificial cases where several decisions are equally good). [This says that the variance of the difference $(f - g)(X)$ goes to zero as g approximates f.]

Kůrková (1991, 1992) gives uniform approximation results for networks with two hidden layers for which only the output weights are varied, including bounds on the numbers of units needed.

Hornik *et al.* (1990) give approximation results for continuous functions and their derivatives; these would take us too far into approximation theory even to state. Stinchcombe & White (1990) give approximation results for networks with bounded weights (for a fixed bound but using very many hidden units).

Barron (1993) gives results on the rate of convergence in $L_2(\mu)$ for some (but not all) functions $f : \mathbb{R}^n \to \mathbb{R}$; this overlaps work by L. K. Jones (1992). A typical result is

Proposition 5.3 (Barron, 1993, Proposition 1) *Suppose* $f : \mathbb{R}^n \to \mathbb{R}$ *has a Fourier representation of the form*

$$f(x) = \int_{\mathbb{R}^n} e^{i\omega^T x} \tilde{f}(\omega) \, d\omega$$

with

$$C_f = \int_{\mathbb{R}^n} \|\omega\| |\tilde{f}(\omega)| \, d\omega < \infty.$$

Then for each N, f can be approximated by a function of the form (5.1)
*with N hidden units (of logistic, threshold or ramp form), in $L_2(\mu)$ on
$B_r = \{\|x\| < r\}$ with error at most $(2rC_f)/\sqrt{N}$. Further, we can take
$\alpha_o = f(0)$ and $\sum_j |w_{jo}| \leqslant 2rC_f$ for the hidden-to-output weights.*

Proof: For rigour, a lot of details are needed in the calculation, for
which we refer the reader to Barron's paper. However, the main ideas
are quite simple. By the choice of α_o we can assume $f(0) = \alpha_o = 0$.
Since f is real, it can be represented in the form

$$f(x) = \int_{\{\omega \neq 0\}} \left[\cos(\omega^T x + b(\omega)) - \cos(b(\omega))\right] |\tilde{f}(\omega)| \, d\omega$$

and so is a probability integral of functions of the class

$$G_{\cos} = \left\{ \frac{\gamma}{\|\omega\|}[\cos(\omega^T x + b) - \cos(b)] \ \middle| \ |\gamma| < rC_f \right\}$$

(with pdf $|\tilde{f}(\omega)|\|\omega\|/C_f$). Thus f is in the closure of the convex hull of
such functions. Each of these is approximated by a convex combination
of single-step functions of height less than $2rC_f$, and then the step
may be approximated by a logistic or ramp unit. This shows that the
approximation is possible.

The rate of convergence then comes for free, by a lemma attributed
to Maurey. Our class of approximating functions is the convex hull of

$$G_\ell = \left\{ \beta\ell(a + b^T x) \ \middle| \ |\beta| < 2rC_f \right\}$$

and each member has norm at most $(2rC_f)^2$. Fix N, and suppose
$\hat{f} = \sum_{i=1}^m p_i g_i$ is a function within the convex hull of G_ℓ within distance
$\sqrt{\delta/N}$ of f. We consider randomly selecting (with replacement) N
members (say g^*) from $\{g_i\}$ with probabilities $\{p_i\}$, and take their
(unweighted) average f^*. Then $Ef^* = \hat{f}$ and

$$E\|\hat{f} - f^*\|^2 = \frac{1}{N}E\|\hat{f} - g^*\|^2 \leqslant \frac{1}{N}E\|g^*\|^2 \leqslant \frac{(2rC_f)^2}{N}.$$

This shows that there must be a realization f^* with

$$\|f - f^*\|^2 \leqslant \|f - \hat{f}\|^2 + \|\hat{f} - f^*\|^2 \leqslant \frac{\delta}{N} + \frac{(2rC_f)^2}{N}$$

and δ was arbitrary. L. K. Jones (1992) gives a constructive version of
this lemma. □

This result has often been regarded as surprising in that the rate of convergence does not vary with n; typically the rate of convergence in approximation theorems is $O(N^{-c/n})$, with c depending on how smooth f is assumed to be. It has been said to break the 'curse of dimensionality'. This is not correct; the conditions do depend on n and impose increasingly strict smoothness on the class of functions as n increases. First, B_r becomes much smaller as n increases; the radius needed to include the unit hypercube is $\sqrt{n}$. (Other forms of the result involve L_1 norms, but with the same rate of increase.) Second, the integral for C_f is dimension-dependent; indeed considering radially symmetric functions $f(\mathbf{x}) = f(\|\mathbf{x}\|)$ suggests that normally C_f grows exponentially fast in n. Results of DeVore *et al.* (1989) show that to approximate all functions with r bounded derivatives to accuracy $1/\sqrt{N}$ in $L_2(\mu)$, at least $\Omega(N^{n/2r})$ units are needed, so the 'curse of dimensionality' cannot be broken by neural networks (nor by any similar non-linear method).

> Often c is the number of continuous derivatives of the function.

> Here $y = \Omega(x)$ means y/x is bounded below for all $x > 0$.

The condition $C_f < \infty$ is not immediately interpretable. It does imply that f is continuously differentiable, with a gradient whose Fourier transform is integrable. Inspection of $G_{\cos}$ shows how the functions are actually restricted; Girosi & Anzellotti (1993) point out that functions with $C_f < \infty$ are precisely those which can be expressed as a convolution with $\|\mathbf{x}\|^{1-n}$, an increasingly restrictive constraint as n increases, and one which Barron (1993, p. 941) shows is satisfied if f has $\lfloor n/2 \rfloor + 2$ continuous derivatives.

Mhaskar & Micchelli (1992) showed that $\Omega\big(m^{n/r+(n+2r)/r^2}\big)$ units are needed for $n \geqslant 2$ to approximate all continuous functions with r bounded derivatives on the unit cube in $\mathbb{R}^n$ uniformly to within distance $1/m$.

It is easy to extend the approximation results to functions with a bound on $b = (w_{ij})$ for each hidden unit j, just by checking how well such functions can approximate step functions. If we impose the limit $\|b\| \leqslant B$, we increase the approximation error by less than

$$\frac{1 + 2\log(rB)}{rB}$$

(Barron, 1993, pp. 936–937). This is an increasingly stringent restriction on b as n increases.

Inverse functions

The universal approximation results for single-hidden-layer networks apply to continuous functions f. In some applications we need to approximate the one-sided inverse of such a function, that is given $\epsilon > 0$ and an compact set K within the range of f, we need to find a function $\phi : \mathbb{R}^p \to : \mathbb{R}^n$ such that $\|f(\phi(\mathbf{x})) - \mathbf{x}\| < \epsilon$ for all $\mathbf{x} \in K$. This arises in control theory, where f describes how the plant responds to control inputs, and ϕ is the control mapping needed to achieve (approximately) a feasible output of the plant.

Sontag (1992) points out that not only may ϕ need to be discontinuous, but it can be outside the class of functions which can be approximated by single-hidden-layer networks, even those with threshold units. However, networks with two hidden layers and threshold units do suffice, as they can approximate the indicator function of any polyhedral region, which cannot be achieved with only one hidden layer.

Dimension bounds

To use the results of Section 2.8 we need to know (or bound) the 'dimension' of families of neural networks. A few results are known. First consider threshold units and one output (so we can consider the VC dimension). Suppose there are M computational units and W weights *in toto*. Baum & Haussler (1989) showed that for any number of hidden layers the VC dimension is bounded by

$$d \leqslant 2W \log_2 eM$$

and for a single hidden layer of H units we have

$$d \geqslant 2p\lfloor H/2 \rfloor \approx pH = W \frac{p}{p+2}$$

where p remains the number of inputs. The upper bound follows from

$$\Delta(m) \leqslant \prod_{i=1}^{M} \Delta_i(m),$$

where $\Delta_i(m) \leqslant (em/d_i)^{d_i}$ refers to unit i with k_i inputs and VC dimension $d_i = k_i + 1$. Since $W = \sum d_i$, if $m = 2W \log_2 eM$

$$\Delta(m) \leqslant \prod_{i=1}^{M} (em/d_i)^{d_i} \leqslant (Mem/W)^W < 2^m.$$

The lower bound comes from a construction of Baum (1988) which shows that a single-hidden-layer net with $2j$ hidden units can separate $2jp$ vectors in $\mathbb{R}^p$.

Shawe-Taylor & Anthony (1991) extend the bound $d \leqslant 2W \log_2 eM$ to multiple (threshold) output units, where d is now the VC dimension defined via graphs of functions. Other bounds are given by Bartlett (1993). Maass (1994a, b) shows that VC dimensions of order $\Omega(W \log W)$ can be achieved with two hidden layers, but it is not known if the true order is $O(W)$ or $O(W \log W)$ for a single hidden layer. (These results apply to networks where both the number of input units and the number of hidden units are allowed to increase.) Sakurai (1993) has $\Omega(W \log W)$ results for networks with one hidden layer and real inputs (whereas Maass considered binary inputs).

If we allow the units to be logistic, the VC dimension increases, but its value is not known; indeed only recently has it been shown to be finite (Macintyre & Sontag, 1993), by decidedly advanced methods from mathematical logic. Indeed, this result is subtle, since with cos units (and hence projection pursuit regression) the VC dimension is infinite. For sigmoidal neural networks, Karpinski & Macintyre (1995a, b) showed that the VC-dimension is $O(W^4)$, and Koiran & Sontag (1996) showed $\Omega(W^2)$.

Karpinski & Macintyre give an explicit bound for a single-output network which has a leading term of $(WM)^2/2$ where M is the number of sigmoidal units

For more than one output and for non-threshold units, the pseudo-dimension is more appropriate. Haussler (1992) gives bounds for networks with bounded weights. A specialization of his Theorem 11 to logistic units (including output units) is:

Proposition 5.4 *Suppose there are $d \geqslant 0$ hidden layers and a total of W adjustable weights. Suppose that the average of the input (non-bias) weights to units in layer i is bounded by b_i. Then*

$$\mathcal{N}(\epsilon) \leqslant \left[\frac{2e(d+1) \prod_{i=1}^{d} b_i}{\epsilon} \right]^W$$

Proof: Haussler (1992, Theorem 11).   □

Bartlett & Williamson (1996) bound the VC-dimension for a single-hidden-layer network by $2W \log_2(24e\,WD)$ if the inputs are restricted to $\{-D, \ldots, D\}$.

6

Non-parametric Methods

We have seen that the Bayes rule is based on $p(k \mid \mathbf{x}) \propto \pi_k p(\mathbf{x} \mid C = k)$. We usually estimate the π_k from the proportions in the training set, and we have considered parametric models for $p_k(\mathbf{x})$ such as the multivariate normal distribution, In this chapter we consider non-parametric estimates of the class distributions. We have also seen that classification can be done by choosing the largest of $f_k(\mathbf{x}) = p(k \mid \mathbf{x})$, and that these can be estimated by regression methods, so in this section we also consider, briefly, non-parametric regression methods. We end with a discussion of the use of mixture distributions which, while parametric, is designed to approximate arbitrary class distributions.

The methods of this chapter are illustrated on small problems. They can be applied to larger problems in many dimensions (and nearest neighbour methods are often very successful) but by their very nature there is nothing to illustrate except performance figures. Nearest neighbour methods are within the diagnostic paradigm; other methods which work within the sampling paradigm and aim to model the class-conditional densities in many dimensions need a very large training set to be successful.

6.1 Non-parametric estimation of class densities

Most non-parametric methods are based on the idea that a function is locally constant, and much of the difficulty in their use is deciding what is meant by 'local' in the high-dimensional space $\mathcal{X}$. We will start by considering *kernel methods*. A kernel K is a bounded function on $\mathcal{X}$ with integral one. Suitable examples include probability density functions such as the multivariate normal. We assume that K is in some sense peaked about $\mathbf{0}$. We then use $K(\mathbf{x} - \mathbf{y})$ as a measure of the proximity of $\mathbf{x}$ and $\mathbf{y}$. (This suggests that we should take $K(-\mathbf{x}) = K(\mathbf{x})$

and this requirement is commonly imposed.) The empirical distribution of $\mathbf{x}$ within a group k gives mass $1/n_k$ to each of the examples. This suggests that a local estimate of the density $p_k(\mathbf{x})$ can be found by summing each of these contributions with weight $K(\mathbf{x} - \mathbf{x}_i)$, that is

$$\widehat{p}_j(\mathbf{x}) = \frac{1}{n_j} \sum_i K(\mathbf{x} - \mathbf{x}_i) \tag{6.1}$$

and this can also be interpreted as an average of kernel functions centred on each example from the class. We have

$$\widehat{p}(k \mid \mathbf{x}) = \frac{\pi_k \widehat{p}_j(\mathbf{x})}{\sum_k \pi_j \widehat{p}_j(\mathbf{x})} = \frac{\frac{\pi_k}{n_k} \sum_{[i]=k} K(\mathbf{x} - \mathbf{x}_i)}{\sum_i \frac{\pi_{[i]}}{n_{[i]}} K(\mathbf{x} - \mathbf{x}_i)}. \tag{6.2}$$

Remember that $[i]$ is the group of training case i.

When the prior probabilities are estimated by n_k/n, (6.2) simplifies to

$$\widehat{p}(k \mid \mathbf{x}) = \frac{\sum_{[i]=k} K(\mathbf{x} - \mathbf{x}_i)}{\sum_i K(\mathbf{x} - \mathbf{x}_i)}, \tag{6.3}$$

the weighted proportion of points around $\mathbf{x}$ which have class k.

The difficulty with kernel methods is the choice of K. Suppose we make the very reasonable choice of a multivariate normal density. Clearly the mean should be zero. How do we choose the covariance matrix? We have seen in discriminant analysis the importance of choosing the right metric via the within-group covariance. The case of univariate $\mathbf{x}$ has been studied in most detail (Silverman, 1986; Härdle, 1990; Wand & Jones, 1995), and Scott (1992) also considers the multivariate case. Even in one dimension it seems that adaptive methods are needed, that is those which change the spread of the kernel over the space $\mathcal{X}$. In our context we want to identify correctly those regions in which $p(j \mid \mathbf{x})$ is maximal, and near-equality will normally occur in the tails of the densities, so it is particularly important to estimate the tails correctly. This seems almost completely ignored in the density estimation literature. For example, plots of the estimated densities on log scale show how rough they are in the tails (e.g. Duda & Hart, 1973, Figures 4.1 and 4.2), and discrimination is based on differences in log densities. In the two-class case, Hall & Wand (1988) suggest estimating $\pi_2 p_2(\mathbf{x}) - \pi_1 p_1(\mathbf{x})$ directly by a kernel estimate. (With estimated prior probabilities this amounts to a kernel estimate for the whole sample, but counting samples from class 1 with a negative kernel.)

Figure 6.1 shows the estimated class densities for the data on Cushing's syndrome, using a normal kernel with bandwidth chosen by a standard reference (Venables & Ripley, 1994, Chapter 5). The decision regions of the resulting classifier are shown in Figure 6.2.

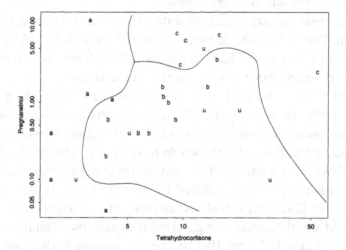

Figure 6.1: The class densities for the data on Cushing's syndrome, estimated by kernel methods with a bivariate normal kernel.

Figure 6.2: The decision regions for the data on Cushing's syndrome, using the estimated class densities shown in Figure 6.1.

Note that if we consider (6.3) with a normal kernel with covariance matrix $\kappa\Sigma$, as $\kappa \to 0$ the posterior probabilities concentrate on the class of the training-set example nearest to $\mathbf{x}$ in Mahalanobis distance. On the other hand, the tail behaviour of the kernels is critical in determining the relative balance of the prior proportions of the classes and the effect of the training data at points $\mathbf{x}$ well outside the training set. It is quite common practice to use kernels with bounded support, so the density estimate at $\mathbf{x}$ could be zero for one or even all classes.

Kernel discriminant analysis is the subject of monographs by Hand (1982) and Coomans & Broeckaert (1986). The ALLOC80 computer package (Hermans *et al.*, 1982) for kernel discriminant analysis is widely used.

Kernel methods can be used to estimate regression surfaces, by averaging the values of y attached to the nearby data points. We obtain the Nadaraya–Watson non-parametric kernel regression

$$\widehat{y}(\mathbf{x}) = \frac{\sum_i y_i K(\mathbf{x} - \mathbf{x}_i)}{\sum_i K(\mathbf{x} - \mathbf{x}_i)}. \tag{6.4}$$

Now suppose we use (6.4) to estimate $f_k(\mathbf{x}) = p(k\,|\,\mathbf{x})$. Then y is the indicator function for class k, and (6.4) reduces to (6.3). This form of classification fits into the framework of Chapter 4. (The non-parametric kernel regression can also be derived by taking a kernel density estimator in the space $\mathscr{X} \times \mathbb{R}$ and evaluating $\mathsf{E}[Y\,|\,\mathbf{X} = \mathbf{x}]$.)

Kernel methods are known in the pattern recognition literature as *Parzen windows* after Parzen (1962). There was earlier work on the method by M. Rosenblatt (1956) and, in passing, Fix & Hodges (1951). The extension to more than one dimension is usually attributed to Cacoullos (1966) and Murthy (1966).

Kernel methods are readily updated for use in leave-one-out cross-validation if we retain the numerator and denominator of (6.3). To find $\widehat{p}(k\,|\,\mathbf{x}_j, \mathscr{T} \setminus \{\mathbf{x}_j\})$ we only have to subtract $K(\mathbf{0})$ from the denominator, and from the numerator when k is the true class of $\mathbf{x}_j$.

Specht (1990a, b, 1991) has re-labelled these methods as neural networks (without any apparent biological motivation); (6.3) he calls a *probabilistic neural network* and (6.4) a *general regression neural network*.

Kernel methods require the whole training set to be retained. Methods which select a smaller set of centres are considered in Section 6.4. An alternative approach advocated by Specht (1967a, b) is to approximate $\widehat{p}(\mathbf{x})$ for Gaussian kernels as the product of a multivariate normal density and a polynomial, using a Taylor expansion. (Details are given by Duda & Hart, 1973, pp. 106–107, and Wasserman, 1993, pp. 46–51.) This is more in the spirit of of the next subsection.

A closer connection to neural networks is made for rotationally symmetric normal kernels and features of unit length, for then $K(\mathbf{x} - \mathbf{x}_i) = f(\mathbf{x}_i^T \mathbf{x})$ where $f(t) \propto \exp(1 - t)/2\sigma^2$.

We stress in Section A.1 that a density is defined with respect to an underlying measure, and can be changed by changing that measure. Thus if we know that p_j is near some density p_0, we should choose that density as the underlying measure, and so estimate p_j by an estimate of the ratio p_j/p_0 times p_0. For a kernel estimator this becomes

$$\widehat{p}_j(\mathbf{x}) = p_0(\mathbf{x})\frac{1}{n_j} \sum_i K(\mathbf{x} - \mathbf{x}_i)/p_0(\mathbf{x}_i).$$

We call p_0 a *fixed start* density; p_j will not integrate to exactly one. We can apply the same procedure with a density p_0 estimated from a parametric family (for example the family of normal distributions) by using a *parametric start*

$$\widehat{p}_j(\mathbf{x}) = \frac{1}{n_j} \sum_i K(\mathbf{x} - \mathbf{x}_i)\frac{p_0(\mathbf{x}; \widehat{\theta})}{p_0(\mathbf{x}_i; \widehat{\theta})}. \tag{6.5}$$

However, the standard theory no longer applies. Hjort & Glad (1995) show that the bias and variance are (asymptotically) essentially unchanged from those of the fixed start estimator using the (unknown) best possible fixed start within the parametric class.

Hjort & Jones (1996) reverse the roles of the parametric and non-parametric parts of (6.5) by estimating the parameters of the parametric family locally (as defined by the kernel). More precisely, the density at $\mathbf{x}$ is estimated by $f(\mathbf{x}; \widehat{\theta}(\mathbf{x}))$ for a parametric family $f(\cdot; \theta)$, where $\theta(\mathbf{x})$ is chosen to maximize a local log-likelihood

$$\frac{1}{n} \sum K(\mathbf{x} - \mathbf{x}_i) \log f(\mathbf{x}_i; \theta) - \int K(\mathbf{t} - \mathbf{x}) f(\mathbf{t}; \theta) \, d\mathbf{t}.$$

In both versions the hope is that the parametric family will capture the broad features of the density, and allow a kernel with a much larger spread to be used. This may make the methods feasible in more dimensions than the simple kernel method, but the issues of the choice of kernel remain. These methods will be most effective in a small number of dimensions. A more constrained form of correction to a parametric start is discussed later under 'projection pursuit density estimation'.

Orthogonal expansion estimators

A general approach towards non-parametric density estimation is via expansions in orthogonal basis functions, estimation of necessary coefficients, and a rule to decide when to stop including terms in the expansion. The expansion approach has some advantages over kernel methods in statistical pattern recognition problems. This approach often yields a compact representation of the estimates of class densities, with a low number of coefficients describing the estimate. Most texts on density estimation mention the approach, whereas Tarter & Lock (1993) consider only orthogonal expansions.

We start with a general orthonormal set of basis functions $\psi_k(x)$ on $\mathscr{X}$ with respect to a suitable weight function w, that is,

$$\int \psi_j(y) \psi_k(y) w(y) \, dy = I\{j = k\}.$$

Examples of such structures abound, see for example Abramowitz & Stegun (1965, Chapter 22) or Thisted (1988, §5.3.2). We will exemplify the method by Fourier series, for $\mathscr{X} = [0, 1]$ and $\psi_k(x) = \exp 2\pi i k x$. (It is always worth bearing in mind that the feature space may profitably be transformed before density estimation; for example the transformation

$\Phi(x)$ will map $\mathbb{R} \to [0, 1]$ and turns near standard normal densities into near uniform ones.) Other examples we shall meet in Section 9.1 are polynomial expansions (Hermite polynomials) orthogonal with respect to a uniform weight on $[-1, 1]$ and with respect to $\phi(x)$ (the normal density) on $\mathbb{R}$.

Suppose we wish to approximate a density function f by a series expansion $\widehat{f}(x) = \sum_k c_k \psi_k(x)$. The best choice of coefficients c_k is given by

$$c_k = \int \psi_k(y) f(y) w(y) \, dy = \mathsf{E} \, \psi_j(X) w(X)$$

in the sense of minimizing the mean integrated squared error $MISE = \mathsf{E} \int (f - \widehat{f})^2 w$. The infinite series $\sum_{k=0}^{\infty} c_k \psi_k(x)$ defined in this way converges pointwise to $f(x)$ if f is continuous. We estimate c_k by $\widehat{c}_k = \sum \psi_k(X_i) w(X_i)/n$. With estimated coefficients we do have to stop the expansion at some term m. More generally, we could use

$$\widehat{f}(x) = \sum_{k=1}^{\infty} b_k \widehat{c}_k \psi_k(x)$$

for some suitable tapering sequence (b_k).

Let us explore Fourier series a little. Since the Fourier coefficients are defined for both positive and negative integers, our sums should extend infinitely in both directions (or be truncated at $\pm m$). We can always write a Fourier series estimator as a kernel estimator on taking

$$K(x) = \sum_{|k| \leqslant m} b_k \exp 2\pi i k x = \frac{\sin \pi (2m + 1) x}{\sin \pi x}$$

if we truncate at $\pm m$. This is the Dirichlet kernel shown in Figure 6.3; it has a narrow peak and slowly decaying oscillations about zero. Thus with this tapering $\widehat{f}$ can take negative values, but for the tapering sequence $b_k = \max[0, 1 - k/(m + 1)]$, the kernel is the Fejér kernel

$$K(x) = \frac{1}{m + 1} \left[\frac{\sin \pi (m + 1) x}{\sin \pi x} \right]^2$$

and so $\widehat{f}$ is non-negative (Figure 6.3).

A number of stopping rules have been proposed. It is fairly easy to show that including term k will decrease $MISE$ if $|c_k|^2 > 1/(n + 1)$ (Tarter & Lock, 1993, §4.2) and with a bias-correction argument this suggests including term k only if $|\widehat{c}_k|^2 > 2/(n + 1)$. Unfortunately this will lead to the inclusion of an infinite number of terms, so it is

Figure 6.3: Kernels for Fourier series estimator. **Left:** Dirichlet kernel. **Right:** Fejér kernel.

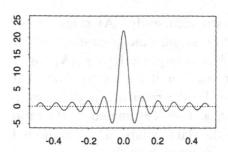 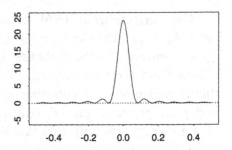

preferable to include terms until this test is failed for 2–4 consecutive terms. Diggle & Hall (1986) give a similar but more complex alternative.

An alternative approach is to allow the data to choose the tapering sequence. Tarter & Lock report good results with the choice

$$b_k = \frac{n}{[n - 1 + |\widehat{c}_k|^{-2}]}.$$

Orthogonal series estimators meet similar difficulties to kernel density estimators once we move to $\mathbb{R}^d, d > 1$. We can, for example, use a multidimensional Fourier series, but these are tied to a particular coordinate system, and the stopping rule has to be defined for a d-tuple index. For Hermite polynomials invariance to rotations can be achieved by including all terms up to degree m, but there are many of these for moderate m and we will need moderate m to approximate all but the simplest functions. Series estimators are particularly useful for low-dimensional projections.

Parametric starts can be used with orthogonal series as well as kernel density estimates; for some univariate examples using a normal start and Hermite polynomials see Buckland (1992a, b).

Projection pursuit density estimation

Projection pursuit density estimation (Friedman *et al.*, 1984) is the application of projection-pursuit ideas (Section 9.1) to density estimation, and so is appropriate when the variation in the densities is concentrated in a linear subspace of $\mathscr{X}$. It estimates a density by the formula

$$p_M(\mathbf{x}) = p_0(\mathbf{x}) \prod_{m=1}^{M} q_m(\mathbf{a}_m^T \mathbf{x}) \tag{6.6}$$

Ridge functions are constant orthogonally to one direction.

where p_0 is an initial density (perhaps an appropriate multivariate normal distribution) and the q_m are multiplicative corrections which are ridge functions. We will consider its application to n samples $\mathbf{x}_i$, perhaps the training samples for a single class.

The corrections in (6.6) are chosen recursively. At stage m we have p_{m-1} and choose $\mathbf{a}_m$ and q_m to maximize the goodness-of-fit of p_m as measured by the Kullback–Leibler divergence $\mathsf{E} \log p_m(\mathbf{X})$ for $\mathbf{X}$ drawn from the true density. Given $\mathbf{a}_m = \mathbf{a}$ it is easy to show that this is maximized by choosing q_m to be the ratio of the densities p and p_{m-1} *projected* onto the direction $\mathbf{a}$. This is estimated by the ratio of a univariate density estimate for the data points $(\mathbf{a}^T \mathbf{x}_i)$ projected on $\mathbf{a}$ to the projection of p_{m-1} (the integration to find the marginal distribution along $\mathbf{a}^T \mathbf{x}$ being done by Monte Carlo methods). Rather than retain the full density estimates, Friedman *et al.* approximated the estimate of q_m by a cubic spline.

Friedman *et al.* (1984) use crude histogram estimators of the densities, but kernel methods could be used.

This method gives us an estimate of p_m given $\mathbf{a}$. Since

$$\mathsf{E} \log p_m(\mathbf{X}) = \mathsf{E} \log q_m(\mathbf{a}^T \mathbf{X}) + \mathsf{E} \log p_{m-1}(\mathbf{X}),$$

we choose $\mathbf{a}_m = \mathbf{a}$ to maximize $\mathsf{E} \log q_m(\mathbf{a}^T \mathbf{X})$ and estimate this by its sample version $\frac{1}{n} \sum \log \widehat{q}_m(\mathbf{a}^T \mathbf{x}_i)$. This is maximized numerically.

It remains to choose the number of terms M. Standard methods such as cross-validation can be used. Often examining q_m (which shows the ratio of the two densities) against the projected data $(\mathbf{a}^T \mathbf{x}_i)$ will indicate if a worthwhile improvement can be made.

The density estimation strategy of Friedman (1987, §4) is 'backward' rather than 'forward' in flavour. Exploratory projection pursuit is used to find a direction $\mathbf{a}$ such that $\mathbf{a}^T \mathbf{X}$ is maximally non-normal. We then remove the marginal structure in direction $\mathbf{a}$ by

$$p(\mathbf{x}) \leftarrow p(\mathbf{x}) \frac{\phi(\mathbf{a}^T \mathbf{x})}{p_{\mathbf{a}^T \mathbf{X}}(\mathbf{a}^T \mathbf{x})},$$

that is by adjusting the marginal density to be standard normal (see page 297), and repeat the process. Eventually the exploratory process will be unable to find an interesting projection, and the remaining density can be fitted by a normal distribution. Reversing the process reveals a density for $\mathbf{X}$ which is a normal density times a series of correction terms, and the corrections are ridge functions. The marginal densities can be fitted by any one-dimensional estimation method, including splines, kernel and orthogonal series methods, but the compact representation of orthogonal series will be especially useful.

It would be possible to use q-dimensional correction terms, for small q.

Discrete distributions

Thus far we have implicitly assumed that we have continuous features, so $\mathscr{X} \subset \mathbb{R}^p$. We can consider non-parametric estimation of class distributions in the discrete case too. For simplicity let us assume that we have p discrete features. Then a distribution is specified by a p-way table of probabilities of all possible combinations, and the natural non-parametric estimator is to use the frequencies of the cells of the table in the training set $\mathcal{T}$. Of course, if p is large and $\mathcal{T}$ is not enormous, most of the cells will be given zero probability. When we come to classify future cases, we are likely to find that the estimated class probabilities are zero under all classes. Thus it is essential to smooth the observed frequencies before using them in a plug-in classifier. We will consider how to do so, but point out that it will normally be better to use a logistic regression than the methods considered here, as that uses the data to model the quantities of direct interest, the posterior probabilities.

The natural idea for statisticians would be to build a contingency table model of the joint distribution of the features (McCullagh & Nelder, 1989). The most common choice would be a log-linear model, in which the log probability

x_C denotes the values of x_i for $i \in C$

$$\log \Pr\{X_1 = x_1, \ldots, X_p = x_p\} = \sum_C \lambda_C(x_C) \qquad (6.7)$$

The empty set $\emptyset$ is included to give a constant term.

is expanded over subsets $C \subset \{1, \ldots, p\}$. When the terms of (6.7) are restricted, for example by omitting $\lambda_C(x_C)$ for large C, the family of probability distributions is restricted, but the coefficients λ_C may be fitted by maximum likelihood. Choosing the appropriate restricted model is an art, and can be considered within the graphical framework of Chapter 8.

The most extreme restriction of (6.7) is to omit all sets C of two or more features, so

$$\Pr\{X_1 = x_1, \ldots, X_p = x_p\} = e^{\lambda_\emptyset} \prod_i e^{\lambda_i(x_i)}$$

which amounts to assuming independence of the features. Thus (6.7) can be seen as an expansion away from an independence model.

Other expansions have been used for binary data. Suppose each feature takes the values 0 and 1, and that feature i takes the value 1 with probability p_i. Let

$$y_i = \frac{x_i - p_i}{\sqrt{p_i(1 - p_i)}}$$

and consider the 2^p polynomials $y_1^{i_1} \cdots y_p^{i_p}$. These are orthogonal with respect to the independence model, so we can take an orthogonal series expansion. If we estimate p_i by the frequency $\hat{p}_i$ with which feature i takes value 1 in the training set, we have the *Bahadur–Lazarsfeld* expansion (Bahadur, 1961a, b; Lazarsfeld, 1961)

$$\Pr\{X_1 = x_1, \ldots, X_p = x_p\}$$
$$= \prod_{i=1}^{p} \hat{p}_i^{x_i}(1 - \hat{p}_i)^{1-x_i}\left[1 + \sum_{i<j} \hat{c}_{ij}\, y_i y_j + \sum_{i<j<k} \hat{c}_{ijk}\, y_i y_j y_k + \cdots\right]. \quad (6.8)$$

The estimates $\hat{c}_{\cdots}$ are just the sample moments of y_i for the appropriate indices. Often just the first correction term is used, which is a correction for correlations only.

Kernel methods

For binary features, Aitchison & Aitken (1976) proposed the kernel smoothing

$$\Pr\{\mathbf{X} = \mathbf{x}\} = \frac{1}{n}\sum_{j} K(\mathbf{x}, \mathbf{x}_i)$$

where $K(\mathbf{x}, \mathbf{y}) = h^{p-d}(1 - h)^d$ and $d = \|\mathbf{x} - \mathbf{y}\|^2$ is the number of disagreements between $\mathbf{x}$ and $\mathbf{y}$. Here $1/2 < h \leqslant 1$ is a smoothing constant; for $h = 1$ there is no smoothing. The kernel gives weight h^p to cell $\mathbf{x}$ and weight $h^{p-k}(1-h)^k$ to cells that differ in k features. Note that the kernel is a product over the features, since d is a sum over features.

This product kernel can be extended naturally to categorical data with $k_i \geqslant 2$ possible outcomes (give probability h to the observed outcome, $(1 - h)/(k_i - 1)$ to all others), to ordered categorical data (spread the probability $1 - h$ over adjacent outcomes) and to features which are counts (Aitken, 1983). For mixed continuous and discrete features we can take a product of an appropriate kernel for each feature. For binary and categorical features the kernel estimator takes a convex combination of the frequencies with a uniform distribution. This has been considered in its own right as a method of smoothing (for example, Fienberg & Holland, 1973).

The smoothing constant h can be chosen separately for each feature. Choosing an appropriate degree of smoothing remains a difficult problem (P. Hall, 1981; Tutz, 1986, 1988, 1989). Averaging with a uniform distribution gives a different perspective which suggests other ways to choose the degree of smoothing (Fienberg & Holland, 1973; Titterington, 1980; Brown & Rundell, 1985).

6.2   Nearest neighbour methods

Choosing the neighbourhood using all the classes may not be a good idea if the class densities have very different structure.

One simple adaptive kernel method is to choose K to be constant over the nearest k examples and zero elsewhere. This does not in fact define a density as the estimate of $p(\mathbf{x})$ (its integral is infinite) but (6.2) suggests a simple estimate of the posterior distribution as the proportions of the classes amongst the nearest k data points. This is a piecewise constant function over $\mathscr{X}$, and gives a classifier known as the *k-nearest neighbour rule*. This differs from using the k-nearest neighbour density estimate for each class, as we choose the neighbourhood from the k nearest points of any class. If the prior probabilities are known and the proportions of the classes in the training set are not proportional to π_k, the proportions amongst the neighbours need to be weighted (Brown & Koplowitz, 1979). (We ignore this in the theory below.)

See the glossary.

The version with $k = 1$ is often rather successful. This divides the space $\mathscr{X}$ into the cells of the Dirichlet tessellation of the data points, and labels each by the class of the data point it contains. We can also consider the analogue of (6.4) which gives a locally constant non-parametric regression surface, and once again corresponds to the k-nearest neighbour estimate of the posterior probabilities.

This was reproduced in Agrawala (1977), Dasarathy (1991) and with a commentary by Silverman & Jones (1989).

Both the k-nn method and kernel discrimination were first given in an unpublished report by Fix & Hodges (1951). There is a very extensive literature on nearest neighbour classifiers, much of which is reviewed or reprinted in Dasarathy (1991).

Ties in the distances can occur with finite-precision data (or if the underlying distribution has a discrete part). One solution is to include all patterns as near as or nearer than the k-nearest neighbour, and take a majority vote amongst them.

Nearest neighbour rules can readily be extended to allow a 'doubt' option by the so-called (k, ℓ)-rules (Hellman, 1970), called in this field a 'reject option'. These take a vote amongst the classes of the k nearest patterns in $\mathscr{X}$, but only declare the class with the majority if it has ℓ or more votes, otherwise declare 'doubt'. Indeed, if there are different error costs, we may want to allow the minimum majority to depend on the class to be declared. Properties of this class of rules are discussed by Devijver & Kittler (1982) and Loizou & Maybank (1987), but for large samples.

The k-nn rule can be critically dependent on the distance used in the space $\mathscr{X}$, especially if there are few examples or k is large (Figure 6.4).

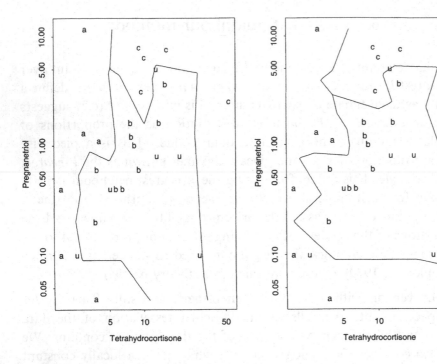

Figure 6.4: Decision boundaries of the 1-nn rule for the Cushing's syndrome data. The left plot uses Euclidean distance on our usual plot, the right Euclidean distance on $\log_{10}$ excretion rates.

Large-sample results

Cover & Hart (1967) gave a famous result on the large-sample behaviour of the nearest neighbour rule. Note that the expected error rate is always bounded below by E^*, by the optimality of the Bayes rule.

Proposition 6.1 *Let E^* denote the error rate of the Bayes rule in a K-class problem. Then the error rate of the nearest neighbour rule averaged over training sets converges in L_1 as the size of the training set increases, to a value E_1 bounded above by*

$$E^* \left(2 - \frac{K}{K-1} E^* \right)$$

Proof: Let X_1 be the nearest neighbour to X, a randomly sampled pattern with class C. The (rather technical) arguments of C. J. Stone (1977) and Devroye (1981a) show that

$$\mathsf{E}\big|p(k \mid X_1) - p(k \mid X)\big| \to 0.$$

Now

$$\Pr(C_1 \neq C \mid X = \mathbf{x}) = \sum_{i \neq j} p(i \mid \mathbf{x}) \, \mathsf{E}\big[p(j \mid X_1) \mid X = \mathbf{x}\big]$$

and so

$$\mathsf{E}\left|\Pr(C_1 \neq C \mid X) - \sum_{i \neq j} p(i \mid X)\, p(j \mid X)\right| \to 0.$$

Thus $E_1 = \mathsf{E}\, e_1(X)$ where $e_1(\mathbf{x}) = \sum_{i \neq j} p(i \mid \mathbf{x})p(j \mid \mathbf{x}) = 1 - \sum_i p(i \mid \mathbf{x})^2$.

The conditional Bayes risk $r^*(\mathbf{x})$ is $1 - \max_i p(i \mid \mathbf{x}) = 1 - p(k \mid \mathbf{x})$, say, so by the Cauchy–Schwarz inequality

$$(K-1)\sum_{i \neq k} p(i \mid \mathbf{x})^2 \geqslant \left[\sum_{i \neq k} p(i \mid \mathbf{x})\right]^2 = r^*(\mathbf{x})^2$$

$$(K-1)\sum_i p(i \mid \mathbf{x})^2 \geqslant r^*(\mathbf{x})^2 + (K-1)\left(1 - r^*(\mathbf{x})^2\right)$$

and

$$1 - \sum_i p(i \mid \mathbf{x})^2 \leqslant 2r^*(\mathbf{x}) - \frac{K}{K-1}r^*(\mathbf{x})^2.$$

On taking expectations we obtain

$$E_1 \leqslant 2E^* - \frac{K}{K-1}\mathsf{E}\left[r^*(X)^2\right] \leqslant 2E^* - \frac{K}{K-1}(E^*)^2$$

using $\mathsf{E}(Y^2) = \mathsf{Var}\, Y + (\mathsf{E}\, Y)^2 \geqslant (\mathsf{E}\, Y)^2$ $\square$

It is easy to see that the upper bound is attained if the densities $p_k(\mathbf{x})$ are identical and so the conditional risks are independent of $\mathbf{x}$.

For the k-th nearest neighbour rule detailed results are only available for two classes. Intuitively one would expect the 2-nn rule to be no improvement over the 1-nn rule, since it will achieve either a majority of two or a tie, which we will suppose is broken at random. The following result supports that intuition. On the other hand, we could report 'doubt' in the case of ties (the $(2,2)$-rule).

Proposition 6.2 *Suppose there are two classes, and let E_k denote the asymptotic error rate of the k-nn rule with ties broken at random and E'_k if ties are reported as 'doubt'. Then*

↗ and ↙ denote
monotone convergence
from below and above,
respectively.

$$E'_2 \leqslant E'_4 \leqslant \cdots \leqslant E'_{2k} \nearrow E^* \swarrow E_{2k} = E_{2k-1} \leqslant \cdots \leqslant E_2 = E_1 = 2E'_2$$

Proof: We rely on L_1 convergence results such as those quoted in the proof of Proposition 6.1 to show the existence of the asymptotic results and to replace $p(k \mid X_r)$ by $p(k \mid X)$ in deriving the asymptotic results. Let $r_k(\mathbf{x})$ denote the (limit of the) probability of misclassifying $\mathbf{x}$. We can think of this as taking $k+1$ samples from $p(\ \mid \mathbf{x})$ and finding that the first belongs to the minority group.

Temporarily denote the posterior probabilities of the two classes by $p = p(1|\mathbf{x})$ and $q = p(2|\mathbf{x})$, and let $\xi = pq$. Consider a_k, the probability that the first of k samples is in a strict minority, and let $b_i = \binom{2i}{i}\xi^i$ be the probability of exactly i examples from each class. First consider k odd. On adding a point, we can create a tie but not a minority, so

$$a_{2k-1} - a_{2k} = p\frac{k-1}{2k-1}\binom{2k-1}{k}p^{k-1}q^k + q\frac{k-1}{2k-1}\binom{2k-1}{k}p^k q^{k-1}$$

$$= 2\frac{k-1}{2k-1}\binom{2k-1}{k}\xi^k = \frac{k-1}{2k-1}b_k.$$

If k is even, adding a point can create a minority but not a tie so

$$a_{2k+1} = a_{2k} + \tfrac{1}{2}b_k$$

$$a_{2k+1} = a_{2k-1} + \tfrac{1}{2}b_k - \frac{k-1}{2k-1}b_k = a_{2k-1} + \frac{1}{2(2k-1)}b_k.$$

Clearly $a_1 = 0$, so by induction

$$a_{2k+1} = \sum_{i=1}^{k}\frac{1}{2(2i-1)}b_i.$$

In terms of these quantities, we have

$$r'_{2k}(\mathbf{x}) = a_{2k+1}$$

$$r_{2k-1}(\mathbf{x}) = a_{2k} + b_k$$

$$r_{2k}(\mathbf{x}) = r'_{2k}(\mathbf{x}) + \tfrac{1}{2}b_k = a_{2k} + \tfrac{1}{2}b_k$$

From these formulae $r_{2k+1} - r_{2k-1} = (2\xi - \tfrac{1}{2})b_k \leqslant 0$. Direct calculation shows that $r_1(\mathbf{x}) = 2\xi$ and $r'_2(\mathbf{x}) = \xi = \tfrac{1}{2}r_1(\mathbf{x})$. On taking expectations this establishes the hierarchy on each side. Now

$$\lim_{k\to\infty} r_{2k-1}(\mathbf{x}) = \lim_{k\to\infty} r'_{2k}(\mathbf{x}) = \sum_{i=1}^{\infty}\frac{1}{2(2i-1)}b_i.$$

Note that

$$r^*(\mathbf{x}) = \min[p(1|\mathbf{x}), p(2|\mathbf{x})] = \tfrac{1}{2}\left[1 - \sqrt{1 - 4\xi(\mathbf{x})}\right]$$

has the Taylor series expansion

$$r^*(\mathbf{x}) = \sum_{i=1}^{\infty}\frac{1}{i}\binom{2i-2}{i-1}\xi^i = \sum_{i=1}^{\infty}\frac{1}{2(2i-1)}b_i.$$

The proof is completed by taking expectations and using the monotone convergence theorem.   □

Figure 6.5:
Large-sample risk r_k
(k odd) or r'_k (k even)
of k-nn rules against
the Bayes risk r^* in a
two-class problem.

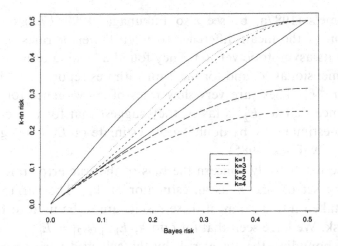

This result is from Cover & Hart (1967); the formulae are given without proof by Devijver & Kittler (1982) and Fukunaga (1990). Figure 6.5 shows $r_k(\mathbf{x})$ as a function of $r^*(\mathbf{x})$; this shows the agreement is excellent for moderate $r^*(\mathbf{x})$ even for small k (but not $k = 1$).

Comparable results for the (k, ℓ)-nn rule were first considered by Tomek (1976a). Later Loizou & Maybank (1987) showed that for $\ell > k/2$ we have

$$E_{k,\ell+1} \leqslant E^*(d) \leqslant E_{k,\ell} \leqslant c(k,\ell,K)E^*(d)$$

where $E^*(d)$ refers to the Bayes rule (if any) with doubt cost d such that its doubt rate is equal to the asymptotic doubt rate of the (k, ℓ)-nn rule, and $c(k, \ell, K)$ is a computable constant. Since $E^*(d) \leqslant E^*$, this can be used to give asymptotic lower bounds for E^*. Clearly the left-hand side increases with ℓ.

Another version of the asymptotics is to allow k to increase with the size n of the training set. C. J. Stone (1977) showed that provided $k \to \infty$ and $k/n \to 0$, the risk for the k-nn rule (*not* averaged over training sets) converges in probability to the Bayes risk. For $k/\log n \to \infty$, Devroye (1981b) strengthened this to almost sure convergence. The same methods allow the variance over training sets for fixed k to be estimated.

All these performance measures are asymptotic and they do not apply to finite samples; for example the error rate does not decrease monotonically with odd k. Notice that the results do not depend on the metric, whereas in practice the choice of metric is often very important. There are few finite-sample results. Cover (1968) is widely quoted as showing that the risk of the 1-nn rule converges to E_1 at rate $O(n^{-2})$, but his results only apply to the bias for one-dimensional $\mathscr{X}$. Fukunaga

& Hummels (1987a, b; see also Fukunaga, 1990, §7.3) consider an expansion of the mean difference between the error rates of the k-nn rule and its asymptotic version. They found a leading term of $O(n^{-2/p})$ for p-dimensional $\mathscr{X}$, and for the 2-nn with ties reported as 'doubt', the rate $O(n^{-4/p})$. These are very slow rates of convergence for moderate p, but since that to E_2' is faster, they suggest that for two classes it is better to estimate E_1 by doubling the estimate of E_2' (as suggested by Fukunaga & Flick, 1985).

These results only concern the bias of the true error rate based on a training set of size n as an estimator of E_k. The variability must also be taken into account if these results are to be used to bound the Bayes risk. We have seen that for odd k, $E_{k-1,\lceil k/2 \rceil} = E_{k-1}' \leqslant E^*$, which suggests bounding the Bayes risk by the achieved performance of the $(k-1, \lceil k/2 \rceil)$-nn rule. However, recall from Section 2.7 that we can achieve lower variability by not using the class labels, but averaging $1 - \max p(k \mid \mathbf{x})$ over a test or training set. If we observe k_i neighbours of class i, this suggests estimating the error rate by $\widehat{E}$, the average of $\min(k_1, k_2)/k$. In Section 2.7 we viewed $\widehat{E}$ as a biased estimate of E_k. Under our large-sample assumptions for the training set but averaging over a test set, it can be shown that (for k odd) $\mathbb{E}\widehat{E} = E_{k-1,\lceil k/2 \rceil}$, but that $\widehat{E}$ has a lower variance than the test-set error rate of the $(k-1, \lceil k/2 \rceil)$-nn rule (Devijver & Kittler, 1982, §10.8).

These results are confined to two classes. For $K \geqslant 2$ we have

Proposition 6.3 *In the large-sample theory the means of the risk-averaged $(3, 2)$-nn rule and the error rate of the $(2, 2)$-nn rule are equal and provide a lower bound for the Bayes risk. The risk-averaged estimator has smaller variance.*

Proof: Condition on $\mathbf{x}$, and draw three samples from $\eta = p(\cdot \mid \mathbf{x})$, as the observed point and its two neighbours. Let ζ be the probability that two are from one class, one from another; unless this occurs both rules score zero. The $(2, 2)$-nn rule scores one if the observed point is in the minority, so has conditional mean and mean square of $\zeta/3$. The risk-averaged 3-nn scores $1/3$ under all assignments of the three samples, so its conditional mean and mean square are $\zeta/3$ and $\zeta/9$. Conditionally and hence unconditionally the means are the same and the risk-averaged estimator has smaller variance.

Now $\zeta/3 = \sum_k (\eta_k^2 - \eta_k^3) = \Delta$, say. We will show that $\Delta \leqslant 1 - \max \eta_k$ and average over $\mathbf{x}$ to bound the Bayes risk. Suppose $\eta_1 \geqslant \eta_k$ for $k > 1$. Consider increasing η_i by δ and decreasing η_j by δ for $i, j > 1$. If $\eta_i + \eta_j < 2/3$ we can increase Δ by zeroing the smaller of η_i and

⌈ ⌉ is defined on page xii.

Check the derivative.

η_j; thus we can zero the smallest $\eta_j > 0$ in turn, transferring the mass to the next smallest. If $\eta_1 > 1/3$, we can make just one $\eta_j > 0$, which shows that $\Delta \leqslant \eta_1^2(1 - \eta_1) + (1 - \eta_1)^2\eta_1 = \eta_1(1 - \eta_1) < 1 - \eta_1$. If $\eta_1 \leqslant 1/3$ we can make just two $\eta_i, \eta_j > 0$; if $\eta_i > 1/3$ the case already proved applied to (η_i, η_1, η_j) shows $\Delta < 1 - \eta_i < 1 - \eta_1$. We can show that the result holds for three $\eta_i = 1/3$ by direct calculation. $\square$

This suggests estimating a lower bound for the Bayes risk by running the 3-nn classifier on the training set and reporting 1/3 the number of occasions on which the neighbours are two of one class, one of another (and of course one of the neighbours will be the training-set example itself). If the distances are tied, we can average over ways of breaking the tie, since this will be equivalent to averaging over perturbations of the points.

Choice of metric

We have not avoided the choice of a metric on $\mathscr{X}$, and this can again cause difficulties. In practice Euclidean distance is normally used, but after a suitable scaling of the variables. It may make sense to use an (estimated) Mahalanobis distance if the within-class distributions are roughly normal and of similar covariance matrix. For two classes, Short & Fukunaga (1980, 1981) looked at a *local* metric with the aim of minimizing the mean-square error between the finite-sample risk and the asymptotic risk (which we have seen does not depend on the metric used). They show the metric should be of the form

$$d(\mathbf{x}, \mathbf{y}) = |p(1 \mid \mathbf{x}) - p(1 \mid \mathbf{y})|,$$

expand this about $\mathbf{x}$ and estimate the coefficients of the expansion. Short & Fukunaga (1980) and Myles & Hand (1990) experiment with extensions to several classes.

Fukunaga & Flick (1984) suggest choosing a *global* quadratic metric with the same aim (and again with two classes). Their metric is a Mahalanobis distance and so they choose the inverse covariance A to obtain

$$d(\mathbf{x}, \mathbf{z}) = \sqrt{(\mathbf{x} - \mathbf{z})^T A (\mathbf{x} - \mathbf{z})}.$$

They compute the value of A which approximately minimizes the error rate, and then estimate the quantities involved (the underlying densities) by k-nn density estimation.

One of the most important steps in choosing a metric may be to exclude features which have little or no relevance. The features selected by classification trees (Chapter 7) can be a very useful guide (Ripley, 1993).

An appealing idea is to combine the features of kernel methods and k-nearest neighbour methods by distance-weighting the classes of the neighbours in reaching a decision. This has proved controversial (Dudani, 1976; Bailey & Jain, 1978; Morin & Raeside, 1981; MacLeod *et al.*, 1987; Parthasarthy & Chatterji, 1990) in that asymptotically for fixed k the distance-weighting does not help. However, this is not the correct basis for the comparison, since using distance-weighting may allow a larger value of k to be used for a given size of training set.

Data editing

One common complaint about both kernel and k-nn methods is that they can take too long to compute and need too much storage for the whole training set. The difficulties are sometimes exaggerated, as there are fast ways to find near neighbours (for example Friedman *et al.*, 1975, 1977; Fukunaga & Narendra, 1975; Kalantari & McDonald, 1983; Kamgar-Parsi & Kanal, 1985; Preparata & Shamos, 1985; Ruiz, 1986; Kim & Park, 1986; Niemann & Goppert, 1988; Bryant, 1989; Jiang & Zhang, 1993) and fast approximate ways to find kernel density estimates by binning (Härdle, 1991; Scott, 1992). However, in many problems it is only necessary to retain a small proportion of the training set to approximate very well the decision boundary of the k-nn classifier. This concept is known as *data editing*. It can also be used to improve the performance of the classifier by removing apparent outliers.

There are many editing algorithms: the literature on data editing is extensive but contains few comparisons. (It is surveyed in Dasarathy, 1991.) The *multiedit* algorithm of Devijver & Kittler (1982) can be specified as follows (with parameters I and V):

1 Put all patterns in the current set.

2 Divide the current set more or less equally into $V \geqslant 3$ sets. Use pairs cyclically as test and training sets.

3 For each pair classify the test set using the k-nn rule from the training set.

4 Delete from the current set all those patterns in the test set which were incorrectly classified.

5 If any patterns were deleted in the last I passes return to step 2.

The edited set is then used with the 1-nn rule (not the original value of k). Devijver & Kittler indicate that (for two classes) asymptotically the

Figure 6.6: Reduction algorithms applied to Figure 1.3. The known decision boundary of the Bayes rule is shown with a solid line; the decision boundary for the 1-nn rule is shown dashed.
(a) *multiedit*.
(b) The result of retaining only those points whose posterior probability of the actual class exceeds 90% when estimated from the remaining points.
(c) *condense* after *multiedit*.
(d) *reduced nn* applied after *condense* to (a).

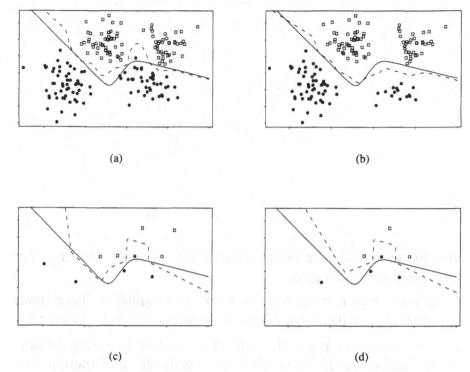

(a)

(b)

(c)

(d)

1-nn rule on the edited set out-performs the k-nn rule on the original set and approaches the performance of the Bayes rule. (The idea is that each edit biases the retained points near $\mathbf{x}$ in favour of the class given by the Bayes rule at $\mathbf{x}$, so eventually this class dominates the nearest neighbours. This applies to any number of classes.)

Figure 6.6(a) illustrates the *multiedit* algorithm applied to the synthetic dataset shown in Figure 1.3 on page 12. The Bayes rule is known in this example (since it is synthetic). In practice *multiediting* can perform much less well and drop whole classes when applied to moderately sized training sets with more dimensions and classes. Another idea (Hand & Batchelor, 1978) is to retain only points whose likelihood ratio $p_y(\mathbf{x})/p_i(\mathbf{x})$ against every class $i \neq y$ exceeds some threshold t. (The densities are estimated non-parametrically.) It make more sense to retain points for which $p(y \mid \mathbf{x})$ is high, for example those which attain a majority ℓ in a (k,ℓ)-rule for a larger value of k. This is illustrated in Figure 6.6(b) for the synthetic example using the (10,9)-nn.

Earlier editing algorithms were given by Wilson (1972), Wagner (1973), Tomek (1976b) and Penrod & Wagner (1977).

The *multiedit* algorithm aims to form homogeneous clusters in $\mathscr{X}$. However, only the points on the boundaries of the clusters are really effective in defining the classifier boundaries. *Condensing* algorithms

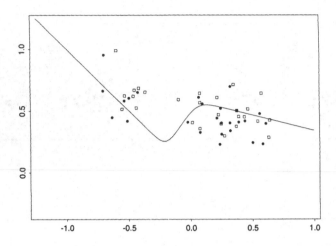

Figure 6.7: The result of the *reduced nearest neighbour rule* of Gates (1972) applied after *condense* to the unedited data of Figure 1.3.

aim to retain only the crucial exterior points in the clusters. For example, Hart (1968) gives:

1 Divide the current patterns into a store and a grabbag. One possible partition is to put the first point in the store, the rest in the grabbag.

2 Classify each sample in the grabbag by the 1-nn rule using the store as training set. If the sample is incorrectly classified transfer it to the store.

3 Return to 2 unless no transfers occurred or the grabbag is empty.

4 Return the store.

This is illustrated in Figure 6.6(c).

A refinement, the *reduced nearest neighbour rule* of Gates (1972), is to go back over the condensed training set and drop any patterns (one at a time) which are not needed to correctly classify the rest of the (edited) training set. As Figure 6.6(d) shows, this can easily go too far and drop whole regions of a class. Other attempts at condensing and reducing are discussed by Swonger (1972), Ullmann (1974), Ritter *et al.* (1975), Tomek (1976c), Chidananda Gowda & Krishna (1979) and Fukunaga & Mantock (1984).

Nearest neighbour methods will give low apparent error rate (zero for 1-nn) so it is essential to use other forms of performance assessment. Fortunately the leave-one-out cross-validated error rate can be computed as easily as the apparent error rate by finding the k neighbours of $\mathbf{x}$ excluding $\mathbf{x}$ itself.

Essentially the same ideas have been considered within the field of machine learning, known as 'memory-based learning' (for example, Stanfill & Waltz, 1986). In its simplest form this is just a 1-nn classifier based on storing all the examples. Editing and condensing techniques have also been considered (E. E. Smith & Medin, 1981; Kibler & Aha, 1987). These seem re-discoveries of the simplest such algorithms.

Examples

The variables in the Pima Indians diabetes example were on very different scales, and so were each scaled to have range about one over the training set. The initial choice of k was made by both leave-one-out and 10-fold cross validation. For the latter the error rates were 57, 65, 57, 54, 51 and 55 (out of 200) for $k = 1, 3, 5, 7, 9, 11$; for leave-one-out we obtained 58, 65, 60, 55, 49 and 41. On the test set the numbers of errors were 98 for $k = 1$ and 82 for $k = 9$, out of 332. This is an example in which there is considerable mixing between the classes, and local methods will be unable to pick up the broad trends. Using the 3-nn rule on the training set suggested a lower bound of about 15% for the Bayes error in this problem; linear methods achieve about 20%.

For the forensic glass data, there are many rational ways to choose the metric, since the features are eight proportions plus the refractive index. We chose to rescale the refractive index to about ± 10 but not to scale the compositional features. With this metric, risk-averaging the 3-nn rule suggests a lower bound of 14% for the Bayes risk. Each of the 1-nn, 3-nn and 5-nn rules had a similar level of performance; the 1-nn rule had a cross-validated error rate of 23.4% and confusion matrix

	WinF	WinNF	Veh	Con	Tabl	Head
WinF	59	7	4	0	0	0
WinNF	12	59	3	2	0	0
Veh	2	5	10	0	0	0
Con	0	2	0	8	1	2
Tabl	1	0	0	2	6	0
Head	1	3	1	1	1	22

Comparing this with the confusion matrices for larger values of k shows the effect of very different proportions for the six classes; as k increases fewer errors are made on the more abundant classes, whereas more are made on the rare classes.

6.3 Learning vector quantization

The refinements of the k-nn rule aim to choose a subset of the training set in such a way that the 1-nn rule based on this subset approximates the Bayes classifier. It is not necessary that the modified training set is a subset of the original and an early step to combine examples to form prototypes was taken by Chang (1974). The approach taken in Kohonen's (1988a, b, 1990a, b, 1995) *learning vector quantization* is to construct a modified training set iteratively. Following Kohonen, we call the modified training set the *codebook*. This procedure tries to represent the decision boundaries rather than the class distributions. Once again the metric in the space $\mathcal{X}$ is crucial, so we assume the variables have been scaled in such a way that Euclidean distance is appropriate (at least locally).

Vector quantization

The use of 'vector quantization' is potentially misleading, since it has a different aim, but as it motivated Kohonen's algorithm we will digress for a brief description.

Vector quantization is a classical method in signal processing to produce an approximation to the distribution of a single class by a codebook. Each incoming signal is mapped to the nearest codebook vector, and that vector sent instead of the original signal. Of course, this can be coded more compactly by first sending the codebook, then just the indices in the codebook rather than the whole vectors. One way to choose the codebook is to minimize some measure of the approximation error averaged over the distribution of the signals (and in practice over the training patterns of that class). Taking the measure as the squared distance from the signal to the nearest codebook vector leads to the *k*-means algorithm which aims to minimize the sum-of-squares of distances within clusters (Section 9.3). An 'on-line' iterative algorithm for this criterion is to present each pattern $\mathbf{x}$ in turn, and update the codebook by

Gersho & Gray (1992) is a reference text on vector quantization; a short introduction is given by Gray (1984).

$$\mathbf{m}_c \leftarrow \mathbf{m}_c + \alpha(t)[\mathbf{x} - \mathbf{m}_c] \qquad \text{if } \mathbf{m}_c \text{ is closest to } \mathbf{x} \qquad (6.9)$$
$$\mathbf{m}_i \leftarrow \mathbf{m}_i \qquad\qquad \text{for the rest of the codebook.}$$

Update rule (6.9) motivated Kohonen's iterative algorithms. Note that this is not a good algorithm for *k*-means; better algorithms are discussed in Section 9.3.

Max (1960) and Zador (1982) have pointed out that choosing the average rth power of the distance as the measure of approximation error amounted to choosing the density of the codebook vectors to approximate the $d/(d + r)$th power of the true probability density in d dimensions. Thus for d large the *k*-means procedure codes an approximation to $p(\mathbf{x})$. This is an asymptotic result for large codebooks, but does indicate that a codebook produced by vector quantization for each class might be a good initial reduction of a very large training set.

Iterative algorithms

Kohonen (1990a) advocated a series of iterative procedures which has since been modified; our description follows the implementation known as LVQ_PAK documented in Kohonen *et al.* (1992). A initial set of codebook vectors is chosen from the training set. (We discuss later precisely how this might be done.) Each of the procedures moves codebook vectors to try to achieve better classification of the training set by the 1-nn rule based on the codebook. The examples from the training set are presented one at a time, and the codebook is updated after each presentation. In our experiments the examples were chosen randomly from the training set, but one might cycle through the training set in some pre-specified order.

The original procedure LVQ1 uses the following update rule. A example $\mathbf{x}$ is presented. The nearest codebook vector to $\mathbf{x}$, $\mathbf{m}_c$, is updated by

$$\mathbf{m}_c \leftarrow \mathbf{m}_c + \alpha(t)[\mathbf{x} - \mathbf{m}_c] \qquad \text{if } \mathbf{x} \text{ is classified correctly by } \mathbf{m}_c$$
$$\mathbf{m}_c \leftarrow \mathbf{m}_c - \alpha(t)[\mathbf{x} - \mathbf{m}_c] \qquad \text{if } \mathbf{x} \text{ is classified incorrectly} \qquad (6.10)$$

and all other codebook vectors are unchanged. Initially $\alpha(t)$ is chosen smaller than 0.1 (0.03 by default in LVQ_PAK) and it is reduced linearly to zero during the fixed number of iterations. The effect of the updating rule is to move a codebook vector towards nearby examples of its own class, and away from ones of other classes. 'Nearby' here can cover quite large regions, as the codebook will typically be small and in any case will cover $\mathcal{X}$ rather sparsely. Kohonen (1990a) motivates this as applying vector quantization to the function $|\pi_1 p_1(\mathbf{x}) - \pi_2 p_2(\mathbf{x})|$ for two classes (or the two classes which are locally most relevant).

A variant, OLVQ1, provides learning rates $\alpha_c(t)$ for each codebook vector, with an updating rule for the learning rates of

$$\alpha_c(t) = \frac{\alpha_c(t-1)}{1 + (-1)^{I(\text{classification is incorrect})} \alpha_c(t-1)}. \qquad (6.11)$$

This decreases the learning rate if the example is correctly classified, and increases it otherwise. Thus codebook vectors in the centre of classes will have rapidly decreasing learning rates, and those near class boundaries will have increasing rates (and so be moved away from the boundary quite rapidly). As the learning rates may increase, they

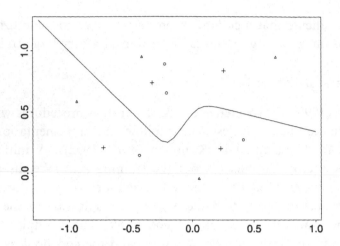

Figure 6.8: Results of learning vector quantization applied to Figure 1.3. The initially chosen codebook is shown by small circles, the result of OLVQ1 by + and subsequently applying 25,000 passes of LVQ2.1 by triangles. The known decision boundary of the Bayes rule is also shown.

are constrained not to exceed an upper bound, often 0.3. Practical experience shows that the convergence is usually rapid, and LVQ_PAK uses 40 times as many iterations as codebook vectors.

An explanation of this rule is given by Kohonen *et al.* (1992) and Kohonen (1995, p. 180) which we interpret as follows. At all times the codebook vectors are a linear combination of the training set vectors (and their initializers, if these are not in the training set). Let $s(t) = (-1)^{I(\text{classification is incorrect})}$, so we can rewrite (6.10) as

$$
\begin{aligned}
\mathbf{m}_c(t+1) &= \mathbf{m}_c(t) + s(t)\alpha(t)[\mathbf{x}(t) - \mathbf{m}_c] \\
&= [1 - s(t)\alpha(t)]\mathbf{m}_c(t) + s(t)\alpha(t)\mathbf{x}(t) \\
&= [1 - s(t)\alpha(t)][1 - s(t-1)\alpha(t-1)]\mathbf{m}_c(t-1) \\
&\quad + [1 - s(t)\alpha(t)]s(t-1)\alpha(t-1)\mathbf{x}(t-1) + s(t)\alpha(t)\mathbf{x}(t).
\end{aligned}
$$

Now suppose $\mathbf{x}(t-1) = \mathbf{x}(t)$ and the same codebook vector is closest at both times (so $s(t-1) = s(t)$). If we ask that the multiplier of $\mathbf{x}(t)$ is the same in both terms, we find

$$
[1 - s(t)\alpha(t)]\alpha(t-1) = \alpha(t)
$$

which gives (6.11). This adaptive choice of rate seems to work well, as in our examples.

The procedure LVQ2.1 (Kohonen, 1990b) tries harder to approximate the Bayes rule by pairwise adjustments of the codebook vectors. Suppose $\mathbf{m}_s$, $\mathbf{m}_t$ are the two nearest neighbours to $\mathbf{x}$. They are updated simultaneously provided that $\mathbf{m}_s$ is of the same class as $\mathbf{x}$ and the class of $\mathbf{m}_t$ is different, and $\mathbf{x}$ falls into a 'window' near the mid-point of $\mathbf{m}_s$ and $\mathbf{m}_t$. Specifically, we must have

$$
\min\left(\frac{d(\mathbf{x}, \mathbf{m}_s)}{d(\mathbf{x}, \mathbf{m}_t)}, \frac{d(\mathbf{x}, \mathbf{m}_t)}{d(\mathbf{x}, \mathbf{m}_s)}\right) > \frac{1-w}{1+w}
$$

for $w \approx 0.25$. (We can interpret this condition geometrically. If $\mathbf{x}$ is projected onto the vector joining $\mathbf{m}_s$ and $\mathbf{m}_t$, it must fall at least $(1 - w)/2$ of the distance from each end.) If all these conditions are satisfied the two vectors are updated by

$$\mathbf{m}_s \leftarrow \mathbf{m}_s + \alpha(t)[\mathbf{x} - \mathbf{m}_s], \qquad (6.12)$$
$$\mathbf{m}_t \leftarrow \mathbf{m}_t - \alpha(t)[\mathbf{x} - \mathbf{m}_t].$$

This rule may update the codebook only infrequently. It tends to over-correct, as can be seen in Figure 6.8, where the result of iterating LVQ2.1 is to push the codebook vectors away from the decision boundary, and eventually off the figure. Thus it is recommended that LVQ2.1 only be used for a small number of iterations (30–200 times the number of codebook vectors).

The rule LVQ3 tries to overcome over-correction by using LVQ2.1 if the two closest codebook vectors to $\mathbf{x}$ are of different classes, and

$$\mathbf{m}_i \leftarrow \mathbf{m}_i + \epsilon\alpha(t)[\mathbf{x} - \mathbf{m}_i] \qquad (6.13)$$

for ϵ around 0.1–0.5, for each of the two nearest codebook vectors if they are of the same class as $\mathbf{x}$. (The window is only used if the two codebook vectors are of different classes.) This introduces a second element into the iteration, of ensuring that the codebook vectors do not become too unrepresentative of their class distribution. It does still allow the codebooks to drift to the centre of the class distributions and even beyond, as Figure 6.9 shows.

The recommended procedure is to run OLVQ1 until convergence (usually rapid) and then a moderate number of further steps of LVQ1 and/or LVQ3.

de Sa & Ballard (1993) motivate a variant of LVQ2.1 by applying stochastic approximation to a kernel regression estimator of $|\pi_1 p_1(\mathbf{x}) - \pi_2 p_2(\mathbf{x})|$ for two classes. This normalizes the step size (replacing $[\mathbf{x} - \mathbf{m}_s]$ by $[\mathbf{x} - \mathbf{m}_s]/\|\mathbf{x} - \mathbf{m}_s\|$) and reduces the window size as well as $\alpha(t)$.

Initialization

The package LVQ_PAK chooses the initial codebook vectors from amongst the training set vectors. It is desirable that the initial vectors lie inside the Bayes boundary for their class, as otherwise they end up representing 'islands' in the classification induced by the 1-nn rule based on the training set which are merely noise. Thus the candidates for initialization are screened to ensure that they are correctly classified by a k-nn rule for moderate k (default 7).

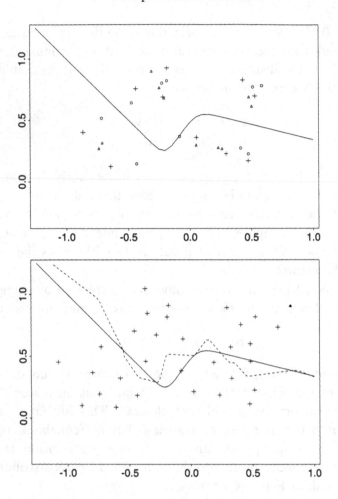

Figure 6.9: Further results of LVQ with a larger codebook. This time the triangles show the results from LVQ3.

Figure 6.10: Result of OLVQ1 with codebook of size 30. The decision boundary is shown as a dashed line.

Kohonen (1990a) advocates a more constructive approach to the initial codebook, for example using a vector quantization of the whole training set (ignoring class) or his own 'self-organizing map' method (Section 9.4).

A further question is how many codebook vectors to use, and how they should be partitioned amongst the classes. The literature seems to assume that the larger number the better, as this will enable a better piecewise linear approximation to the decision boundaries of the Bayes classifier. However, this argument assumes that the iterative algorithms can make a good job of distributing the codebook, and as Figures 6.9 and 6.10 show, this is not necessarily so. We may do better to start with a better-designed codebook such as the result of editing procedures.

It is unclear how many codebooks vectors should be selected for each class, since the number needed depends as much on how well they are employed as on the proportions of the class. Kohonen *et al.* (1992) suggest using equal numbers per class initially, and altering the proportions so that the median of the nearest neighbour distances between vectors within a class is roughly similar across classes.

Examples

For the Pima Indians data we used OLVQ1 plus LVQ2 or LVQ3 with 2–4 vectors per class, and found a test-set error rate of around 70/332. Increasing this to 10 vectors per class increased the error rate to about 75/332.

A procedure consisting of initializing, running OLVQ1 to convergence then 10,000 iterations of LVQ3 was run under the standard 10-fold cross-validation for the forensic glass data. The estimated error rate was 30.4%. Assigning equal numbers of vectors to each class (4 each) reduced this slightly, to 29.9%, making less errors in the containers and tableware groups, and more for non-float window glass.

6.4   Mixture representations

Another way of looking at kernel density estimation with a non-negative kernel is that it represents a probability density by

$$\widehat{p}(\mathbf{x}) = \sum_i w_i f_i(\mathbf{x})$$

where the densities $f_i(\mathbf{x}) = K(\mathbf{x} - \mathbf{x}_i)$ and the weights are uniform (or $n\pi_{[i]}/n_{[i]}$ for known class probabilities). Then $\widehat{p}_j(\mathbf{x})$ is of the same form, but giving weights $1/n_j$ to points from class j and zero to the others. Vector quantization uses a somewhat more general uniformly-weighted mixture. This suggests representing densities by a mixture of a fixed set of densities, so

$$\widehat{p}_j(\mathbf{x}) = \sum_i w_{ij} f_i(\mathbf{x}), \qquad \widehat{p}(\mathbf{x}) = \sum_i \left[\sum_j \pi_j w_{ij} \right] f_i(\mathbf{x}) \tag{6.14}$$

and

$$\widehat{p}(j \mid \mathbf{x}) = \frac{\pi_j \sum_i w_{ij} f_i(\mathbf{x})}{\sum_i \left[\sum_j \pi_j w_{ij} \right] f_i(\mathbf{x})}. \tag{6.15}$$

We may also want to allow parameters within f_i, for example the means and in the covariance matrix of a normal density, so we will write $f_i(\mathbf{x}; \theta_i)$. We considered the case of a two-component normal mixture in Chapter 2 on pages 41–42.

Mixture densities of this sort have been considered occasionally as general models for density estimation (for example by Roeder, 1990).

We could estimate the class-conditional densities p_j by choosing the w_{ij} on the basis of the training samples labelled by class j and then use (6.15) for classifying future samples. It is also possible to use unclassified cases to assist in the estimation process, as proposed in this context by Tråvén (1991). Sebestyen (1962) proposed an iterative process of approximating the class-conditional densities by Gaussian mixtures; two recent accounts also using Gaussian mixtures are Chou & Chen (1992) and Streit & Luginbuhl (1994), both of which regard this as a 'neural network' method.

There is an extensive literature on the estimation of mixture distributions surveyed by Redner & Walker (1984), Titterington *et al.* (1985) and McLachlan & Basford (1988). It is straightforward to write down the likelihood for the observed training patterns $(\mathbf{x}^p, y^p)$ and any unclassified patterns $\mathbf{z}^u$ as

$$\prod_p \pi_{y^p} \sum_i w_{iy^p} f_i(\mathbf{x}^p; \theta_i) \prod_u \sum_{i,j} \pi_j w_{ij} f_i(\mathbf{z}^u; \theta_i)$$

but finding a maximum may be another matter. Indeed, it is possible that the maximum may occur with the components $f_i(\mathbf{x}; \theta_i)$ degenerating around observed points. This can be avoided by suitably constraining the parameters θ_i. One favourite device for numerically finding a maximum is the EM algorithm (Section A.2 has the details). This pretends the example really did come from one of the components $f_i(\mathbf{x}; \theta_i)$, say I, but this is unobserved. The posterior probabilities of $I = i$ given $\mathbf{x}^p$ are then used to weight the various components in the mixture; for example for a general Gaussian mixture we use weighted mean and covariance estimators for each mixture component. Of course, the posterior probabilities depend on the parameters, so the process must be iterated.

The convergence of the EM algorithm is notoriously slow (Redner & Walker, 1984), and it may be better to use a conventional numerical optimization technique. It is unclear whether this reputation is justified, as it may be much easier to find a nearly-optimal solution (in the sense of high log-likelihood) than to find the maximizing parameters precisely. (For pattern recognition purposes, only a good approximation to the mixture density is needed.) There are a number of ways to find good starting points for the optimization, for example by *ad hoc* partition of the space $\mathcal{X}$ and fitting a component to the patterns falling in each partition. To emphasize the importance, note that if the EM algorithm is applied to each class of the synthetic data of Figure 1.3 for a two-component mixture with equal covariance matrices (the truth), it becomes trapped in a poor local minimum from starting points which do not separate the group means sufficiently on the x axis. Ingrassia (1992) demonstrates that both the standard EM algorithm and his implementation of simulated annealing can find an inappropriate local minimum in univariate multimodal normal mixture problems with quite high probability.

A Bayesian approach will have a prior on the parameters (θ_i) and on the mixing proportions w_{ij}. These are generally taken to be independent, and the proportions for each class given a Dirichlet distribution. The EM algorithm can be used to find a posterior mode for the parameters, but the predictive distributions $\widetilde{p}(j\,|\,\mathbf{x})$ can only be found by the iterative simulation methods discussed in Section A.3 (Diebolt & Robert, 1994; Gelman *et al.*, 1995). The simplest way to apply the Gibbs sampler will be in its 'blocked' form, simulating alternately all the components I^p for the examples given the component parameters, then the component parameters given (I^p).

In the spirit of neural networks, an 'on-line' approach has been sought, in which the parameter estimates are adjusted whenever an example is presented (and the training set will be presented many times). Tråvén (1991) considers an on-line approximation to the EM algorithm for general Gaussian mixtures (with separate covariance matrices). The current estimates of the means and variance are found as weighted means and variances. Those weights depend on the current parameter estimates, so the estimates cannot be updated exactly when a new example is presented unless the data are retained. When a new example $\mathbf{x}$ is observed, Tråvén uses

$$\widehat{\mu}_i \leftarrow \widehat{\mu}_i + \eta_i(\mathbf{x} - \widehat{\mu}_i)$$
$$\widehat{\Sigma}_i \leftarrow \widehat{\Sigma}_i + \eta_i(\mathbf{x} - \widehat{\mu}_i)(\mathbf{x} - \widehat{\mu}_i)^T$$

This update formula is not quite correct, as $\widehat{\Sigma}_i$ depends on the new $\widehat{\mu}_i$.

for each component, where $(\eta_i) = p(i\,|\,\mathbf{x})/[p(i\,|\,\mathbf{x})+\sum_{j=1}^{N} p(i\,|\,\mathbf{x}^p)]$. This is unavailable, and is approximated by $p(i\,|\,\mathbf{x})/(N+1)p_i$ on the assumption that the $\mathbf{x}^p$ were a random sample from the class and N is large. Using a constant rather than N will allow 'forgetting' of examples. The only advantage of this procedure seems to be to avoid storing the data, and if the dataset is very large it may well be sufficient to use a smaller sample to estimate the parameters.

An alternative way to estimate $p(j\,|\,\mathbf{x})$ would be to regress the class indicator on the variables $f_i(\mathbf{x})$, which will differ from (6.15) and which for radially symmetric component functions has been discussed in Section 4.2. As there, we have avoided the question of choosing the number of component densities which can in general only be done using knowledge of the complexity of the densities $p_i(\mathbf{x})$ or by cross-validation.

Vector quantization can be seen as a special case of a finite mixture, in which the components are constant densities over the tiles of the Dirichlet tessellation formed by the codebook. There have also been methods to design modified training sets for use with kernel methods, and these can be seen as using finite normal mixtures (or normalized Gaussian radial basis functions) with equally weighted components. For example, Specht (1991) describes a simple clumping method to select cluster centres, and Burrascano (1991) uses LVQ to select a set of centres for the normals.

Hastie & Tibshirani (1996) explore normal mixtures with a common covariance matrix Σ for all components in all classes. This is a rather restrictive assumption, but does allow all the components to be rescaled simultaneously to $\mathsf{N}_p\{\mu, I\}$ as in linear discriminant analysis. Each class is then represented by a distribution over component means rather than a single mean, but the between-groups covariance matrix can still be decomposed to find 'canonical variates' on which the data may be displayed. This model is similar to using LVQ with the Mahalanobis distance for Σ, as the latter can be considered to be using equally-weighted mixtures. The LVQ model has the advantage of choosing the codebook vectors for good discrimination, and mixture models for the classes suffer from modelling the class populations accurately in regions where this is not needed.

Choosing the number of mixture components is notoriously difficult (McLachlan & Basford, 1988; Peck *et al.*, 1989; Furman & Lindsay, 1994).

Examples

Finding a good local maximum for maximum likelihood fitting of a mixture of normals is difficult in practice, and we found a wide range of fits from different starting points. The k-means algorithm of Section 9.3 was used to initialize the means, with the covariance matrices started at the within-cluster covariance matrix. However, k-means is itself a random algorithm subject to local minima, so the whole procedure was run several times and the best fit selected. The k-means procedure depends on the distance used in $\mathscr{X}$; Euclidean distance was used after careful scaling of the features.

Figure 6.11 shows the plug-in classifiers for normal mixtures fitted to the synthetic dataset (which was generated from a normal mixture). In this case unequal mixtures were used (to avoid biasing the fit too

Figure 6.11: The decision boundaries for the plug-in Bayes classifier for mixture models fitted to the synthetic dataset. The dashed line corresponds to fitting two normals with unequal covariances to each class; the dotted line to fitting five components to each class, whose means are shown by +.

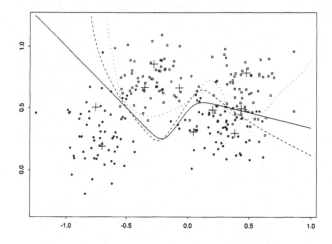

much to the correct model). Fitting five rather than two components per class decreased the deviance by about 20 but used 30 extra parameters. The use of AIC strongly suggested two components per class, but the presence of many local minima makes this theoretically dubious (and we know that the global maximum has infinite log-likelihood).

For the Pima Indians data we saw a suspicion of bimodal distributions on page 99. Using two normal components per group with a common covariance matrix (between as well as within groups) gave a test-set error of 64/332, a negligible improvement over linear discriminant analysis. Allowing different common covariance matrices within classes achieved 84/332, comparable with quadratic discriminant analysis.

For the forensic glass data some of the classes are too small to fit even a single normal density, so we need to use a common covariance matrix over all the classes and components. Some trials suggested that three components per class was a reasonable compromise, for which the cross-validated error rate was 30.8%, more than for LVQ.

7

Tree-structured Classifiers

The use of tree-based methods for classification is relatively unfamiliar in both statistics and pattern recognition, yet they are widely used in some applications such as botany (Figure 7.1) and medical diagnosis as being extremely easy to comprehend (and hence have confidence in).

The automatic construction of decision trees dates from work in the social sciences by Morgan & Sonquist (1963) and Morgan & Messenger (1973). (Later work such as Doyle, 1973, and Doyle & Fenwick, 1975, commented on the pitfalls of such automated procedures.) In statistics Breiman *et al.* (1984) had a seminal influence both in bringing the work to the attention of statisticians and in proposing new algorithms for constructing trees. At around the same time decision tree induction was beginning to be used in the field of *machine learning*, which we review in Section 7.4, and in engineering (for example, Sethi & Sarvarayudu, 1982).

The terminology of trees is graphic, although conventionally trees such as Figure 7.2 are shown growing down the page. The *root* is the top node, and examples are passed down the tree, with decisions being made at each *node* until a terminal node or *leaf* is reached. Each non-terminal node contains a question on which a split is based. Each leaf contains the label of a classification. A *subtree* of T is a tree with root a node of T; it is a *rooted subtree* if its root is the root of T.

A classification tree partitions the space $\mathcal{X}$ of possible observations into sub-regions corresponding to the leaves, since each example will be classified by the label of the leaf it reaches. Thus decision trees can be seen as a hierarchical way to describe a partition of $\mathcal{X}$. We could give the botanist a description of each species and ask for the description which matches the current specimen. Even in small domains this can be too difficult, and a decision tree provides a structured description of the knowledge base. Often the same information can be structured in

1.	Leaves subterete to slightly flattened, plant with bulb	2.
	Leaves flat, plant with rhizome	4.
2.	Perianth-tube > 10mm	**I. × hollandica**
	Perianth-tube < 10mm	3.
3.	Leaves evergreen	**I. xiphium**
	Leaves dying in winter	**I. latifolia**
4.	Outer tepals bearded	**I. germanica**
	Outer tepals not bearded	5.
5.	Tepals predominately yellow	6.
	Tepals blue, purple, mauve or violet	8.
6.	Leaves evergreen	**I. foetidissima**
	Leaves dying in winter	7.
7.	Inner tepals white	**I. orientalis**
	Tepals yellow all over	**I. pseudocorus**
8.	Leaves evergreen	**I. foetidissima**
	Leaves dying in winter	9.
9.	Stems hollow, perianth-tube 4–7mm	**I. sibirica**
	Stems solid, perianth-tube 7–20mm	10.
10.	Upper part of ovary sterile	11.
	Ovary without sterile apical part	12.
11.	Capsule beak 5–8mm, 1 rib	**I. enstata**
	Capsule beak 8–16mm, 2 ridges	**I. spuria**
12.	Outer tepals glabrous, many seeds	**I. versicolor**
	Outer tepals pubescent, 0–few seeds	**I. × robusta**

Figure 7.1: Key to British species of the genus *Iris*. Simplified from Stace (1991) p. 1140, by omitting parts of his descriptions.

other ways. Many botanical trees amount to a set of rules describing one class, so each class is eliminated in turn. Another research area in machine learning has been to induce sets of rules from a training set, either directly or via an induced tree (e.g. Michalski, 1980; Quinlan, 1987a, b, 1993).

The idea of tree induction is to construct a decision tree from a set of examples, which is how humans construct trees. It is usual to do so by growing the tree, that is by successively splitting leaves. Tree construction is easiest when there is an exact partition of $\mathscr{X}$, that is one which classifies every example correctly. The alternative, in which the distributions of observations from the classes overlap, is often called a *noisy* classification problem. For the exact case, we need to continue to grow the tree until every example is classified correctly. In a noisy problem to do so would over-fit the examples at hand, and the two possible strategies are to stop growing the tree early, or to prune the tree after constructing, closely analogous to forwards and backwards selection in regression.

Confusingly, these strategies are sometimes called pre- and post-pruning.

Table 7.1: Example decisions for the space shuttle autolander problem, from Michie (1989).

stability	error	sign	wind	magnitude	visibility	decision
any	any	any	any	any	no	auto
xstab	any	any	any	any	yes	noauto
stab	LX	any	any	any	yes	noauto
stab	XL	any	any	any	yes	noauto
stab	MM	nn	tail	any	yes	noauto
any	any	any	any	Out of range	yes	noauto
stab	SS	any	any	Light	yes	auto
stab	SS	any	any	Medium	yes	auto
stab	SS	any	any	Strong	yes	auto
stab	MM	pp	head	Light	yes	auto
stab	MM	pp	head	Medium	yes	auto
stab	MM	pp	tail	Light	yes	auto
stab	MM	pp	tail	Medium	yes	auto
stab	MM	pp	head	Strong	yes	noauto
stab	MM	pp	tail	Strong	yes	auto

Figure 7.2: Decision tree for shuttle autolander problem. The numbers m/n denote the proportion of training examples reaching that node which are misclassified.

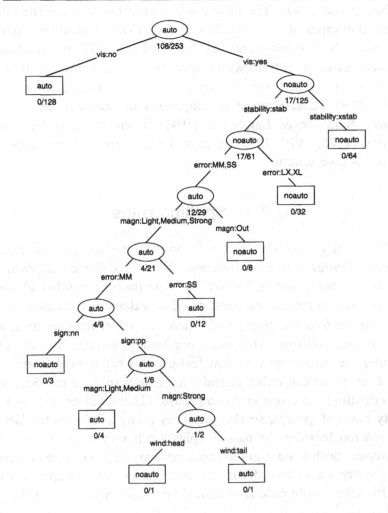

The main differences between algorithms for tree construction are the pruning strategy used and the exact rule for splitting nodes. Many algorithms only allow binary splits, that is to divide a node into two; a few allow multi-way splits (for example by flower colour). Note that these are just algorithms; there are only very simple models and no deep theorems in this field.

There are two types of optimality to be considered. One is optimality of the partition of $\mathscr{X}$, which can be judged by the error rate achieved. In principle we could seek an optimal partition amongst all prescribed partitions of $\mathscr{X}$, for example those representable by a set of decision rules splitting on a single feature. This is a computationally infeasible procedure for all but the smallest problems, but the step-wise construction of a partition by a decision tree can be seen as an approximation to finding the optimal partition.

The other sense of optimality is to represent a partition by a tree in the best possible way. The most obvious criterion is to use the minimal expected number of tests. Hyafil & Rivest (1976) showed this particular problem to be NP-complete; Payne & Meisel (1977) give an algorithm to construct optimal trees with respect to fairly general cost functions.

There are a number of partial surveys of the literature. Dietterich (1990) covers 'recent developments in practical learning algorithms'. Safavian & Landgrebe (1991) is wide-ranging but shallow. Quinlan (1986, 1990, 1993) surveys the machine-learning approaches within his own school.

7.1   Splitting rules

In this section and the next we consider the component pieces of currently favoured tree-construction algorithms. Some historical alternatives are mentioned in Section 7.4. Note that the number of possible trees is vast, so there is no question of an exhaustive search over trees.

Consider first splitting a leaf. There is a set of features from which to construct splitting attributes. For binary features we will clearly consider the binary split on that feature. For categorical features with $L > 2$ levels we can either consider an L-way split, or consider binary splits dividing the levels into two groups. (There will be $2^{L-1} - 1$ non-empty pairs of groups, so this generates many attributes for large L.) For ordered features the natural splits are binary of the form $x \leqslant x_c$; this applies both to continuous measurements and to ordered categories. Some systems also consider linear combinations of continuous features and Boolean combinations of logical ones. (See Section 7.5.)

Each leaf will have a set of *attributes* A on which it might be split. How should we consider the value of the split? There have been many suggestions from several different viewpoints. Consider first a population viewpoint. That is, there is a known probability distribution over $\mathscr{X} \times \mathscr{C}$ of examples which would reach that leaf. This gives a marginal probability distribution p_k over $\mathscr{C}$. Consider splitting on attribute A which has levels $a_1, \ldots, a_m$. There is then a probability distribution p_{ik} over attributes and classes, and the child leaf corresponding to $A = a_i$ would have probability distribution $p(k \mid a_i) = p_{ik}/p_{i\cdot}$ over classes k.

$\mathscr{C}$ is the set of classes.

$A\cdot$ denotes summation over that index.

We can then ask if the child nodes are on average 'purer' than their parent. A measure of impurity should according to Breiman *et al.* (1984, p. 24) be zero if p_j is concentrated on one class, and maximal if p_j is uniform. Two commonly used measures of impurity are the *entropy*

$$i(p) = -\sum_j p_j \log p_j$$

(where $0 \log 0 = 0$) and the *Gini index*

$$i(p) = \sum_{i \neq j} p_i p_j = 1 - \sum_j p_j^2.$$

One interpretation of the Gini index is the expected error rate if the label is chosen randomly from the class distribution at the node. (It may be better to use this than the error rate from the Bayes rule at the node since it gives an element of 'look ahead'. Quite often no feasible split reduces the error rate, yet after two or three splits large reductions in error rate emerge; see the right-hand branch of Figure 7.2.)

The decrease in average impurity on splitting by attribute A is then

$$i(p_c) - \sum_{i=1}^{m} p_{i\cdot} \times i(p(c \mid a_i)).$$

A common approach is to choose the split that maximizes this. Since this will in general favour many-valued attributes, Breiman *et al.* and many others confine attention to binary attributes. (See Section 7.4 for adjustments for multi-way splits.)

Breiman *et al.* preferred the Gini index. The entropy index has been used widely, for example by Sethi & Sarvarayudu (1982) and Quinlan (1983) in the engineering and machine learning literature respectively.

The premise of the following proposition holds for both the entropy and Gini measures of impurity. Part (ii) reduces the number of attributes which need consideration for two classes from $2^{L-1} - 1$ to $L - 1$, but

it has no simple extension to three or more classes. (The result is due
to Breiman *et al.*, but the very much shorter proof is original.)

Proposition 7.1 *Suppose $i(p)$ is strictly concave.*

(i) *The decrease in impurity is non-negative, and zero if and only if the
the distributions are the same in all children.*

(ii) *Suppose there are two classes. For a categorical feature, order
the levels in increasing $p(1 \mid x = x_i)$. Then a split of the form
$\{x_1, \ldots x_\ell\}, \{x_{\ell+1}, \ldots, x_L\}$ maximizes the reduction in average impurity.*

Proof: (i) We have by Jensen's inequality   See the glossary.

$$\sum_i p_i \cdot i(p(c \mid a_i)) \leqslant i(\sum_i p_i \cdot p(c \mid a_i)) = i(p_c)$$

with equality if and only if $p(c \mid a_i) = p(c)$ for all i and c.

(ii) With just two classes we can regard $i(p)$ as a function of p_1
only; it remains strictly concave. Consider dividing into two groups by
allocating to group 1 with probability a_i when $x = x_i$. Then

(a) the average impurity of the two groups is minimized by taking
$a_i = 0$ or 1 by concavity, and

(b) the partial right derivative of the average impurity with respect to
a_i (which exists by concavity) at $a_i = 0$ is of the form

$$p(X = x_i)[Ap(1 \mid x = x_i) + B]$$

for constants A and B, and so is positive (when the optimal solution
is to allocate x_i to group 1) for all $i \leqslant \ell$ or all $i > \ell$ for some ℓ.

and both examples lead to the postulated form of split since which
group is labelled 1 is arbitrary. $\square$

   Another way to look at this approach is to define the average
impurity of the tree as

$$I(T) = \sum_{\text{leaves } t} q_t i(p(c \mid t))$$

where q_t is the probability an example reaches node t. The decrease in
I on splitting the node is then q_t times the decrease in node impurity
we considered before, so that strategy is equivalent to splitting the node
to minimize the average tree impurity.

Of course, to use this population approach we estimate all the probabilities by frequencies in the training-set examples reaching the node. Ciampi *et al.* (1987) and Clark & Pregibon (1992) take another approach, viewing the tree as a probability model for the training set. For each node t there is a probability π_{tc} that an example reaches that node and is of class c, which can in principle be computed from the distribution over $\mathcal{X} \times \mathcal{C}$ by partitioning. Suppose we condition on the features for all the examples in the training set. We then know the number n_t of examples which will reach leaf t, and the numbers n_{tc} of each class at that node will have a multinomial distribution with probabilities $\pi_{c|t} = \pi_{tc}/\pi_{t\cdot}$. The conditional likelihood is then proportional to

$$\prod_{\text{leaves } t} \prod_{\text{classes } c} \pi_{c|t}^{n_{tc}}$$

and this allows us to write a deviance for the tree probability model as

$$D(T) = \sum_{\text{leaves } t} D_t, \qquad D_t = -2 \sum_{\text{classes } c} n_{tc} \log \pi_{c|t}.$$

(This *is* a deviance, since in the perfect model $\pi_{c|t} = 1$ whenever $n_{tc} > 0$ at a leaf t.) If we estimate π_{tc} by the maximum likelihood estimate $\hat{\pi}_{c|t} = n_{tc}/n_t$, the maximized deviance is

$$D(T) = 2 \left[\sum_t n_t \log n_t - \sum_{t,c} n_{tc} \log n_{tc} \right].$$

The splitting strategy is to choose the attribute which maximizes the deviance.

Now consider the average tree impurity for the entropy measure. When the probabilities are estimated this is

$$I(T) = \sum_t \frac{n_t}{n} i(n_{tc}/n_t) = -\sum_{t,c} \frac{n_t}{n} \frac{n_{ct}}{n_t} \log \frac{n_{ct}}{n_t}$$

$$= -\sum_{t,c} \frac{n_{ct}}{n} \log \hat{\pi}_{c|t} = D(T)/2n$$

so the splitting strategies for deviances and for entropy-based impurity are identical.

We can take this duality of approaches further. Many impurity measures can be written as a sum over examples: for example the Gini index is the sum of $(1 - p_c)/n$ where c is the class of the example. Suppose there are n_c examples of class c. Then

$$I = \sum_c n_c (1 - p_c)/n$$

This paragraph is technical and not needed elsewhere.

and this is minimized over (p_c) by taking $\widehat{p}_c = n_c/n$. Thus the use of the Gini index can be considered as a probability model with a different measure of goodness-of-fit. Chou (1991) considers a larger class of measures of the form

$$I(T) = \sum_{\text{leaves } t} i(t) = \sum_t \frac{n_t}{n} \mathsf{E}\left[\ell(Y, \widehat{\mathbf{p}}(t)) \mid t\right]$$

where Y is the class of an example, ℓ is a loss function and $\widehat{\mathbf{p}}(t)$ is chosen to minimize the conditional expectation. The Gini index then arises from $\ell(Y, \mathbf{p}) = \|\text{ind}(Y) - \mathbf{p}\|^2 = 1 + \|\mathbf{p}\|^2 - 2p_Y$ (and $\text{ind}(Y)$ is the K-tuple of indicators $Y = k$) and the entropy from $\ell(Y, \mathbf{p}) = -\log p_Y$. Clearly $I(T)$ is always reduced by a split (since there is a $\mathbf{p}(t)$ for each child to minimize over). Further, let

$$d(t, \mathbf{p}(t)) = \mathsf{E}\left[\ell(Y, \mathbf{p}(t)) \mid t\right] - i(t).$$

Then if we consider a binary split over values of a categorical split, it is optimal only if for each category x assigned to the left child t_L $d(x, \widehat{\mathbf{p}}(t_L)) \leqslant d(x, \widehat{\mathbf{p}}(t_R))$, and conversely for the right child t_R (Chou, 1991). This extends Proposition 7.1, at least for impurity indices of Chou's form.

Priors, weights and costs

There are several assumptions made so far which we may wish to relax. Quite often the training set is not a random sample from the whole population, but chosen to disproportionally represent the classes, especially to over-represent rare classes. Suppose that the classes are known to have probabilities (π_k) in the population, but have n_k representatives out of n in the training set. The population approach works with (p_k), the population distribution of classes within the node t. Clearly n_{tk}/n_t is no longer an appropriate estimate of p_k, and we would use the probability vector proportional to $n_{tk}/n_t \times n\pi_k/n_k$.

Another extension is to allow *weights* to be attached to the examples. The most obvious reason is that we have an integer number w_i of examples like this one, and wish to avoid the overhead of computing with many copies. This suggests interpreting n_{tk} and n_t as the sum of weights, not merely counts of examples. Note that we can incorporate priors for the classes via weights, by giving all examples in class k a weight $n\pi_k/n_k$ (or multiplying the current weight by this factor).

Suppose we wish to attach costs to different misclassifications, say the cost C_{ij} of misclassifying examples of class i as class j. One approach is to say that the tree construction is merely modelling the posterior probabilities $p(k \mid \mathbf{x})$, and the costs should be used to choose the classification at each node, but not otherwise. However, differential costs suggest that we would like a more accurate model for some classes than for others. Breiman *et al.* (1984, §4.4) suggest that this can sometimes be incorporated into the impurity index. For example, the interpretation given for the Gini index suggests the modified form

$$i(p) = \sum_{i \neq j} C_{ij}\, p_i p_j.$$

Unfortunately, this effectively symmetrizes the costs (since the coefficient of $p_i p_j$ is $C_{ij} + C_{ji}$) and so is completely ineffective in two-class problems. It can also fail to be concave, and so give rise to splits with negative 'decreases' in impurity.

For two classes there is a simple approach to misclassification costs. Each example in class 2 costs a factor C_{21}/C_{12} more to misclassify than an example in class 1, which suggests weighting the examples in class i by C_{ij} for $j \neq i$. This will also be appropriate for more classes if the misclassification costs depends only on i and not on $j \neq i$.

How about the deviance approach? We use the weights for each example to weight the log-likelihood; the deviance becomes

$$D(T) = \sum_{\text{leaves } t} D_t, \qquad D_t = -2 \sum_{\text{classes } c} n_{tc} \log \pi_{c\mid t}.$$

where n_{tc} now represents the sum of the weights of examples reaching leaf t of class c. (This is certainly appropriate if weights represent multiple examples.) We will once again estimate $\pi_{c\mid t}$ by n_{tc}/n_t, but this is an estimate of the biased posteriors, and will be adjusted to be proportional to $n_{tc}/n_t \times n\pi_k/n_k$ to estimate the posteriors in the population. Note that the latter is what we get if we weight examples in class k by $n\pi_k/n_k$.

7.2 Pruning rules

There are $\lfloor 1.5028369^\ell \rfloor$ rooted subtrees of a binary tree with ℓ leaves; Breiman *et al.* (1984, p. 284).

The number of rooted subtrees of a binary tree is very large so we need a way to navigate this family efficiently.

Cost-complexity pruning

The best-known procedure for tree pruning is that proposed by Breiman *et al.* (1984). Let $R(T)$ be a measure of a tree formed by adding the contributions from the leaves. One obvious candidate is the number of misclassifications on the training set or a test set; another is the entropy or deviance of the partition. Let the *size* of a tree be the number of leaves. (For a binary tree the total number of nodes is twice the size minus one.) Then Breiman *et al.* (1984) proposed choosing a rooted subtree T of the full tree T_0 which minimizes

$$R_\alpha(T) = R(T) + \alpha \, \text{size}(T).$$

We can also consider $R_\alpha(T)$ as the sum of $R(t) + \alpha$ over the leaves of T. This can be seen as using a Lagrange multiplier for size, so finding the minimizing trees for all α is equivalent to finding the trees with minimum $R(T)$ for each size. (Our results are equally valid for other measures of size such as the total costs of the tests at the nodes.)

When using the apparent error rate on the training set we will want to choose a positive α to penalize size, but our results also apply to $\alpha = 0$ which would be appropriate with the error rate on a test set. Ciampi *et al.* (1987) consider pruning with the Akaike Information Criterion which corresponds to taking $R(T)$ as the deviance and $\alpha = 2(K-1)$. (The AIC penalizes minus the log-likelihood by the number of parameters. Estimating the probability distribution within a leaf takes $K-1$ parameters. This count ignores parameters in the splitting attribute and the selection of the attribute itself; it is unclear how these should be counted.)

Breiman *et al.* showed that there is a nested family of subtrees T_k of $T_0 = T$ such that each is optimal for a range of α, and so there are values

$$-\infty = \alpha_0 < \alpha_1 < \cdots \infty$$

such that T_i is an optimal tree for $\alpha \in [\alpha_i, \alpha_{i+1})$. Further, they gave an algorithm to construct the tree sequence (T_k). Often $\alpha_1 \geqslant 0$, for example if $R(T)$ is the measure used to grow the tree (such as deviance or Gini) or the error rate on the training set (from Proposition 7.5). However, $\alpha_1 = 0$ is quite common.

We will now prove these results. There can be a number of trees with the same value of $R_\alpha(T)$; we will consider only one which is a subtree of all to be optimal, and if this exists we call it $T(\alpha)$. Consider a non-trivial tree T, and for any non-terminal node t let T_t be the subtree rooted at that node. Let

$$g(t, T) = \frac{R(t) - R(T_t)}{\text{size}(T_t) - \text{size}(t)}$$

If we have weights, we would use these in calculating $R(T)$. To handle missing values we will need a modest extension, in which $R(T)$ also contains contributions from all nodes. Precisely, $R(T)$ is assumed to be a sum over leaves plus a sum over non-leaves, the summands being different in the two cases.

which compares the reduction in R by including the subtree with the increase in size. Note that $g(t, T) > \alpha$ if and only if $R_\alpha(t) > R_\alpha(T_t)$. The effect of pruning at node t is to replace T_t by t.

Proposition 7.2 *Suppose we number the nodes of a tree T so that each node precedes its parent. If we visit the nodes in this order (bottom-up) and prune at node t if $R_\alpha(t) \leqslant R_\alpha(T'_t)$ for the current tree T', the result is $T(\alpha)$.*

This was proposed as a new algorithm by Gelfand & Delp (1991), Gelfand *et al.* (1991) and Guo & Gelfand (1992), the latter including a more complex proof. However, the algorithm is implicit in earlier work, and explicit, without proof, in Quinlan (1987a), under the name of *reduced error pruning*. It follows immediately from Theorems 10.7 and 10.10 of Breiman *et al.* (1984).

Proof: We will establish by induction that when node t is considered all the branches at t are optimally pruned. This is clearly true for the leaves. At node t we either prune with value $R_\alpha(t)$ or not with value $R_\alpha(T'_t) = \sum_{\text{branches } B} R_\alpha(T'_B)$ if this is strictly smaller. If there is a subtree T'' rooted at t with a smaller value of R_α it must be non-trivial, and there must be a branch B with $R_\alpha(T''_B) < R_\alpha(T'_B)$ and so T'_B is not optimally pruned, a contradiction. Now suppose there is another subtree with the same value of R_α. Then each of its branches (it must have some) will have the same value of R_α as the corresponding branch of T'_t and so include that branch. Thus after node t is considered, the current T'_t is optimally pruned. When the root is reached the current tree is optimally pruned, so is $T(\alpha)$. □

This gives an algorithm to find $T(\alpha)$ for a single α. We now show how to find (α_k) and the tree sequence T_k. From now on we assume that size is increasing, that is adding nodes increases (weakly) the size.

Proposition 7.3 *Let α_1 be the smallest value of $g(t, T)$ for any non-terminal node t of T. The optimally pruned tree is T for $\alpha < \alpha_1$, and $T_1 = T(\alpha_1)$ is obtained by pruning at all nodes t with $g(t, T) = \alpha_1$. Further, $g(t, T_1) > \alpha_1$ for all non-terminal nodes of T_1.*

Proof: The optimality of T for $\alpha < \alpha_1$ is immediate from $R_\alpha(t) > R_\alpha(T_t)$ and Proposition 7.2. Consider $\alpha = \alpha_1$, and pruning by Proposition 7.2. Whenever the tree is pruned, $R_\alpha(T_s)$ is unchanged for all nodes s of the new tree. Thus $R_\alpha(t) \leqslant R_\alpha(T'_t)$ for the current tree T' if and only if $R_\alpha(t) \leqslant R_\alpha(T_t)$ if and only if $g(t, T) \leqslant \alpha_1$. Then for a retained node t,

$$\begin{aligned} R_{\alpha_1}(t) - R_{\alpha_1}(T_{1t}) &= R_{\alpha_1}(t) - R_{\alpha_1}(T_t) + [R_{\alpha_1}(T_t) - R_{\alpha_1}((T_1)_t)] \\ &= R_{\alpha_1}(t) - R_{\alpha_1}(T_t) = g(t, T)[\text{size}(T_t) - \text{size}(t)] \\ &> \alpha_1[\text{size}(T_t) - \text{size}(t)] \geqslant \alpha_1[\text{size}((T_1)_t) - \text{size}(t)] \end{aligned}$$

so $g(t, T_1) > \alpha_1$. □

Proposition 7.4 *For $\beta > \alpha$ $T(\beta)$ is a subtree of $T(\alpha)$ and is the result of β-pruning of $T(\alpha)$.*

Proof: We will show by induction that $T_t(\beta)$ is a subtree of $T_t(\alpha)$ and conclude that $T(\beta)$ is a subtree of $T(\alpha)$. This is true at the leaves. At node t we compare $R_\alpha(t)$ to $R_\alpha(T_t(\alpha))$ and $R_\beta(t)$ to $R_\beta(T_t(\beta))$ and in each example prune if the first is (weakly) smaller. We must show that if $R_\alpha(t) \leqslant R_\alpha(T_t(\alpha))$ then $R_\beta(t) \leqslant R_\beta(T_t(\beta))$. Now since $T_t(\beta)$ is a candidate for α-pruning of the tree rooted at t, we have

$$R_\beta(t) = R_\alpha(t) + (\beta - \alpha)\text{size}(t) \leqslant R_\alpha(T_t(\alpha)) + (\beta - \alpha)\text{size}(t)$$
$$\leqslant R_\alpha(T_t(\beta)) + (\beta - \alpha)\text{size}(t)$$
$$= R_\beta(T_t(\beta)) - (\beta - \alpha)[\text{size}(T_t(\beta)) - \text{size}(t)]$$
$$\leqslant R_\beta(T_t(\beta)).$$

Since $T(\beta)$ minimizes $R_\beta(T')$ over all rooted subtrees T' of T and is a subtree of $T(\alpha)$, it also minimizes $R_\beta(T')$ over rooted subtrees of $T(\alpha)$.   □

The algorithm of Proposition 7.3 can be applied to the new tree $T_1 = T(\alpha_1)$ to find $\alpha_2 > \alpha_1$ (since $g(t, T_1) > \alpha_1$ for all non-terminal nodes of T_1) and $T_2 = T(\alpha_2)$ and so on until T_k is the trivial tree, the root of $T_0 = T$. From Propositions 7.3 and 7.4, $T(\alpha) = T_1$ for $\alpha_1 \leqslant \alpha < \alpha_2$ and $T(\alpha_2) = T_2$. Repeating the process gives $T(\alpha)$ for all $\alpha \leqslant \alpha_k$, and Proposition 7.4 shows that the trivial tree is optimal for $\alpha \geqslant \alpha_k$. This completes the algorithm to find the tree sequence:

1   Set $k = 0$ and write out $T_0 = T$.
2   Set $\alpha = \infty$.
3   Visit the non-terminal nodes t in bottom-up order and calculate $R(T_t)$ and size (T_t) by summing over the descendants (and including any contribution at t). Set

$$g(t) = \frac{R(t) - R(T_t)}{\text{size}(T_t) - \text{size}(t)}$$

and $\alpha = \min(\alpha, g(t))$.
4   Visit the nodes in top-down order and prune whenever $g(t) = \alpha$.
5   Set $k = k + 1$ and write out $\alpha_k = \alpha$ and $T_k = T$.
6   If T is non-trivial go to 2.

We could visit the nodes in any order at step 4, but a top-down order avoids considering nodes which themselves will be pruned away.

It remains to choose the particular tree within this sequence. If we have a *validation* set we can use its error rate with $\alpha = 0$. Otherwise Breiman *et al.* propose selecting the value of α using *cross-validation*. The training set is split into V parts; Breiman *et al.* (1984, pp. 11–12) seem to prefer 3 but later users (e.g. Clark & Pregibon, 1992) recommend 10. For each of the parts, a tree sequence is constructed from the remaining $V - 1$ parts of the training set, and its measures $R(T_k)$ calculated, to give a piecewise constant function for $R(T(\alpha))$. This is averaged over all V parts, and α chosen to minimize the function.

In fact Breiman *et al.* averaged only at the values $(\sqrt{\alpha_k \alpha_{k+1}})$ for the sequence α_k for the original tree, but this saves but little effort.

Because the training set is disjoint from the test set in each of the V cross-validation experiments, we can expect to form a reasonably unbiased estimate of $R(T(\alpha))$. If V is small we have used a considerably smaller training set, and so might expect to overestimate the error rates, but this does not necessarily mean that the relative values of $R(T(\alpha))$ for different α are seriously biased. In practice the estimates of $R(T(\alpha))$ are highly variable over the choice of parts of the training set, and the estimated function may have no minimum within the range of α considered, or a very broad minimum. Breiman *et al.* suggest choosing the largest value of α with the cross-validated $R(T(\alpha))$ just above the minimum (the 'one SE' rule). The standard error can be estimated from a binomial distribution for error-count pruning, or a chi-squared distribution for deviance pruning.

There is a difficulty with cross-validating deviance measures $R(T)$ not found with error rates nor the Gini measure. Suppose that at some leaf t a class c occurs in the test set but not in the training set. Then the fitted probability $\widehat{\pi}_{tc} = 0$ and so the deviance at that leaf is infinite. (The other measures give a unit penalty.) This might be thought appropriate, and will certainly lead to that leaf being pruned, but makes it difficult to average $R(T(\alpha))$. There are several *ad hoc* solutions, all of which involve altering the fitted probabilities. One we have used successfully is to give a prior of one example per class at each node so $\widehat{\pi}_{tc} = (n_{tc} + 1)/(n_t + K)$ for K classes, which is never zero, but can approach zero if a class does not occur in a large number of examples.

This is one of a family of *shrinking* approaches. Bahl *et al.* (1989), Chou (1991) and Buntine (1992) each smooth at all splits, not just the leaves, taking the fitted probabilities to be a convex combination of those of the parent and the frequencies in the child node. (Clark & Pregibon, 1992, also propose this.) It remains to choose the convex combination, and indeed to decide if it should be the same at each node. Chou uses leave-one-out cross-validation at each node; Buntine uses a combination which depends on the sample size (see Section 7.7). Clark & Pregibon use a constant factor over the tree, chosen by cross-validation, and see this as an alternative to pruning.

Another approach to pruning

Gelfand *et al.* (1991) and Gelfand & Delp (1991) point out that the optimally pruned tree with respect to the true misclassification rate $R(T)$, were this available, need not be within the family $T(\alpha)$ pruned with respect to the apparent error rate. We have already seen how to prune with respect to an honest measure of error rate. Gelfand *et al.* propose a pruning algorithm based on dividing the training set into two and alternating the role of the halves. Initially the tree is grown (and nodes labelled) using one half and pruned using the error rate on the other half. The tree is then re-grown from the pruned tree using the previous test half and pruned using the previous training half to estimate the error rate. This is repeated until the tree size is unchanged. The pruned subtrees are nested and increasing, and if a node is terminal at two successive steps growth from that node can be stopped.

Long formal proofs are given in Gelfand *et al.* (1991) for the Gini measure of impurity. We can give a short and general argument. It is important here that ties are broken consistently when labelling leaves; we need to choose the class of the parent node if this is a contender.

Proposition 7.5 *Suppose a training set is partitioned, and the whole set and each cell of the partition are labelled by a class with the highest frequency within it. Then the apparent error rate is decreased by partitioning, strictly so unless the whole partition is given the same class.*

Proof: Let the frequency of class c within cell i be n_{ci}. Then the success count before division is $\max_c n_{c\cdot}$, maximized by k, say, and after division is $\sum_i \max_c n_{ci} \geq \max_c \sum_i n_{ci} = n_{k\cdot}$ with equality only if k maximizes n_{ci} for each class. If the error rate is the same, the class of the parent is a contender for the class of each cell.   □

Suppose the two halves of the training set are $\mathcal{T}_1$ and $\mathcal{T}_2$, and let $R^{(i)}(T)$ denote the number of errors for test set $\mathcal{T}_i$, using the labels assigned when the tree was grown. Let T^* denote the tree grown using $\mathcal{T}_1$ and optimally pruned using $R^{(2)}$.

Proposition 7.6 *The tree T^* is unchanged under optimal pruning using $R^{(1)}$*

Proof: From Proposition 7.2 it suffices to show that $R^{(1)}(T_t) < R^{(1)}(t)$ for each interior node. Now T_t corresponds to a partition of the examples of $\mathcal{T}_1$ reaching node t, so by Proposition 7.5 $R^{(1)}(T_t) \leqslant R^{(1)}(t)$ with equality only if T_t gives the same partition as t and hence $R^{(2)}(T_t) = R^{(2)}(t)$ which would contradict the optimality of the pruning of T^*. □

Now suppose a tree S is grown starting from T^* using $\mathcal{T}_2$ and optimally pruned to S^* using $R^{(1)}$. Since S contains T^*, Proposition 7.6 shows that S^* contains T^* (since pruning is monotone on trees). If a leaf in T^* remains a leaf in S^* the growing and pruning process to form T^* will be repeated at the next step, and so that node will always remain a leaf. As there are only a finite number of examples and empty leaves will never be generated, the process must stop.

Gelfand *et al.* (1991) propose reporting an error rate based on classifying at each leaf the examples from the half of the training set other than the one on which that leaf was labelled (when it was grown). They recommend this procedure only for large datasets (since it works with half the data at a time), and we have found it unsatisfactory for moderately sized datasets, in which T^* is often just the root subtree.

'Pessimistic' and 'error-based' pruning

Quinlan (1987a, 1993) introduced two much cruder ideas for pruning. In the first approach, he proposes a continuity correction for cost-complexity pruning, so that the number of errors on the training set at each node is increased by one half. The idea was to better estimate the true rather than apparent error rate. (This idea is exactly equivalent to taking $\alpha = 0.5$, since $R(t)$ is increased by one half.) He compares the error rate of the tree T_t with the error rate at node t (after a 'continuity correction' adding one half to each error count). Rather than prune only if the adjusted error rate for node t is smaller, he proposes to prune unless it is somewhat larger, specifically when

error rate for t < error rate for T_t + std. dev.(error rate for T_t).

(Quinlan is vague about how to calculate the last term; his example appears to use a binomial formula with a common probability, but it would be better to calculate the standard deviation within each leaf.) This looks like an approximation to a significance test, except that the variability of the left-hand side is not taken into account. Again the details are vague, but Quinlan states that all subtrees are considered as candidates for pruning (unlike Proposition 7.2). As there are very many such subtrees, this seems unlikely.

A much larger adjustment of the apparent error rate is proposed in his 1993 book. Suppose a leaf t covers N examples, J of which are misclassified. Then $R(t)$ is taken to be the 87.5% point of a binomial$(N, J/N)$ distribution. This could be calculated exactly (and is in the C4.5 program) but given the approximate nature of the justification, using a normal approximation to the binomial to give

$$R(t) = J + 1.15 \times \sqrt{J(1 - J/N)}$$

seems perfectly adequate. Our understanding is that the algorithm of Proposition 7.2 is used.

Examples

The data on diabetes amongst Pima Indians have provided a difficult example for many methods, and is typical of the difficulty of using tree-based methods. It is easy to grow a tree (using the deviance/entropy approach) with many nodes: our initial tree has 22 nodes, shown in Figure 7.3. (One of the splits has another attribute with exactly the same split.) Four nodes can be pruned without changing the classifications at all; the error rate on the test set is 81/332. This is just about significantly worse than the logistic regression with 66/332, as the McNemar statistic is 1.96.

Figure 7.4 shows the difficulty in choosing the size by 10-fold cross-validation of error-rate pruning. There is little variation with the size of tree down to size 3, which suggests the latter should be adopted. This gives the rule that diabetes should be predicted if the plasma glucose level exceeds 123.5 and the diabetes pedigree function exceeds 0.31. This rule has a test-set error rate of 90/332, worse than the unpruned tree. However, the difference is not statistically significant, for McNemar's test statistic is $[|29 - 20| - 1]/\sqrt{29 + 20} \approx 1.14$. In this circumstance we should not use risk averaging, as most examples are predicted with $\max \widehat{p}(c \,|\, x) = 1$, and so the true test-set error rate is dramatically underestimated (11% instead of 24.4%).

Quinlan's 'pessimistic' pruning removes just 3 nodes, and AIC removes just one. Indeed, if we grow an even larger tree by allowing smaller populations within the leaves, both Quinlan and AIC select a tree with about 30 nodes. Such large trees have a test-set error rate of about 99/332.

Using the Gini index rather than entropy to grow the tree produced a similar but not identical tree, with splits occurring in a slightly different

Figure 7.3: The
classification tree grown
for the Pima Indians
diabetes data based on
a training set of size
200. Growth was
stopped at leaves with
10 or fewer examples.

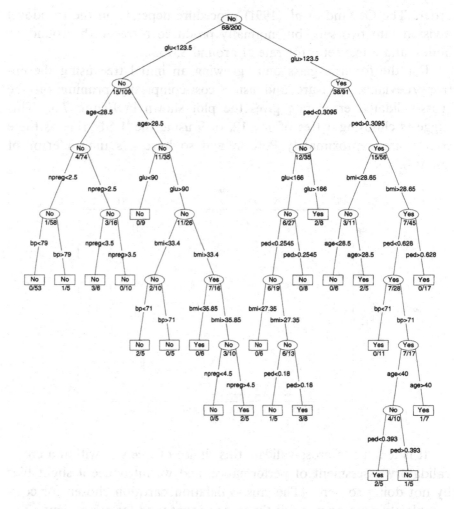

Figure 7.4: Number of
errors *vs* size for
error-rate pruning of
the tree of Figure 7.3.

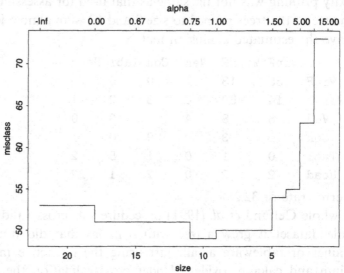

order. The Gelfand *et al.* (1991) procedure depends on the (random) division into two sets, but normally produced a tree with around 15 nodes, and a test-set error rate of around 85/332.

For the forensic glass data, growing an initial tree using the entropy/deviance measure and using cost-complexity pruning on the cross-validated error rate gives the plot shown in Figure 7.5. This suggests choosing a tree of size 12, or 9 using the '1 SE' rule as these counts are approximately Poisson and so have a standard error of about 8.

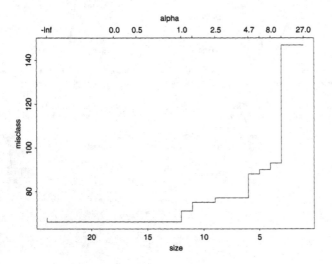

Figure 7.5: Cross-validated error count *vs* tree size for the forensic glass data.

It is tedious to cross-validate this choice of tree size within a cross-validatory assessment of performance, and we introduce a slight bias by not doing so here. (The cross-validation partition chosen for cost-complexity pruning was not the same as that used for assessment.) The performance of the trees pruned to size 9 and 12 were almost identical; size 9 gave the estimated confusion matrix

	WinF	WinNF	Veh	Con	Tabl	Head
WinF	55	13	2	0	0	0
WinNF	14	50	6	3	2	1
Veh	5	8	4	0	0	0
Con	0	3	0	9	0	1
Tabl	0	1	0	1	5	2
Head	2	2	0	2	1	22

and an error rate of 32.2%.

The whole Gelfand *et al.* (1991) procedure was cross-validated. On the whole dataset it grew a tree with 6 nodes that did not classify as container or tableware at all, just using the refractive index and magnesium and calcium oxides. Under cross-validation the tree size

Figure 7.6: Pruned classification tree for the forensic glass data.

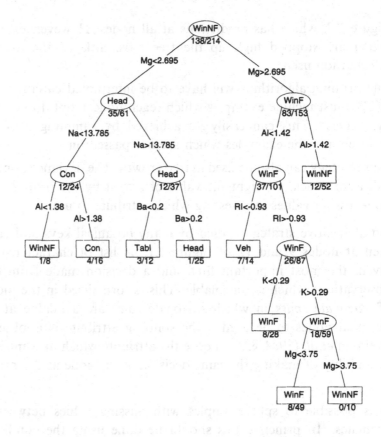

varied widely between subsets; the assessment of the error rate was 42%.

The Quinlan pruning procedure tended to prune lightly; its cross-validated error rate was 31%.

7.3 Missing values

One attraction of tree-based methods is the ease with which missing values can be handled. Consider the botanical key of Figure 7.1. We only need to know about a small subset of the 10 observations to classify any example, and part of the art of constructing such trees is to avoid observations which will be difficult or missing in some of the species (or, as in the case of capsules, for some of the examples). However, missing values may be unavoidable, and there are several approaches to handling them.

1 A general strategy is to 'drop' an example down the tree as far as it will go. If it reaches a leaf we can predict y for it. Otherwise we use the distribution at the node reached to predict y, as shown in

Figure 7.2, which has predictions at all nodes. However, examples which are stopped high up the tree have little of the available information used.

The pruning algorithms will have to be interpreted carefully, since $R(T_t)$ must include examples which reach node t but do not reach the leaves. This can easily be achieved by summing over both children and the examples which are not passed on.

This strategy can also be used in tree growth. The deviance approach will automatically weight the value of a split by the proportion of non-missing values in assessing which attribute to use.

2 An alternative strategy is used by many botanical keys and can be seen at nodes 9 and 12 of Figure 7.1. A list of characteristics is given, the most important first, and a decision made from those observations which are available. This is formalized in the method of *surrogate splits* in which surrogate rules are available at non-terminal nodes to be used if the splitting attribute is unobserved. Breiman *et al.* (1984, §5.3) choose the attribute which maximizes the probability of making the same decision at the node as the primary split.

3 It is possible to split examples with missing values between the branches. In principle this should be done using the conditional probabilities of left and right splits given all the observed information. In general that probability is unavailable. What we can estimate easily is the probability of going left or right given the attributes used in earlier splits, from the frequencies of complete examples at the node.

Quinlan (1986) attributes to Alen Shapiro the idea of building a tree to estimate the conditional distribution of the missing value.

In this approach each example is split into a probability distribution over leaves; each time a missing value is encountered the current fractional example is subdivided. (There is a potential problem here, especially if it is the same feature under consideration as a higher split, since different conditional distributions will be used each time the example is split. If an example has already been split, the imputed values at earlier splits also have to go into the conditioning.)

The obvious way to produce a classification for a split example is to combine the posterior probabilities in the leaves reached by its fractions using the probabilities assigned to leaves, and then assign the class with the highest overall posterior probability. However, Quinlan's (1993) C4.5 system takes the simpler but less rational approach of weighting the leaf classifications, and choosing the class with the overall highest probability.

4   Another possibility is to take 'missing' as a further level of the attribute (e.g. Kass, 1980). For methods which allow multi-way splits this has the disadvantage of increasing the number of levels, so that making some values missing can increase the gain in impurity (Quinlan, 1986). This can be circumvented by allowing only binary splits, or by penalizing multi-way splits.

In most approaches tree construction is based on the examples without any missing observations. Where missing values are very frequent this may be unacceptable or even impossible. Quinlan (1986) suggests replacing missing values by the distribution within the class at that node when computing the expected value of a split. On the other hand, Quinlan (1993) multiplies the impurity gain calculated on known examples by the proportion of missing values (as implied by the deviance approach) and his C4.5 system uses fractional examples throughout tree construction. However, note that in the deviance approach if example splitting is used the partitioning is no longer recursive (as the fitted probability for such examples depends on each branch).

All of these ideas have merits and demerits, depending on how common missing values are and whether they are missing at random. For example, in medical diagnosis the absence of a test might well carry information, and examples with the value of an attribute missing could be very different from those with a recorded value. On the other hand, if missing values are rare, there will not be enough information to usefully treat 'missing' as a separate attribute value.

Sometimes features are not missing but also not known exactly; for example a continuous feature may only be known to lie within an interval, or a test may indicate a 80% chance of being positive. Such information is best handled by splitting examples.

Example

The Pima Indians diabetes data has many missing values, so we tried out the value of splitting examples, with a training set that had 200 complete examples and 100 partially missing examples. Growing and pruning a tree on this augmented training set led to a slightly larger tree shown in Figure 7.7, which makes 74/332 errors on the test set.

This highlights the role of the plasma glucose level and the body mass index. Figure 7.8 shows the training-set examples on those two features, which indicates that the separation is rather weak.

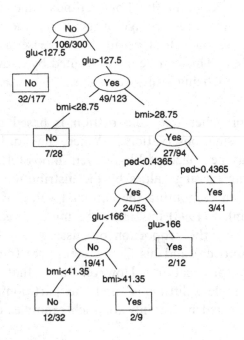

Figure 7.7: The classification tree grown and pruned for the Pima Indians diabetes data based on a training set of size 300.

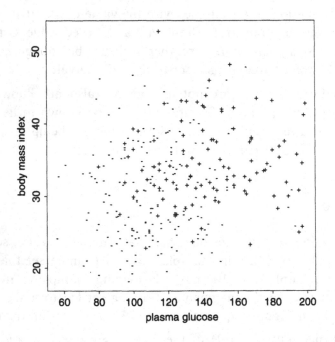

Figure 7.8: The presence or absence of diabetes against plasma glucose and body mass index for the training set of the Pima Indians data.

7.4   Earlier approaches

Quinlan (1986) provides an historical overview of developments in the field of machine learning. He considers the *TDIDT* family of algorithms (Top Down Induction of Decision Trees) to stem from Hunt's Concept Learning System (Hunt *et al.*, 1966), via his own ID3 (Quinlan, 1979) and the ACLS system of Patterson & Niblett (1983). Another branch is the ASSISTANT systems of Kononenko *et al.* (1984) and Cestnik *et al.* (1987). Quinlan calls his own descendant of ID3 C4.5. Many of the later systems are commercial and so not documented in the scientific literature. There was a family of CLS systems; the last, CLS-9, chose the split which maximized the number of examples correctly classified over the new leaves.

All of these systems (except CLS) are based on the entropy measure of impurity. ID3 examines all candidate attributes and chooses that with the largest 'information gain', which is what we called the reduction in average impurity at the node. All the probabilities are estimated from frequencies in the training set. Most of these systems allowed only two classes.

What attributes are allowed? In ID3 only categorical features were considered and the split is into all levels of the feature. Both ACLS and ASSISTANT use a binary division of the feature. Continuous features could in principle be divided into ranges or split as in Figure 7.1.

If multi-way splits are allowed, they would be expected to have greater information gain; indeed they may have a large information gain even if the attribute has no predictive power. Quinlan (1986, 1988) suggests guarding against this by comparing the information gain to what he terms the 'information value' IV of the attribute A, that is the entropy of the distribution of attribute values at the node. He suggests choosing the attribute which maximizes

$$\text{gain}\,(A)/IV$$

over attributes 'with average-or-better gain amongst all tests examined' (itself a size-biased selection criterion).

One feature of the original ID3 was that it works with a 'window', that is a subset of the training data. This is initially chosen as a random subset and the tree grown on it. The rest of the training set is tested, and a selection of incorrectly classified examples is added to the window. The tree is extended and the process repeated. In a logical domain this can reduce the computation, but has not been used in the descendants of ID3 (except C4.5).

The main differences between these algorithms come in their stopping rules. The original ID3 had no stopping rule. Quinlan (1983) proposed a chi-square test for the value of the split on the chosen A, that is a test of independence in the 'A cross class' table. Niblett (1987) suggested Fisher's exact test for the same purpose. The ASSISTANT system compared a cross-validation estimate of the error after splitting at the node with the apparent error at the node, and stopped if the former was worse. Its successor, ASSISTANT86, computed the node size times the information gain divided by the entropy and stopped if this was smaller than a preset threshold (for example 4%).

So far we have assumed that no attributes have missing values. ASSISTANT either used the proportions of the attribute amongst examples of the same class at that node to fill in the most frequent value (as used by CN2, Clark & Niblett, 1989) or used fractional examples to express the distribution over values. Many of the general schemes discussed above have been used.

Suppose there are K classes. The Laplacian error estimate replaces the error rate e_i/n_i at a leaf by $[e_i+K-1]/[n_i+K]$, in a crude attempt to compensate for the optimistic bias of the re-substitution estimator of error. Niblett & Bratko (1986) pruned the tree at node t if the naive Laplacian error estimate at that node is less than that for the subtree rooted at t. (This is described in detail in Niblett, 1987, and also used by Casey & Nagy, 1984.)

Bratko & Kononenko (1987) give a number of comparisons for domains of medical diagnosis. Their results show that binary trees are generally smaller and have slightly lower error rates than multi-way trees, and that stopping early slightly improves the error rate but markedly improves comprehension.

One early strand of work in statistics was given by Kendall & Stuart (1966, §44.30–32) and Richards (1972). They consider continuous-valued attributes and two classes. Suppose for a feature X that one class, say class 1, has generally smaller values than the other. The split is then $(-\infty, \min_2 X_i)$ to class 1, $(\max_1 X_i, \infty)$ to class 2 and the overlap $[\min_2 X_i, \max_1 X_i]$ is passed to the next level. The attribute is chosen for which this rule decides the most examples, and the process repeated at the next level.

The system THAID of Morgan & Messenger (1973) was one of the first statistical applications of decision trees. Its splitting criterion was the error rate. A descendant, CHAID (Kass, 1980) chooses the split with the highest significance in the A cross class table. However, it found the (approximately) most significant table including amalgamating categories of multi-level attributes. The stopping rule was again based on significance, with some allowance for selection. Mingers (1987) also based the choice of split on the highest statistical significance of the contingency table of A cross classes.

Ciampi *et al.* (1987) *merge* leaves in a post-processing step if their populations are sufficiently similar. An agglomerative clustering algorithm is applied to the leaves with a dissimilarity measure computed from a log-likelihood-ratio test of the difference in within-leaf class distributions. We do not see this as preferable to pruning methods.

see Section 9.3.

The work in the engineering literature is diverse and wide-ranging; we will only highlight a few ideas. (Safavian & Landgrebe, 1991, catalogue many more.) Henrichon & Fu (1969) set up a tree with a linear combination at each node whose range was partitioned into positive, negative and undecided; the undecided examples are passed to the next layer. The partitioning criterion was to approximately minimize the error rate. Swain & Hauska (1977) suggested minimizing the sum of measurement and error costs using a 1-step lookahead. Again for a two-class problem, Friedman (1977) and Rounds (1980) used the Kolmogorov–Smirnov distance between the distributions of the two classes to choose the feature, and split at a maximum of the distance. Multi-class problems can be considered by building a tree contrasting each class with the first, and combining information in the leaves of the $K - 1$ trees to decide between the K classes. Sethi & Sarvarayudu (1982) took an information-based approach identical to that which was emerging in machine learning.

Engineers have continued to be active in this field, for example Argentiero *et al.* (1982), Casey & Nagy (1984), Dattatreya & Sarma (1981, 1985), Goodman & Smyth (1988), Kurzynski (1983a, b), Li & Dubes (1986), Schuermann & Doster (1984) and Wang & Suen (1984, 1987).

7.5   Refinements

A modest amount of progress towards more efficient algorithms has been made since Breiman *et al.* (1984).

Chan & Bao (1991) and Fayyad & Irani (1992) noticed that we can restrict the set of cut-points tried for a continuously-valued attribute A with the entropy measure of impurity or equivalently using deviances. The empirical distribution of A jumps only at observed values, so clearly the optimal cut-point will be one of the observed values. What these authors proved is that the cut-point always occurs on the boundary between two classes, so we do not need to consider observed values if those to the immediate left and right correspond to examples of the same class. How much of a saving this produces depends on the number of classes (clearly it is best if this is small) and on the degree of overlap of class-conditional distributions of A. The results of Fayyad & Irani showed a very modest speed-up (less than two overall), and our experiments showed even less.

Chou (1991) provided a partial extension to part (ii) of Proposition 7.1 based on the ideas discussed earlier. This result corresponds to finding a locally optimal partition of an attribute with L levels in the sense of reduction in average impurity or deviance, in linear expected time in the number of examples. It is only *locally* optimal, a point Chou glosses over in his title and description.

Crawford (1989) considered alternative estimators of the error rate $R(T)$ to be used in cost-complexity pruning, based on the *bootstrap* (Section 2.7). The idea of the bootstrap is to resample with replacement a sample of size n (the original size) from the training set. (Clearly each of the original examples will occur an integer number of times in the bootstrap sample.) A tree sequence can be grown from each bootstrap sample, and the *bias* in the error rates for the bootstrap samples used to estimate the bias of $R(T(\alpha))$. That is, for each of B bootstrap samples we grow and prune a tree to find $T^b(\alpha)$ and evaluate the difference between the error rate for the real training set and the bootstrap sample. The average of this quantity over the B samples, $\widehat{\omega}(\alpha)$, is the bootstrap estimate of the bias of $R(T(\alpha))$, so finally α is chosen to minimize

$$R(T(\alpha)) + \widehat{\omega}(\alpha).$$

Breiman *et al.* (1984, p. 312) give some calculations which suggest that the bootstrap estimator of the bias $R(T)$ will systematically underestimate the bias. This property is not shared by Efron's (1983) .632 bootstrap (Section 2.7). Crawford (1989) reports experiments on pruning via both bootstrapped and the .632 bootstrap estimators of the error rate, generally preferring the .632 bootstrap to both the ordinary bootstrap and cross-validation.

Incremental learning

Thus far we have assumed that the whole training set is available initially. It is easy to envisage situations in which the training set becomes available from an on-line process, and it is desired to maintain an up-to-date decision tree.

Quinlan (1979) originally envisaged building a decision tree by ID3 incrementally, but this was as a computational shortcut in a noisefree problem where it might be hoped that a small subset of examples would induce the broad shape of a suitable tree.

Incremental tree induction has been taken up by Schlimmer & Fisher (1986), Utgoff (1988a, 1989, 1990), Utgoff & Brodley (1990) and Van de Welde (1989, 1990). The difficulty with the incremental growth of trees is that early decisions on which attribute to split were based on few examples and so are likely to be wrong. Utgoff (1988a) allows his procedure ID5 to recover by testing the current optimality of a split, and if it is sub-optimal to re-order the subtree if the optimal split occurs within the sub-tree. Utgoff (1990) gives a modification which is guaranteed to recover the tree grown by ID3. Van de Welde's objective is to grow the smallest possible tree.

All this work was for noiseless problems. Crawford (1989) considers incremental tree growth for noisy problems, carrying out the whole procedure (growth and pruning) on a subtree when a new example shows that the split at the root of that subtree is sub-optimal *and* will affect the path of the new example through the existing subtree. Subsequently he used bootstrap resampling to estimate if the gain by re-growing the subtree was significant.

Hybrid methods

We have mentioned that some systems allow linear combinations of continuous variates or Boolean combinations of binary ones at each node. Some of these have been termed *hybrid* by Utgoff (1988b), and are discussed by Dietterich (1990).

The STAGGER system of Schlimmer & Granger (1986) combines a 'Bayesian weight-learning algorithm with a method for constructing Boolean expressions'. The FRINGE algorithm of Pagallo (1989), Pagallo & Haussler (1989, 1990) builds on STAGGER, and post-processes trees constructed by ID3 to include new attributes constructed as Boolean combinations of existing ones.

Soft splits

A classification tree makes hard splits; for example in Figure 7.1 completely different paths are taken if we measure the perianth tube as longer or shorter than 10 mm. We might be worried if we measured 9.9 mm, and test both possibilities. An automated system will not do that unless it is enhanced by soft splits, of the form 'branch right with probability $\sigma(x)$'. Hitherto $\sigma(x) = I(x > x_0)$, but we can envisage a smoother transition, and average the predictions by splitting examples as for missing values. Carter & Catlett (1987) used a piecewise linear σ, linearly interpolating between $(x_L, 0), (x_0, 0.5)$ and $(x_R, 1)$ for $x_L < x_0 < x_R$. They and Quinlan (1993, §8.1.2) suggested choosing x_L and x_R (in various *ad hoc* ways) after the main cut-point x_0 has been chosen in the usual way. Training set examples can then be divided if their value of x falls in the range $[x_L, x_R]$, and the tree growth continued.

It would not be difficult to choose x_L and x_R as well as x_0 to maximize the reduction in impurity at the split. We could use a logistic σ, but an asymmetric smoothing of the split may be desirable.

7.6   Relationships to neural networks

Two distinct relationships between neural networks and decision trees have been pointed out. The first is that the splitting mechanism invoked at a node is a way to split optimally the examples reaching that node into two (or more), and a neural network could be used to select the attribute to be used. The simplest network would split a linear combination of the variates, and this gives the *perceptron trees* of Utgoff (1988b) and *neural trees* of Sankar & Mammone (1993) (and Strömberg *et al.*, 1991). However, Breiman *et al.* (1984) had already considered allowing linear combinations of variates when setting out the list of attributes at each node, so this gives no added generality. (The idea goes back to at least Henrichon & Fu, 1969.) The growth procedures are different, in that both Utgoff and Sankar & Mammone use the perceptron learning rule (Section 3.6); Utgoff also used incremental induction for a noiseless problem. Breiman *et al.* used a gradient descent method to find a local maximum in the change in average impurity.

An obvious extension is to allow a non-linear discrimination rule at each node. Indeed, we can avoid the combinatorial search over the set of attribute splits at a node by seeking a non-linear combination of the features as a new feature and splitting on that. Many smooth and non-linear regression techniques could be used, including feed-forward neural networks as considered by Guo & Gelfand (1992). They found difficulty in extending minimizing the impurity to non-linear functions, and instead used standard least-squares neural network methods to train the function to discriminate between two groups of classes. With more than two classes they need a rule to choose the partition of the classes; one idea for a moderate number of classes is to compare the impurity change for each partition of the classes.

The other relationship which has been explored is to use the decision tree to guide the design of a neural network. Brent (1991) considers perceptron trees with t splits and so $t + 1$ leaves dividing $\mathscr{X}$ into $t + 1$ regions with piecewise linear boundaries. There is an immediate correspondence between such trees and a neural network with threshold units and two hidden layers of sizes t and $t + 1$, the first hidden layer corresponding to non-terminal nodes of the tree and the second hidden layer to a path from the root to a leaf with weights zero or ± 1. This can be used as a starting point for optimizing the neural network, perhaps with sigmoidal units. Brent minimizes the deviance of a split (in fact using a slightly different form similar to Fisher's exact test), but also uses the fact that the threshold units can be approximated by sigmoidal units of high gain to allow optimization methods on the neural network to approximate finding an optimal split in the sense of deviance reduction.

Sethi's (1990, 1991) *entropy nets* have essentially the same idea. The first hidden layer again computes the splitting functions at the nodes. The second and third layers are of AND and OR nodes respectively, and are wired to produce the same partition as the decision tree. (Note that AND and OR can both be produced by a perceptron.) The second layer corresponds to ANDing conditions down each path, and the OR layer collects paths with the same terminal class. Again, the threshold nodes may be relaxed to have sigmoidal response functions. Notice that although standard neural-network training algorithms could be used, Sethi points out that the derived neural network will be sparsely connected and of a special form. He proposes the use of a simple reinforcement learning rule. Each node in the AND and OR layers is associated with identifying a single class, since it is part of one path through the tree, and only the weights to these layers are trained, reinforcing the strongest signals amongst nodes associated with the same class.

7.7 Bayesian trees

A full Bayesian approach to tree construction will be stymied by the vast number of possible trees, each of which should appear in the posterior average. The best that is possible is to average over a few good trees, which is the approach taken by Buntine (1992).

The prior has to be given in several steps. A tree is fully specified by its topology (hitherto what we have meant by T) and the specification of the conditional probabilities. Buntine chooses independent and identical Dirichlet priors at each node, principally for computational convenience. For topologies, one can put a uniform distribution over all possible trees, or over tree shapes (ignoring the choice of attributes), or code for the complexity of the tree.

Given the prior and a dataset, it is in principle possible to find the posterior distribution over trees, and sum out over the topologies, giving a posterior distribution over classes for any future example. In practice this is impossible, and Buntine uses a variety of heuristics to grow trees which look to be of high posterior probability, and then averages over those trees. Let $B(\alpha_1, \ldots, \alpha_K) = \prod \Gamma(\alpha_j)/\Gamma(\sum \alpha_j)$ be the normalizing constant in the density of the Dirichlet distribution. Then

$$P(T \mid \text{training set}) \propto P(T) \prod_{\text{leaves } t} \frac{B(n_{t1} + \alpha_1, \ldots, n_{tk} + \alpha_K)}{B(\alpha_1, \ldots, \alpha_K)} \quad (7.1)$$

$$P((\pi_{tc}) \mid T, \text{training set}) \propto \prod_{\text{leaves } t} \frac{1}{B(\alpha_1, \ldots, \alpha_K)} \prod_{\text{classes } c} \pi_{tc}^{n_{tc} + \alpha_c - 1}.$$

Then (π_{tc}) has mode at $(n_{tc} + \alpha_c)/(n_t + \sum \alpha_c)$, which is (for $\alpha_c \equiv 1$) the smoothing we proposed earlier.

Buntine uses (7.1) to suggest an heuristic for measuring the quality of a split,

$$P(\text{test}) \times \prod_{\text{children}} \frac{B(n_{t1} + \alpha_1, \ldots, n_{tk} + \alpha_K)}{B(\alpha_1, \ldots, \alpha_K)}.$$

Note that this would be appropriate if (7.1) was a product over paths, not leaves. Similarly the shrinking proposed is towards the parent node, whereas (7.1) suggests a shrinking towards $\alpha_c / \sum \alpha_j$, and this is set from the overall distribution of classes in the training set.

Perhaps the most interesting heuristic is the use of 'option trees', in which more than one attribute at each node can be considered, but not simultaneously. Thus not only the best one-step lookahead split is chosen, but the best few are kept in play, allowing a range of trees with high posterior probability to be generated. An earlier idea along similar lines is that of Kwok & Carter (1990).

8

Belief Networks

The supervised methods considered so far have learnt both the structure of the probability distributions and the numerical values from the training set, or in the case of parametric methods, imposed a conventional structure for convenience. Other methods incorporate non-numerical 'real-world' knowledge about the subject domain into the structure of the probability distributions. Such knowledge is often about causal relationships, or perhaps the lack of causality as expressed by conditional independence.

These ideas have been most explored within the field of *expert systems*. This is a loosely defined area, and definitions vary:

> 'The label "expert system" is, broadly speaking, a program intended to make reasoned judgements or to give assistance in a complex area in which human skills are fallible or scarce . . . .' (Lauritzen & Spiegelhalter, 1988, p. 157)

> 'A program designed to solve problems at a level comparable to that of a human expert in a given domain.' (Cooper, 1989)

> 'An expert system has two parts. The first one is the knowledge base. It usually makes up most of the system. In its simplest form it is a list of IF . . . THEN rules: each specifies what to do, or what conclusions to draw, under a set of well-defined circumstances.
>
> The second part of the expert system often goes under the name of "shell". As the name implies, it acts as a receptacle for the knowledge base and contains instruments for making efficient use of it. These include a short-term memory, tree-searching machinery and a user interface.' (Crevier, 1993, pp. 156–7)

The last definition is the traditional one in AI, but excludes expert systems based on probabilistic knowledge by assuming that the knowledge base is made up of 'if . . . then rules'. Another aspect of expert

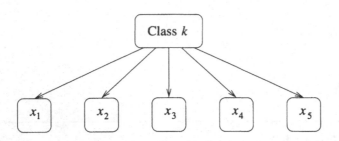

Figure 8.1: The causal graph for the idiot's Bayes rule. The arrows represent causal influence.

systems which is often stressed is the availability of facilities to provide *explanations*, usually in the form of a chain of deductions which lead to the conclusion.

In the development of probabilistic reasoning by Pearl (1986, 1988, 1993a), Lauritzen & Spiegelhalter (1988) and co-workers, the structural division is slightly different. The knowledge base is represented by a qualitative description of the dependencies (more accurately, lack of dependence) between the variables in the system, *and* the quantitative description of the numerical values of those dependencies. The role of the 'shell' is taken by the set of algorithms which manipulate the probabilities in an automatic way to present conclusions, such as the posterior probabilities of the various classes. Systems based on these ideas have many names: they have been called Bayesian expert systems, Bayes(ian) net(work)s, belief net(work)s, causal (probabilistic) networks, probabilistic expert systems and probabilistic reasoning on causal graphs.

This list of names is not exhaustive.

A very simple example may help to fix ideas. Suppose the input $\mathbf{x}$ is a set of m features $x_1, \ldots, x_m$. As usual, we wish to find the posterior probabilities $p(k \mid \mathbf{x})$ to classify a future case. The rule called *naive* or *idiot's Bayes* (Warner *et al.*, 1961; Titterington *et al.*, 1981) takes

$$p(k \mid \mathbf{x}) \propto \pi_k \prod_{i=1}^{m} p(x_i \mid k). \qquad (8.1)$$

One derivation of (8.1) is to assume $p(\mathbf{x} \mid k) = \prod p(x_i \mid k)$, that is that the features are conditionally independent given the class, from which it is immediate that

$$p(k \mid \mathbf{x}) p(\mathbf{x}) = p(\mathbf{x}, k) = p(\mathbf{x} \mid k) \pi_k = \pi_k \prod_{i=1}^{m} p(x_i \mid k).$$

The qualitative part of the knowledge base is then the assumption of conditional independence, and the quantitative part is the specification of the probabilities π_k and $p_k(x_i) = p(x_i \mid k)$. The shell is the set of rules for manipulating probabilities (essentially Bayes' formula) plus a user interface. The qualitative part of naive Bayes can be expressed graphically as in Figure 8.1.

In the rest of this chapter we will develop more complex networks than Figure 8.1 and corresponding stylized ways to apply Bayes' formula to derive the posterior probabilities, and to modify them as more information becomes available.

Overviews of this area with various applications are provided by Spiegelhalter *et al.* (1993), Andreassen *et al.* (1991), Charniak (1991) and Neapolitan (1990), as well as by papers within the collections edited by Oliver & Smith (1990), Shafer & Pearl (1990) and Gammerman (1995). Applications in computer vision are described by Agosta (1990), Binford *et al.* (1989), Levitt *et al.* (1990) and Rimey & Brown (1992). Various commercial and free shells are available, including BAIES (Cowell, 1992, 1995), Hugin (Andersen *et al.*, 1989), IDEAL (Srinvas & Breese, 1990) and PRESS (Gammerman *et al.*, 1995). Although we work with probabilities, the same calculations can be applied to other measures of belief which satisfy certain axioms (see Section A.4)—see Pearl (1988), Dempster & Kong (1988), Shafer & Shenoy (1986), Shenoy *et al.* (1988), Shenoy (1989) and Shenoy & Shafer (1990)—and also to consistency calculations in computer databases (Fagin, 1977).

Recent books include Almond (1995), Jensen (1996) and Shafer (1996).

The methods of this chapter are more complicated than, say, classification trees, so it is worth asking if the ability to feed in qualitative knowledge actually improves the accuracy of classification. Several of the discussants of Spiegelhalter *et al.* (1993) asked this, specifically in the context of medical diagnosis. Their answer (page 280) is equivocal. Belief networks are designed and trained to answer more than just the question of classifying future cases. They are able to give a much higher level of explanation, including exploring what were important input features in reaching the conclusion and whether the input data were in some sense in conflict. To do so they model the whole joint distribution. Although there will be an advantage in using qualitative knowledge (at least if it is a reasonable approximation to reality), the need to model the whole distribution makes more demands on limited data resources.

Two restrictions need to be noted. Most of the development of belief networks has been restricted to categorical variables, that is discrete random variables with a finite (and usually small) number of levels; some developments using continuous variables are being made (such as Lauritzen, 1992, and Gammerman *et al.*, 1995). Second, the problem of conditioning belief networks on observations is in general NP-hard (Cooper, 1990) and so the methods described here are potentially prohibitively slow. Fortunately, in many real systems the networks are sparsely connected, and the shells do seem to work quite fast enough.

There has been a parallel (and until recently completely separate) development of methods within the field of pedigree analysis in genetics (Spiegelhalter, 1990; Cannings *et al.*, 1978; Cannings & Thompson, 1981; Thompson 1985).

Belief networks are often associated with notions of causality. Opinions on the usefulness of this vary from complete scepticism (Speed, 1990) to enthusiasm (Pearl, 1993b, 1995). Since our purpose is prediction of pattern classes, we will avoid discussion of causality except to use known causality to help us specify probability models.

In the past directed graphs were widely used because there was perceived to be a problem with zero probabilities in graphical models on undirected graphs. This is inaccurate; these problems disappear when special distributions or (especially) special graphs are considered. Thus we derive most of the methodology in the context of decomposable (undirected) graphs after considering the simple case of a directed tree.

8.1   Graphical models and networks

Graphs such as Figure 8.1 are used to represent conditional independence properties on a collection of random variables. It will be important to keep a clear distinction between directed graphs which can represent causality, and undirected graphs without arrows which represent dependence without specifying a causal direction; some terms are used for both with subtly different meanings.

Throughout this section we will assume we are given a finite collection of random variables $X_v, v \in V$, and we wish to describe qualitatively the dependencies between these random variables. In our applications these random variables will include the features in the pattern **x** and the class C. However, they may also include unobserved features. For example, an important extension to Figure 8.1 is given in Figure 8.2 where the class is not assumed to be reported accurately, as may be common where diagnosis is difficult.

To describe dependence we will use the language of graph theory. This is usually self-explanatory, but more formal treatments can be found in many basic accounts of theoretical computer science, including Knuth (1968), Cormen *et al.* (1990) and Sedgewick (1990), as well as the specialist books by Berge (1973) and Golombic (1980). Maier (1983) gives a different perspective, that of database theory.

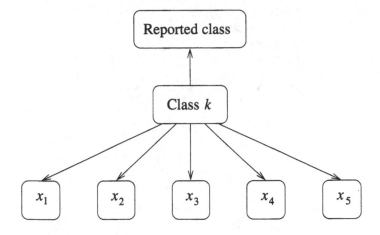

Figure 8.2: The causal graph for the idiot's Bayes rule with inaccurate reporting of classes.

A *graph* is a collection of *vertices* and *edges*. The vertices will represent the set of random variables (hence the use of *V* to denote the set). The set of *edges* is a set of unordered pairs of distinct vertices; if an edge is present it is indicated on a diagram by a line (without an arrow) joining the pair of vertices. A *path* on a graph is list of vertices for which each successive pair is joined by an edge. A *subgraph* is a subset of vertices together with those edges both of whose vertices are in the subset. A subgraph is said to be *connected* if there is a path joining every pair of vertices, and *complete* if every possible edge is present. The maximal complete subgraphs of a graph are called its *cliques*. A *cycle* is a path which returns to its origin and visits no vertex more than once. A connected graph with no cycles is called a *tree*.

Another convention is to call complete subgraphs cliques, when a clique is maximal if no vertex can be added without making the subgraph incomplete.

Where necessary, we will refer to graphs as *undirected graphs*. *Directed graphs* also have a set of vertices and edges, but the edges are ordered pairs of vertices, and are represented on a figure (such as Figure 8.2) by lines with arrows from the first vertex to the second vertex. The first vertex is often called the *parent* and the second (marked by the arrow) the *child*. The notions of paths and cycles extend immediately to directed graphs. We will make frequent use of directed acyclic graphs, *DAGs*, that is directed graphs without cycles. A *directed tree* has the properties that it has one vertex, the *root*, such that a (directed) path leads from the root to any vertex, and any other vertex has precisely one incoming arrow. An *ancestral subgraph* of a directed graph contains all the ancestors of its vertices; for a directed tree ancestral subgraphs are rooted subtrees.

Pedants will call these acyclic directed graphs.

A *polytree* is a singly-connected DAG, that is a DAG in which at most one path exists between any two vertices (see Figure 8.3).

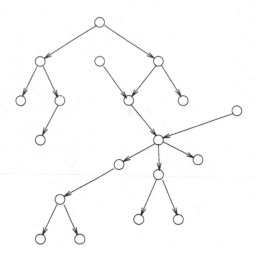

Figure 8.3: An example of a polytree.

Markov networks

The usual way to interpret the (in)dependencies represented by an undirected graph $\mathscr{G}$ is what Pearl (1988) calls an *I-map*. Given three subsets A, B, C of vertices, we say C *separates* A and B in $\mathscr{G}$ if every path from (a vertex in) A to (a vertex in) B goes through (a vertex in) C. For any subset A of V let X_A denote the collection of random variables associated with the vertices in A. Then we consider whether X_A and X_B are conditionally independent given X_C, which we write as

$$X_A \perp\!\!\!\perp X_B \mid X_C \quad \text{or sometimes} \quad A \perp\!\!\!\perp B \mid C.$$

We say the graph $\mathscr{G}$ is an I-map of the distribution if this is true whenever A and B are separated by C. We say the distribution is *global Markov* with respect to $\mathscr{G}$ if separation implies conditional independence.

> The two concepts are the same, but the graph varies for an I-map and the distribution for global Markov.

Note that the complete graph on V will be an I-map of any probability distribution on X_V, since then there will never be any separating C. This shows that there may be subsets A, B and C with the conditional independence property $A \perp\!\!\!\perp B \mid C$, for which not every path from A to B goes through C. If it is also true that all conditional independencies are represented by separation, Pearl calls the representation by this graph a *perfect map*. An I-map $\mathscr{G}$ is called *minimal* if removing any edge from $\mathscr{G}$ makes it no longer an I-map. If there is a unique minimal I-map, it is called the *Markov network* of the distribution.

> Distributions for which there is a Markov network that is perfect are often called *graphical models*, although this term is also used more loosely.

Markov properties of distributions on graphs have been studied in the areas of random fields (Preston, 1974, 1976) and image analysis (Geman & Geman, 1984; Geman, 1990; Isham, 1981). Several Markov properties have been defined. Denote by ∂A the *boundary* of A, the set of vertices in A^c which have a neighbour in A (so all neighbours of points in A are in $A \cup \partial A$). Then the most important Markov properties for a given graph $\mathcal{G}$ are

global For any disjoint subsets A, B and C such that C separates A and B (all paths from A to B contain a member of C) we have $X_A \perp\!\!\!\perp X_B \mid X_C$.

local The conditional distribution of X_a given $X_{V \setminus \{a\}}$ depends only on $X_{\partial \{a\}}$, or equivalently $X_a \perp\!\!\!\perp X_{V \setminus [\{a\} \cup \partial \{a\}]} \mid X_{\partial \{a\}}$. This is probably easier to describe in words: the random variables at a and those at vertices not connected to a by an edge are conditionally independent given those which are so connected.

pairwise X_a and X_b are conditionally independent given all the other random variables if there is no edge from a to b.

These properties allow us to read off successively weaker conditional independence statements from the graph; the global Markov property is equivalent to the graph being an I-map. The three properties can be strictly different. Consider the four discrete random variables with joint distribution

This example is from Pearl (1988, p. 135).

a	b	c	d	Pr
0	0	0	0	1/3
0	1	1	1	1/3
1	1	0	2	1/3

These are taken as the random variables at the vertices of a graph. The pairwise Markov property is satisfied by any graph on the vertices, even that with no edges. On the other hand, the four random variables are far from independent, and the conditional distribution of X_d given X_a, X_b and X_c depends on at least two of the conditioning random variables. For the local Markov property to hold we have two possible minimal graphs:

The set $V \setminus \{a, b\}$ can be any pair of vertices, and knowledge of any pair of random variables determines which of the three outcomes occurs. Thus conditionally X_a and X_a are constant.

and for both the global Markov property holds. Thus in this example there are two distinct minimal I-maps.

An example of a local but not global Markov distribution is given by taking the graph a—b c—d and the same non-constant random variable at the four vertices. This is trivially local Markov, but $\{b\}$ and $\{c\}$ are separated by $\emptyset$, and $X_b \perp\!\!\!\perp X_c$ is false.

Our main example demonstrates the inadequacy of the 'obvious' way to construct a minimal I-map, that is to include the edge $\{a, b\}$ in the graph if and only if $X_a \not\perp\!\!\!\perp X_b \mid X_{V \setminus \{a,b\}}$ holds. This is clearly the minimal graph to satisfy the pairwise Markov property, but need not be global Markov. By the following result, if we confine attention to discrete random variables and strictly positive probability distributions this will be an I-map, and therefore the unique minimal I-map. This result has a confusing history; it is often attributed to Hammersley & Clifford in 1971, although they did not publish for nearly twenty years and gave one of a series of increasingly more general statements.

Proposition 8.1 *Suppose we have a collection of discrete random variables defined on the vertices of a graph.*

(i) *Suppose the joint distribution is strictly positive. Then the pairwise Markov property implies that there are positive functions ϕ_C, symmetric in their arguments, such that*

$$\Pr\{X_V = x_V\} \propto \prod_C \phi_C(x_C) \qquad (8.2)$$

the product being over cliques of the graph.

(ii) *A potential representation (8.2) implies the global Markov property for any distribution.*

Proof: (i) We may assume (by re-labelling if necessary) that each random variable X_s can take the value 0. The proof proceeds by induction on the size of $A = \{s \mid X_s \neq 0\}$, and we prove the existence of functions ϕ_C (with $\phi_C \equiv 1$ for non-complete C) such that

$$\Pr\{X = x\} = \Pr\{X = 0\} \prod_{C \subset A} \phi_C(x_C). \qquad (8.3)$$

This can be reduced to a product over cliques by assigning ϕ_C to a clique which contains C, and multiplying the original clique function by ϕ_C.

Define the functions ϕ_C recursively by

$$\phi_C(x_C) = \Pr\{X_C = x_C, X_{C^c} = 0\} \Big/ \Pr\{X = 0\} \prod_{D \subsetneq C} \phi_D(x_D)$$

The Hammersley–Clifford result relates the *local* Markov property to the global property and to a potential representation. See Clifford (1990) for the published version and historical comment.

The rest of this section is rather technical and may be skipped at first reading.

Defining $\phi_C = 1$ for non-complete C allows us to take products over all subsets.

where the product is over *strict* subsets, and $\phi_C(0) \equiv 1$. (Note that $\phi_C > 0$ which avoids having $0 \times p/0$ in the manipulations that follow.) Clearly (8.3) holds if A is complete, and so holds if A is empty or has one element. Now suppose it holds if A has k or fewer elements. Split a non-complete A with $k+1$ members as $B \cup \{s\} \cup \{t\}$ where B has $k-1$ elements and s and t are not neighbours. Then $X_s \perp\!\!\!\perp X_t \mid X_B, X_{A^c}$, so

$$
\begin{aligned}
\Pr\{X_V = x_v\} &= \Pr\{X_s = x_s, X_t = x_t, X_B = x_B, X_{A^c} = 0\} \\
&= \Pr\{X_B = x_B, X_s = x_s, X_t = 0, X_{A^c} = 0\} \\
&\quad \times \frac{\Pr\{X_t = x_t \mid X_B = x_B, X_s = x_s, X_{A^c} = 0\}}{\Pr\{X_t = 0 \mid X_B = x_B, X_s = x_s, X_{A^c} = 0\}} \\
&= \Pr\{X_B = x_B, X_s = x_s, X_t = 0, X_{A^c} = 0\} \\
&\quad \times \frac{\Pr\{X_t = x_t \mid X_B = x_B, X_s = 0, X_{A^c} = 0\}}{\Pr\{X_t = 0 \mid X_B = x_B, X_s = 0, X_{A^c} = 0\}} \\
&= \Pr\{X_B = x_B, X_s = x_s, X_t = 0, X_{A^c} = 0\} \\
&\quad \times \frac{\Pr\{X_t = x_t, X_B = x_B, X_s = 0, X_{A^c} = 0\}}{\Pr\{X_t = 0, X_B = x_B, X_s = 0, X_{A^c} = 0\}} \\
&= \Pr\{X = 0\} \prod_{C \subset B \cup \{s\}} \phi_C(x_C) \frac{\prod_{C \subset B \cup \{t\}} \phi_C(x_C)}{\prod_{C \subset B} \phi_C(x_C)} \\
&= \Pr\{X = 0\} \prod_{C \subset B \cup \{s\} \cup \{t\}} \phi_C(x_C)
\end{aligned}
$$

where we use conditional independence at step 3, (8.3) for sets of size at most k at step 5 and the fact that a complete subset $C \subset A$ cannot contain both s and t at the last step. This establishes the result for any set A of size $k+1$ and completes the inductive step of the proof.

(ii) Suppose a potential representation (8.2) is given. Then

$$
\begin{aligned}
\Pr\{X_A \mid X_{A^c}\} &= \frac{\Pr\{X_V\}}{\Pr\{X_{A^c}\}} = \left. \prod_{\text{cliques } C} \phi_C(X_C) \middle/ \sum_{X_A} \prod_{\text{cliques } C} \phi_C(X_C) \right. \\
&= \left. \prod_{\substack{\text{cliques } C \\ C \cap A \neq \emptyset}} \phi_C(X_C) \middle/ \sum_{X_A} \prod_{\substack{\text{cliques } C \\ C \cap A \neq \emptyset}} \phi_C(X_C) \right.
\end{aligned}
$$

where we cancel terms for cliques C disjoint from A. A clique with $C \cap A \neq \emptyset$ is contained in $A \cup \partial A$, so the right-hand side is a function of $X_{A \cup \partial A}$ and $\Pr\{X_A \mid X_{A^c}\} = \Pr\{X_A \mid X_{\partial A}\}$. Some of the potentials ϕ_C may take zero values, but these calculations still hold if we take $0/0 = 0$.

Now suppose A and B are separated by C. Let B' be the set of vertices which can be reached by a path from B which does not meet C

and let $D = (B' \cup C)^c \supset A$; by construction D, B' and C are disjoint and C separates D and B', so no neighbour of D is in B'. Then $\Pr\{X_D \mid X_{B'}, X_C\} = \Pr\{X_D \mid X_{D^c}\} = \Pr\{X_D \mid X_{\partial D}\}$ does not depend on $X_{B'}$. Thus $D \perp\!\!\!\perp B' \mid C$ and hence $A \perp\!\!\!\perp B \mid C$ as $A \subset D, B \subset B'$.   □

Some partial relaxation of the positivity condition is possible: see Moussouris (1974), Averintsev (1975) and Ripley & Kelly (1977). Our counter-example on page 249 also shows (with the minimal I-map shown as the right-hand graph) that a distribution can be global Markov but not have a potential representation.

If there were a potential representation, $\Pr\{a = 0, b = 1, c = 0, d = 2\} = \phi_{\{a,b\}}(0,1)\phi_{\{a,c\}}(0,0) \times \phi_{\{b,d\}}(1,2)\phi_{\{c,d\}}(0,2) > 0$.

An alternative to imposing strict positivity on the *distribution* is to impose further conditions on the *graph*. Matúš (1992) shows that all three Markov properties are equivalent for any distribution if and only if every subgraph on three vertices contains two or three edges. A graph is said to be *triangulated* or *chordal* if every cycle of length four or more has a chord (an edge joining two non-consecutive vertices), and we will see in Proposition 8.2 that we can construct a potential representation for a triangulated I-map. Conversely, if a graph is not triangulated, it has a chordless cycle of length four or more, and our counter-example (extended if necessary by copies of X_d along the cycle, and constant variables elsewhere) shows a distribution on the vertices of the graph that is global Markov but does not have a potential representation. Thus being global Markov and having a potential representation are equivalent for all distributions on a graph if and only if it is triangulated.

We could ask if all conditional independence properties entailed by being global Markov can be read from the graph by separation. Geiger & Pearl (1993) show that this *is* so, by constructing a strictly positive distribution such that $X_A \perp\!\!\!\perp X_B \mid X_C$ if and only if C separates A and B on the graph. (The random variables used in this construction do take a finite set of values, but not one that can be specified in advance.)

Markov trees

If an undirected graph is a tree, any vertex can be declared as the root, and a directed tree formed by assigning arrows to point away from the root (along the unique path from the root to a vertex). The simplest possible tree is one with no branches, that is a chain of vertices. The simplest Markov network is a Markov chain, for which the Markov property is usually stated symmetrically

'past and future are independent given the present'

but the usual calculations on a Markov chain depend heavily on the time ordering of the vertices to work either forwards or backwards in time. Most of these methods can be extended to trees, with calculations proceeding up towards the root or downwards away from the root. Although some of the ideas logically belong in the next section, trees are so important an idea that we treat this special case here.

Markov chains are usually specified by the transition probabilities $\Pr\{X_t = j \mid X_{t-1} = i\}$, and Markov trees are also commonly specified by giving $\Pr\{X_v \mid X_a, a \text{ the parent of } v\}$. This *does* specify the joint distribution, since we can label the vertices in increasing order away from the root of the tree so that a vertex is preceded by its parent, and then

$$\Pr\{X_V\} = \prod_i \Pr\{X_i \mid X_j, j < i\} = \prod_i \Pr\{X_i \mid X_a, a \text{ the parent of } i\}$$

(8.4)

where at the second step we use the separation of v from the rest of the graph by its parent. The root has a special role in (8.4): it has no parent and $\Pr\{X_1\}$ appears without conditioning. In a tree the cliques each contain just one edge, so (8.4) is a potential representation of the distribution (without any positivity condition).

Figure 8.2 provides a natural example of a directed tree, with the true class as the root. The calculation we would wish to perform is to condition the distribution by restrictions $E = \bigcap\{X_v \in S_v\}$ on the values at some or all of the vertices (which we think of as the 'evidence'), specifically to find $\Pr\{X_v \mid E\}$ for one or more individual vertices. There are simple ways to do so making use of the tree structure (Pearl, 1982, 1988). Conditioning on the value taken by X_v has two effects. One is to break the tree at v so we can find the probability distribution of the descendants of v independently of all the rest of the tree. The other effect is to change the distribution of the ancestors of v and all their descendants (that are not direct descendants of v). We can consider both effects by first propagating messages up the tree, then down from the root. In the following a vertex is considered to be a descendant of itself, but not its own ancestor.

We suppose that we have calculated the marginal distributions p_v at each vertex: this can be done by

Here S_v is any set of values, possibly all possible values. We are excluding evidence such as $X_a = X_b$.

The rest of this subsection is technical and not needed elsewhere.

$$p_v(x_v) = \sum_{x_a} p_a(x_a)\Pr\{X_v = x_v \mid X_a = x_a\}$$

where a is the parent of v (and we are given the marginal distribution of the root vertex). We want $p_v^*(x_v) = \Pr\{X_v = x_v \mid E\}$. For any vertex v let E_v^- denote the conditioning event (if any) on the descendants of v, and let E_v^+ denote the condition on the remaining random variables.

Then by Bayes' formula

$$p_v^*(X_v) \propto \Pr\{E_v^- \mid X_v, E_v^+\} \Pr\{X_v \mid E_v^+\} = \Pr\{E_v^- \mid X_v\} \Pr\{X_v \mid E_v^+\} \quad (8.5)$$

using the separation by vertex v.

The first term is only non-trivial if v has some children u_i. Partition E_v^- into events concerning the descendants of each child, which are conditionally independent given X_v so

$$\Pr\{E_v^- \mid X_v\} = I(X_v \in S_v) \prod_{\text{child } u} \Pr\{E_u^- \mid X_v\}$$

and

$$\Pr\{E_u^- \mid X_v\} = \sum_{x_u} \Pr\{E_u^- \mid X_u = x_u\} \Pr\{X_u = x_u \mid X_v\} \quad (8.6)$$

which can be computed by a pass over the tree towards the root. Now consider the second term of (8.5). Variables in E_v^+ can only influence X_v through its parent X_a, so

$$\Pr\{X_v \mid E_v^+\} = \sum_{x_a} \Pr\{X_v \mid X_a = x_a\} \Pr\{X_a = x_a \mid E_v^+\}. \quad (8.7)$$

Now E_v^+ includes the restriction on X_a and E_a^+, plus E_b^- for any other children b of a. We have

$$\Pr\{X_a \mid E_v^+\} = \Pr\{X_a \mid X_a \in S_a, E_a^+, E_b^- \text{ for other children } b\}$$
$$\propto I(X_a \in S_a) \Pr\{X_a \mid E_a^+\} \prod_b \Pr\{E_b^- \mid X_a\}. \quad (8.8)$$

The terms in the product will have been found at (8.6) in the inwards pass, and so (8.7) and (8.8) can be computed on a subsequent outwards pass.

These computations may also be organized as asynchronous message-passing. Each vertex keeps a current version of $\lambda(x_v) = \prod_u \Pr\{E_u^- \mid X_v\}$ as a product over its children (empty products being one) and $v(x_v) = \Pr\{X_v = x_v \mid E_v^+\}$. Then $p_v^*(x_v)$ is $I(x_v \in S_v)\lambda(x_v)v(x_v)$ normalized to unit sum. When information becomes available at vertex v, it passes $\Pr\{E_v^- \mid X_v \in S_v\}$ to its parent a, which is converted to $\Pr\{E_v^- \mid X_a\}$ by (8.6) and used to update $\lambda(x_a)$. It also passes to each child $\Pr\{X_v \mid X_v \in S_v, E_v^+, E_b^-\}$ where that child is excluded from the b considered. This message is found from (8.8) by re-normalizing. Whenever another vertex receives a message it passes on messages to its other neighbours.

It will be helpful to consider an example. The graph is shown in Figure 8.4. There are 12 binary random variables which we label

Figure 8.4: An example of a directed tree.

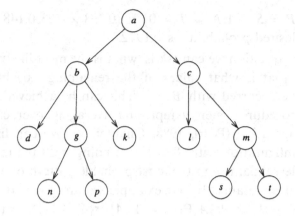

Table 8.1: Specifications of conditional probabilities for the directed tree of Figure 8.4.

$\Pr\{A = 1\} = 0.3$	$\Pr\{L = 1 \mid C = 0\} = 0.1$
	$\Pr\{L = 1 \mid C = 1\} = 1.0$
$\Pr\{B = 1 \mid A = 0\} = 0.9$	$\Pr\{M = 1 \mid C = 0\} = 0.0$
$\Pr\{B = 1 \mid A = 1\} = 0.1$	$\Pr\{M = 1 \mid C = 1\} = 0.7$
$\Pr\{C = 1 \mid A = 0\} = 0.2$	$\Pr\{N = 1 \mid G = 0\} = 0.5$
$\Pr\{C = 1 \mid A = 1\} = 0.8$	$\Pr\{N = 1 \mid G = 1\} = 0.1$
$\Pr\{D = 1 \mid B = 0\} = 0.5$	$\Pr\{P = 1 \mid G = 0\} = 0.1$
$\Pr\{D = 1 \mid B = 1\} = 0.4$	$\Pr\{P = 1 \mid G = 1\} = 0.6$
$\Pr\{G = 1 \mid B = 0\} = 0.7$	$\Pr\{S = 1 \mid M = 0\} = 0.8$
$\Pr\{G = 1 \mid B = 1\} = 0.2$	$\Pr\{S = 1 \mid M = 1\} = 0.5$
$\Pr\{K = 1 \mid B = 0\} = 0.3$	$\Pr\{T = 1 \mid M = 0\} = 0.3$
$\Pr\{K = 1 \mid B = 1\} = 0.9$	$\Pr\{T = 1 \mid M = 1\} = 0.7$

The two values of the probability refer to the values zero and one of the unspecified variable, in that order.

by capital letters. Their joint distribution can be specified via the probabilities of Table 8.1.

We first calculate the marginal probabilities by a downwards pass in alphabetical order of the vertices. We then are told that $C = P = S = 1$ and $N = T = 0$, and asked for the conditional probability that $B = 1$ (it was 0.66 before conditioning). The first step is to pass the information up to the root, then down to B. Knowledge of S and T is actually irrelevant, as we know C which separates B from S, T. Propagating $N = 0, P = 1$ to g gives $\Pr\{E_g^- \mid G\} = (0.05, 0.54)$, and propagating this to b gives $\Pr\{E_b^- \mid B\} = (0.393, 0.148)$. Propagating $C = 1, S = 1, T = 0$ to a gives a term $\Pr\{A \mid E_b^+\} = \Pr\{A \mid C = 1, S = 1, T = 0\} = \Pr\{A \mid C = 1\} = (0.14, 0.24)/0.38$. This is passed down to b, giving $\Pr\{B \mid E_b^+\} = (0.23, 0.15)/0.38$ using (8.7). From (8.4) we have

$\Pr\{B\,|\,C=P=S=1,N=T=0\} \propto (0.393\times0.23, 0.148\times0.15)$, and finally the desired probability is 0.1972.

A further question we can ask is 'what is the most likely explanation of $B=1$', that is what values of the remaining variables are most likely to have occurred with $B=1$. This can be achieved by the same updating procedures, merely replacing averaging over children by a maximum operation (Pearl, 1988, Chapter 5). We can find the most probable configuration with $B=1$ by finding both the most probable pattern of descendants and the most probable pattern of ancestors and their other descendants. In our example we are given $B=C=1$, so $\Pr\{A\,|\,B=C=1\} \propto \Pr\{A\}\Pr\{B=1\,|\,A\}\Pr\{C=1\,|\,A\} = (0.126, 0.024)$, so $A=0$ is the most probable ancestor. Clearly $D=0$ and $K=1$ are most probable. Finally $\Pr\{G\,|\,B=1,N=0,P=1\} \propto \Pr\{G\,|\,B=1\}\Pr\{N=0\,|\,G\}\Pr\{P=1\,|\,G\} = (0.040, 0.108)$ and the most plausible explanation is $(A=0,C=1,D=0,G=1,K=1,N=0,P=1)$.

This mode of reasoning is called *abductive inference*.

Another method to organize these calculations is via the joint marginal distributions of each adjacent pair of vertices, which is just the marginal distribution of the parent times the transition probability to the child. When a condition is imposed at a vertex, this alters the joint distributions of that vertex and each neighbour. The evidence can then be propagated to each neighbour of that neighbour (and hence all over the tree) by updating the joint density of X_a, X_b as

$$\Pr_{\text{new}}\{X_a, X_b\} = \Pr_{\text{old}}\{X_a, X_b\} \times \frac{\Pr_{\text{new}}\{X_a\}}{\Pr_{\text{old}}\{X_a\}}.$$

We illustrate this for our example, considering only the vertices *abcgnp* for compactness. (The distributions of the remaining vertices can be found conditionally on this set quite easily.) Initially the marginal probabilities that $X_v=1$ are

a	b	c	g	n	p
0.3	0.66	0.38	0.37	0.352	0.285

and the joint probabilities on the five edges are

	0 0	0 1	1 0	1 1
a—b	0.07	0.63	0.27	0.03
a—c	0.56	0.14	0.06	0.24
b—g	0.102	0.238	0.528	0.132
g—n	0.315	0.315	0.333	0.037
g—p	0.567	0.063	0.148	0.222

Now we condition on $N = 0$. This changes the marginal probabilities to

	0 0	0 1	1 0	1 1
g—n	0.4861	0	0.5139	0

Vertex g then sends the message $\mathrm{Pr}_{\mathrm{new}}\{G\}/\mathrm{Pr}_{\mathrm{old}}\{G\} = (0.7716, 1.3889)$ to its neighbours b and p and updates $\mathrm{Pr}\{G = 1\} = 0.5139$. Let us consider the effect on p, which is to update the edge marginal to

	0 0	0 1	1 0	1 1
g—p	0.4375	0.0486	0.2056	0.3083

If we now condition on $P = 1$, we find $\mathrm{Pr}\{G = 1\} = 0.8638$.

We now update the b—g edge by rescaling by the G marginal. We have not updated it since $\mathrm{Pr}\{G\} = 0.37$, so the result is

	0 0	0 1	1 0	1 1
b—g	0.0221	0.5556	0.1141	0.3082

which gives $\mathrm{Pr}\{B = 1\} = 0.4223$. Updating the a—b edge we find $\mathrm{Pr}\{A = 1\} = 0.4780$ and

	0 0	0 1	1 0	1 1
a—b	0.1189	0.4031	0.4588	0.0192

Finally, consider conditioning on $C = 1$. First update the a—c edge to produce

	0 0	0 1	1 0	1 1
a—c	0.4176	0.1044	0.0956	0.3824

then set the entries for $C = 0$ to zero and renormalize to give $\mathrm{Pr}\{A = 1\} = 0.7855$. Now we update the a—b edge again to get

	0 0	0 1	1 0	1 1
a—b	0.0489	0.1656	0.7539	0.0316

This gives $\mathrm{Pr}\{B = 1\} = 0.1972$, which is the conditional probability we require. Propagating this ends up with marginal probabilities of

a	b	c	g	n	p
0.7855	0.1972	1	0.916	0	1

and the joint probabilities of the five edges are

	0 0	0 1	1 0	1 1
a—b	0.0489	0.1656	0.7539	0.0316
a—c	0	0.2145	0	0.7855
b—g	0.0307	0.7721	0.0533	0.1439
g—n	0.084	0	0.916	0
g—p	0	0.084	0	0.916

The details of such calculations are tedious to do by hand, but lend themselves to automated recursive calculations, and have been used by Binford *et al.* (1989). Kim & Pearl (1983) (see also Pearl, 1988) extend the principles to polytrees, where a vertex may have more than one parent as well as more than one child. The method fails if applied to DAGs which would have loops if viewed as undirected graphs. There are a few ways to convert such DAGs into polytrees (Pearl, 1988, §4.4):

1 *clustering methods*, in which vertices are joined into a composite node. For example, if in Figure 8.5 we combine vertices b and c the DAG is converted to a chain (which is of course a tree). We will see a systematic version of this idea in the next section.

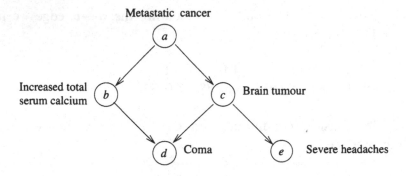

Metastatic cancer

Increased total serum calcium

Brain tumour

Coma

Severe headaches

Figure 8.5: A hypothetical medical belief network (after Cooper, 1984, and Pearl, 1988).

2 *conditioning*, in which the values at one or more vertices are fixed, and the results averaged over the conditioned variables. If in Figure 8.5 we condition on X_a, we have a polytree.

3 *simulation* methods which we discuss in the next section.

Decomposable models

The Markov network of a strictly positive distribution has a parametrization via the clique functions ϕ_C of (8.2). Computations are very much simplified for graphs that have no cycle of four or more vertices without a chord (an edge joining two non-consecutive vertices). Such graphs are called *chordal* or *triangulated*. A probability distribution is said to be *decomposable* with respect to a graph $\mathscr{G}$ if $\mathscr{G}$

This is equivalent to another graph property called decomposability, so these graphs are also called *decomposable*.

is an I-map and triangulated. If $\mathcal{G}$ is omitted, the minimal I-map is assumed (if it is unique). The undirected version of the directed graph in Figure 8.5 is not decomposable, and neither is the right-hand I-map for our counter-example on page 249.

For decomposable distributions there is a natural way to specify the joint distribution by 'local' pieces, which we will call a *marginal representation*. These are the marginal distributions over the cliques, and are much more easily interpreted than general potential representations.

Proposition 8.2 *A distribution which is decomposable with respect to $\mathcal{G}$ can be written as a product of the distributions on the cliques of $\mathcal{G}$ divided by the product of the distributions on their intersections.*

To prove this we need a few more concepts. A *join tree* T of cliques is a tree with the cliques as its vertices, such that if we remove all the cliques containing a vertex of V, the tree stays connected. Thus if two cliques contain a common vertex $v \in V$, so does every clique on the path in the tree between them. Every triangulated graph has a join tree (Beeri *et al.*, 1983), and triangulatedness can be tested and a join tree found by the following procedure (Tarjan & Yannakakis, 1984):

1   Order the vertices by *maximal cardinality search*. Start anywhere, and number next a vertex with the largest number of already numbered neighbours.

2   Starting with the highest numbered vertex, check for each vertex that all its lower-numbered neighbours are themselves neighbours. (By adding missing edges here, a triangulated graph will be produced.)

3   Identify all the cliques, and order them by the highest numbered vertex in the clique.

4   Form the join tree by connecting each clique to a predecessor (in the ordering of step 3) which shares the most vertices.

It can be useful to use the freedom to start anywhere in constructing the join tree, and subsequently to take any clique as its root. Incidentally, since this algorithm labels cliques by a subset of the vertices, there can be no more cliques than vertices in a triangulated graph.

This construction orders the join tree so that the unique path from C_1 to C_j has cliques in increasing order. For $i \geqslant 2$ let $j(i)$ be the number of the predecessor of clique i on the unique path, and let $S_i = C_i \cap (C_1 \cup \cdots \cup C_{i-1})$. Then $S_i \subset C_{j(i)}$, which is called the *running intersection property* (Beeri *et al.*, 1983). Let $H_i = C_1 \cup \cdots \cup C_{i-1}$ and $R_i = C_i \setminus S_i$. The clique sequence is also *perfect*, which means that $\partial R_i \cap H_i$ is a complete subgraph for all $i \geqslant 2$.

A join tree is also called a junction tree. Blair & Peyton (1993) give an introduction to these ideas, and Lauritzen (1996, Chapter 2) gives a self-contained account of the graph-theoretical properties based on decomposability.

$S_i \subset C_{j(i)}$ follows from the definition of a join tree: if $v \in C_j \cap C_i$ for $j < i$ then $v \in C_{j(i)}$.

Proposition 8.3 *The separator S_i separates R_i from $H_i \setminus S_i$.*

This is corollary A.8 of Dawid & Lauritzen (1993), but their proof is skeletal.

Proof: Fix a path from R_i to $H_i \setminus S_i$, and suppose it has a vertex v in R_j for some $j > i$ but none in $\bigcup_{k>j} R_k$. Let a and b be the first vertices before and after v which are not in R_j. Then a and b are both in $\partial R_j \cap H_j$ and hence neighbours. We can construct a connected subset of the path within H_j by omitting the vertices between a and b. By repeating this process we find a connected subset of the original path wholly in H_{i+1}, and this path will have an edge from $a \in R_i$ to $b \in H_i$. We will show that $b \in S_i$. Since a and b are neighbours, they both belong to some clique C_k, and since $a \in R_i$, $k \geqslant i$. Suppose $k > i$. Then $a, b \in C_k \cap H_k \subset C_s$ for some $s < k$, and by repeating if necessary we find $a, b \in C_s$ for some $s \leqslant i$. Thus $b \in C_i \cap H_i = S_i$. $\qquad\square$

Proof of Proposition 8.2:
The sets R_i form an ordered partition of V, with $\bigcup_{j<i} R_i = \bigcup_{j<i} C_i = H_i$, so

$$\Pr\{X_V\} = \prod_i \Pr\{X_{R_i} \mid X_{R_1}, \ldots, X_{R_{i-1}}\} = \prod_i \Pr\{X_{R_i} \mid X_{H_i}\}.$$

The separation property shows that $X_{R_i} \perp\!\!\!\perp X_{H_i} \mid X_{S_i}$, so $\Pr\{X_{R_i} \mid X_{H_i}\} = \Pr\{X_{R_i} \mid X_{S_i}\}$ and

$$\Pr\{X_V\} = \prod_i \Pr\{X_{R_i} \mid X_{S_i}\} = \prod_i \Pr\{X_{C_i}\} / \Pr\{X_{S_i}\} \qquad (8.9)$$

known as the *set-chain* and *marginal* representations. If any denominator is zero, the expression is taken as zero. $\qquad\square$

Note that we have not used a positivity condition here, but (8.9) provides a potential representation. Thus for triangulated graphs we have a potential representation if and only if the global Markov property holds, by Proposition 8.1(ii).

The set-chain representation provides a minimal way to specify the joint distribution. It is not the same as specifying the transition probabilities $\Pr\{C_i \mid C_{j(i)}\} = \Pr\{R_i \mid C_{j(i)}\}$ of the join tree as a Markov tree, since $S_i = C_i \cap C_{j(i)}$ is strictly smaller than $C_{j(i)}$.

The fill-in produced by the Tarjan–Yannakakis algorithm is fast to compute, but it may add more edges than are necessary. Algorithms to produce minimal fill-ins are known (Rose *et al.*, 1976), but are potentially very slow (this is an NP-complete problem: Yannakakis, 1981; Wen, 1990). An approximate algorithm using simulated annealing is described by Kjærulff (1992).

See the glossary.

The next proposition shows that we can read off conditional independence properties from the join tree.

Proposition 8.4 *Suppose C_A and C_B are sets of cliques in the join tree of a decomposable distribution separated by C_C. Then the sets of variables $A = \bigcup C_A$ and $B = \bigcup C_B$ are conditionally independent given the set $C = \bigcup C_C$.*

Proof: It will suffice to show that C separates $A \setminus C$ and $B \setminus C$ on the original graph and then use the global Markov property. We prove this by contradiction. Suppose there is a path in $V \setminus C$ from $\gamma_0 \in A$ to $\gamma_m \in B$ via $\gamma_1, \ldots, \gamma_{m-1}$. Associate γ_0 with a clique $C_0 \in C_A$ and for $s \geqslant 1$ associate γ_s with a clique C_s that contains both it and γ_{s-1}; as the vertices are not in C, these cliques cannot belong to C_C. The cliques C_{s-1} and C_s need not be neighbours in the join tree, but because they have γ_s in common, all cliques on the unique path from C_{s-1} to C_s also contain γ_s and so are not members of C_C. In this way we can assemble a path in the join tree from $C_0 \in C_A$ to C_m avoiding the collection C_C. It is not necessarily the case that $C_m \in C_B$, but if it is not there is another clique in C_B containing γ_m, and we can adjoin the path from C_m to that clique (which again avoids C_C). This contradicts the separation of C_A and C_B by C_C, so there is no such path (γ). □

These results show that for decomposable distributions we can work with the sets of variables in the join tree, and for trees the local specification of the distribution is easy. Further, we can update the marginal distributions in cliques by message-passing when conditioning information becomes available. This makes it attractive to work not with a minimal I-map $\mathscr{G}$ for a distribution, but with a triangulated I-map $\mathscr{G}^\Delta$ formed by triangulating $\mathscr{G}$. The distribution is decomposable with respect to $\mathscr{G}^\Delta$, and this gives us potential and marginal representations and allows us to compute using the join tree of $\mathscr{G}^\Delta$. Of course, some of the *interpretation* is lost, but this does not affect the calculations and the original graph may be used for interpretations. What may make this procedure unattractive is that $\mathscr{G}^\Delta$ may be very different from $\mathscr{G}$; it could even be complete.

We hope the sets are small.

This graph has more edges, so all separation properties on $\mathscr{G}^\Delta$ also hold on $\mathscr{G}$; thus $\mathscr{G}^\Delta$ is an I-map.

Conditioning on evidence

Suppose we wish to introduce the evidence $E = \{X_v \in S_v\}$ which is a restriction on the variable at a single vertex v. We will do so by finding the marginal representation of $\Pr\{\cdot, E\}$. Then summing over any clique gives us $\Pr\{E\}$ and we can divide by this to give the marginal representation of $\Pr\{\cdot \mid E\}$. Select a clique C_i that contains v. Suppose we have a potential representation (ϕ_C) (for example, the set-chain representation). By setting $\phi_C(x_C) = 0$ whenever $x_v \notin S_v$ we obtain a potential representation of $\Pr\{\cdot, E\}$, and we employ the general procedure below to turn a potential representation into a marginal one.

Suppose we have a potential representation (ϕ_C). We can allow a little more generality by having functions ψ_S on the separators S and asking that

$$\Pr\{X_V = x_v\} = \prod_C \phi_C(x_C) \Big/ \prod_S \psi_S(x_S) \qquad (8.10)$$

where $1/0$ is taken as 0 (and so if $\psi_S(x_S) = 0$ we can adjust $\phi_C(x_C)$ to be zero). For any neighbouring pair of cliques C_i, C_j with $S = C_i \cap C_j$ consider the operation of replacing ψ_S by the marginal ψ_S^* of ϕ_{C_i} (formed by summing over the variables in $C_i \setminus S$) and replacing ϕ_{C_j} by $\phi_{C_j} \times \psi_S^*/\psi_S$. This step maintains a potential representation. We describe ψ_S^*/ψ_S as the *message* passed over the edge of the join tree. (When conditioning a marginal representation, this can be thought of as finding the revised marginal distribution of S, and adjusting the marginal distribution of the adjoining clique to $\Pr\{C_j \mid S\}\Pr\{S\}$.)

Now suppose this step is performed for multiple conditioning events and each edge in the join tree until no further progress can be made. The steps can be organized in many ways (Jensen *et al.*, 1990; Shenoy & Shafer, 1990; Dawid, 1992): an attractive order is to pass messages in to the root and then out to the leaves. When this has been done, we will have the marginal representation. (Formal proofs are given by Dawid, 1992.)

By replacing averaging by maximizing in forming ψ_S^* we can find the most plausible explanation just as for Markov trees (Dawid, 1992). A modified averaging (omitting evidence on S) computes the distribution at each variable conditional on the evidence everywhere else (Cowell & Dawid, 1992) which is useful in monitoring consistency of information (Spiegelhalter *et al.*, 1993).

8.2   Causal networks

We now turn to causal networks, which are defined on DAGs (directed graphs without cycles). Such graphs can always be numbered so that the parent(s) of a vertex have a lower number than the vertex itself. (This is called a *topological sort* of the DAG.) It is always true (for discrete random variables) that

$$\Pr\{X_V\} = \prod_v \Pr\{X_v \mid X_a, a < v\} \qquad (8.11)$$

but to relate the distribution to the graph we ask that for a *recursive model*

$$\Pr\{X_V\} = \prod_v \Pr\{X_v \mid X_a, a \text{ a parent of } v\}. \qquad (8.12)$$

(We will use $pa(v)$ as a shorthand for this condition.) By summing over X_v in reverse order of the vertices we find from (8.12) that

$$\Pr\{X_v, v \leqslant j\} = \prod_{v \leqslant j} \Pr\{X_v \mid pa(v)\}$$

To avoid any measure-theoretic difficulties on the existence of conditional probabilities, we will formally define recursive to mean $X_v \perp\!\!\!\perp X_{a<v} \mid pa(v)$.

and so being recursive is equivalent to

$$\Pr\{X_v \mid X_a, a < v\} = \Pr\{X_v \mid pa(v)\}.$$

Conversely, if we are given the conditional probabilities of each variable given its parents, (8.12) can be used to define the joint distribution. This is the main attraction of a causal representation, as the conditional distributions given the parents are often easier to supply than clique potentials.

Given a distribution on a set of vertices, we can ask for what DAGs it is a recursive model. It is certainly recursive for the DAG which makes each vertex a child of all lower-numbered vertices, from (8.11), and we can form a smaller DAG by declaring as parents of vertex v only a subset of earlier vertices on which $\Pr\{X_v \mid X_a, a < v\}$ actually depends. If there are zeroes in the probability distribution then this procedure may not be unique. Consider the counter-example to a potential representation on page 249 in order (a, b, c, d). We find a to be the parent of b, (a, b) to be the parents of c, but for d we have the choice of any two from (a, b, c). If the distribution is strictly positive we can select as parents those vertices b for which $X_v \not\perp\!\!\!\perp X_b \mid X_a, a \neq b, a < v$ (using the equivalence of pairwise and local Markov on $\{a : a \leqslant v\}$).

Pearl (1986, p. 245) incorrectly claims uniqueness.

Note that this construction of a DAG depends on the ordering of the vertices, whereas the notion of a recursive model does not: some orderings may produce much simpler DAGs than others. It is here that causality is used to select a beneficial ordering. (Todd, 1995, illustrates this for the example of Lauritzen & Spiegelhalter, 1988, by choosing a different set of causalities.)

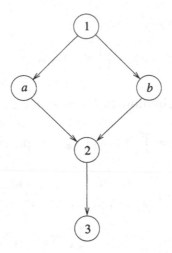

Figure 8.6: An example to illustrate conditional independence on a DAG.

To make use of the machinery for Markov networks (particularly on decomposable graphs) we would like to turn a recursive model on a DAG into a Markov network. It does *not* in general suffice to use the graph formed by dropping directions; Figure 8.6 will help illustrate why. Suppose all the variables are binary, with $\Pr\{X_a = 1\} = \Pr\{X_b = 1\} = 0.3 + 0.4X_1$, $\Pr\{X_2 = 1\} = 0.3 + 0.4I(X_a = X_b)$ and $\Pr\{X_3 = 1\} = 0.3 + 0.4X_2$. Then $X_a \perp\!\!\!\perp X_b \mid X_1$, but $X_a \not\perp\!\!\!\perp X_b \mid X_1, X_2$, whereas on the undirected graph $\{1, 2\}$ separates a and b. Clearly two vertices that have a common child need to be joined in the undirected graph. This is reflected in the idea of the *moral graph* (Lauritzen & Spiegelhalter, 1988). The steps to convert the DAG to its moral (undirected) graph are

1   replace all the directed edges by undirected ones and

2   add edges joining the parents (in the DAG) of each vertex if necessary.

The neighbours of a vertex in the moral graph are its parents and children in the original DAG, plus the other parents of those children. (Pearl calls the other parents of a vertex's children its *mates*.)

Proposition 8.5 *A recursive model on a DAG is global Markov and has a potential representation on the moral graph of its DAG.*

Proof: We will show the existence of a potential representation; by Proposition 8.1(ii) the distribution must be global Markov.

Start by setting the potential for each clique to one. For each vertex v, select a clique which contains it and all its parents and multiply its potential by $\Pr\{X_v \mid X_a, a \text{ a parent of } v\}$. (There may be more than

There must be such a clique: the graph is moral.

one choice of clique, in which case any will do.) This converts (8.12) to a potential representation by grouping terms into cliques.   □

Note once again that no positivity condition is required.

The moral graph is then triangulated by filling in, if necessary, and converted to a join tree of its cliques, to which the methods of Markov trees can be applied.

By Proposition 8.2 the joint distribution can be specified by giving the marginal probability distributions for each clique. Originally we had (8.12), which may seem to be the easiest way to specify the joint distribution. However, having the clique marginals allows us to specify the marginal distribution of each random variable. Further, the distribution will change as information becomes available. Consider Figure 8.2 on page 247. Initially the distribution can be specified conditionally on the true class. However, once some of the variables and/or the reported class are known, we will wish to find the conditional probability of the true class. In this case the moral graph comes from dropping the arrows, and any ordering that has the true class first or second may be used. The cliques are all the edges individually, and can be in any order. In this case the DAG is already a tree and can be used directly.

Markov properties on DAGs

This subsection is technical and not used elsewhere.

We have seen that a recursive model induces a global Markov distribution on the moral graph. Because there can be some in-filling at moralization, the conditional independencies implied by the moral graph may not include all those implied by the original recursive model. Consider Figure 8.6; its moral graph is the same as that of the DAG with a link from a to b, and on the latter DAG $X_a \not\perp\!\!\!\perp X_b \mid X_1$ is allowed. Of course, the conditional independencies still hold but are encoded in the numerical values of the particular representation used. In-filling to triangulate the moral graph can produce further masking of conditional independence properties; this is tolerated for the simplification in computation that results.

This has aroused interest in looking directly at Markov properties on a DAG. Several definitions of *directed Markov* are in use. That of Dawid & Lauritzen (1993) is

Spirtes *et al.* (1993) call this *causal Markov*. We can replace $nd(v)$ by $nd(v) \setminus pa(v)$ to give disjoint sets of variables.

$$X_v \perp\!\!\!\perp nd(v) \mid pa(v) \quad \text{for all } v \in V$$

where $nd(v)$ is the set of variables whose vertices are not descendants of v. (This is called *directed local Markov* by Lauritzen *et al.*, 1990.) The definition of Lauritzen (1989) and Lauritzen *et al.* (1990) (*directed global Markov*) is that $X_A \perp\!\!\!\perp X_B \mid X_C$ whenever C separates A from B in the moral graph of the smallest ancestral DAG containing A, B and C. This resolves the difficulty with Figure 8.6. We know that $X_a \perp\!\!\!\perp X_b \mid X_1$, but a and b are joined in the moral graph. They are not however joined in the moral graph of the subgraph on vertices $\{a, b, 1\}$, which is ancestral. That we have to take an ancestral subgraph is shown by noting that $X_a \not\!\perp\!\!\!\perp X_b \mid X_3$; the dependence arises through the ancestor X_2 of X_3.

Lauritzen *et al.* (1990) show that their two definitions of directed Markov are equivalent to each other and to being a recursive model. It is obvious that directed global Markov implies directed Markov which implies recursive. For any moralized ancestral DAG (8.12) provides a potential factorization, which by Proposition 8.1(ii) implies the distribution is global Markov on the moral graph hence $X_A \perp\!\!\!\perp X_B \mid X_C$. Another proof is given in Propositions 8.6 and 8.7 which also applies to continuous random variables (without assuming a joint density).

> Kiiveri *et al.* (1984) have other definitions of local and global Markov properties on a DAG.

Pearl (1986, 1988) finds conditional independence properties via a more complicated notion of separation for DAGs, called *d*-separation. A DAG is said to be an I-map of a distribution if $X_A \perp\!\!\!\perp X_B \mid X_C$ whenever C *d*-separates A and B. To define *d*-separation, consider a *trail* from A to B on the DAG, following edges in either direction. At each vertex inside the trail there will be two arrows. A trail is *blocked* by C at a vertex v if the two edges at v either

> Spirtes *et al.* (1993) call a vertex with converging arrows a *collider*.

(i) do not have converging arrows and $v \in C$, or

(ii) do have converging arrows and neither v nor any of its descendants are in C,

and A and B are said to be *d*-separated by C if every trail from A to B is blocked by C at some internal vertex. Consider once again the example of Figure 8.6. Then $X_a \perp\!\!\!\perp X_b \mid X_1$ follows from the *d*-separation of $\{a\}$ and $\{b\}$ by $\{1\}$, but $\{a\}$ and $\{b\}$ are not *d*-separated by $\{1, 2\}$ or $\{1, 3\}$, as the trail a—2—b is no longer blocked.

The *d*-separation condition for independence (empty C) is that A and B have disjoint ancestral sets, which is as we would expect.

For a DAG to be an I-map for a distribution is equivalent to the distribution being directed global Markov by the following proposition, and so is equivalent to being a recursive model.

Proposition 8.6 (Lauritzen *et al.*, 1990)
For sets A, B and C of vertices, d-separation of A and B by C is equivalent to the separation of A and B by C in the moral graph of the subgraph of ancestors of $A \cup B \cup C$.

Proof: Let $\mathscr{A}$ be the moralized ancestral subgraph. Suppose A and B are separated by C in $\mathscr{A}$. We will show that A and B are d-separated by C. Consider a trail from A to B in the DAG.

First we show that any trail which is not wholly in $\mathscr{A}$ is blocked. Follow the trail from A, and let a be the first vertex not in $\mathscr{A}$. By the ancestral property, the arrow on the edge leaving $\mathscr{A}$ must point to a. Only descendants of a can be reached by continuing the trail without converging arrows, and these cannot be in $\mathscr{A}$ (or their ancestor a would also be). To reach B the trail must have a pair of converging arrows at a vertex outside $\mathscr{A}$, and will be blocked at that vertex since it is not an ancestor of any vertex in C (as all such vertices are in $\mathscr{A}$).

Second, consider a trail from A to B on the DAG with all vertices in $\mathscr{A}$. If this contains a vertex with converging arrows, the predecessor and successor will have been 'married' in the moral graph, and so by replacing the converging arrows by the added edge we can find a path in the moral graph $\mathscr{A}$ which avoids vertices with converging arrows. This path must contain a vertex from C, and hence the trail on the DAG must be blocked at a vertex without converging arrows.

Conversely, suppose A and B are not separated by C in $\mathscr{A}$. Take a path from A to B which does not meet C, and form a trail on the DAG by replacing any edges formed at moralization by a diversion to their common child (which is in $\mathscr{A}$). Suppose this trail is blocked; this can only happen at a vertex a with converging arrows and no descendants in C. Since the vertex is in $\mathscr{A}$, it must have a descendant in either A or B. Thus we can follow the trail from A to a and then follow its descendants to B, or from B to a and then via descendants to A. This gives us a trail with strictly fewer blocking vertices. By repeating the process we can find an unblocked trail in the DAG from A to B. $\qquad\square$

These results show that for a recursive model on a DAG we can read off conditional independence properties by d-separation in the DAG or separation in the ancestral moral graph. Can all conditional independencies of sets of variables be found in this way? The general answer must be 'no', as conditional independencies could be encoded in the numerical values of the distribution, but Geiger & Pearl (1990) construct a recursive model for which any sets of variables which are not d-separated are not conditionally independent. Thus no other conditional independencies can be found from the DAG alone. If we know that some relations between variables are deterministic (occur with probability one) we can extract more information by D-separation, which differs from *d*-separation by also blocking a trail at a vertex whose variable is determined by X_C (Geiger *et al.*, 1990).

Proposition 8.7 (Verma & Pearl, 1990)
If a distribution is a recursive model on a DAG, that DAG is an I-map for the distribution.

Proof: The proof is by induction on the size of the DAG $\mathscr{D}$: the result is trivial if the DAG has only one element. Let ω be the last vertex in the DAG (and hence has no children), and suppose the result is true for the subgraph $\mathscr{D}'$ omitting ω. Our distribution is still recursive on $\mathscr{D}'$. Let A, B, C be disjoint sets of vertices such that A and B are *d*-separated by C in $\mathscr{D}$; we will show $A \perp\!\!\!\perp B \mid C$. Any trail with ω as an internal vertex necessarily has converging arrows at ω. Consider three exhaustive cases.

(a) The vertex $\omega \neq A \cup B \cup C$. Every trail including ω is blocked, so A and B are *d*-separated by C in $\mathscr{D}'$ and $A \perp\!\!\!\perp B \mid C$.

(b) Suppose that vertex ω is in either A or B; we will assume $\omega \in A$, and let $A' = A \setminus \{\omega\}$. Note that A' is *d*-separated from B by C in $\mathscr{D}'$. Now ω has no parents in B (or there would be a one-step trail from A to B). Let P be the set of parents of ω which are not in C; P is also *d*-separated from B by C in $\mathscr{D}'$. (Any trail from P to B either goes through ω and is blocked, or can be extended to a trail from ω to B which is not blocked at the added internal vertex in P.) Thus $A' \cup P \perp\!\!\!\perp B \mid C$ and for a recursive model $\{\omega\} \perp\!\!\!\perp B \mid A' \cup C \cup P$ (since the conditioning set includes all the parents of ω). These imply $A \cup P \perp\!\!\!\perp B \mid C$ (see (A.4)) hence $A \perp\!\!\!\perp B \mid C$.

(c) We have $\omega \in C$, so no trail can be blocked at ω. It follows that A and B are *d*-separated by $C' = C \setminus \{\omega\}$. Now ω must be *d*-separated from either A or B by C' or there would be an unblocked trail from A to B via ω. Suppose this is true for B, so $A \cup \{\omega\}$ and B are *d*-separated by C'. We have $A \cup \{\omega\} \perp\!\!\!\perp B \mid C'$ by case (b), which implies $A \perp\!\!\!\perp B \mid C' \cup \{\omega\} = C$ (see (A.3)).   □

Calculations on moral graphs

There are several calculations we may wish to perform on the joint distribution, but the most common are to condition on observed variables and to marginalize, especially to find the distribution over the true class. As we have seen, it will be normal to start with the conditional probability tables (8.12), which give a potential representation (see the proof of Proposition 8.5). This may be converted to a marginal representation by the standard message-passing algorithm (page 261), which also allows us to condition a marginal representation on evidence of the form $E = \bigcap \{X_v \in S_v\}$.

In general finding a marginal distribution of a subset of random variables is difficult, but it is immediate from the marginal representation if the subset is contained in some clique. Fortunately we will mostly be interested in marginals of a single variable, and then will have at least one clique to choose from. There is a trick (Pearl, 1988, §3.5.3; Jensen, 1991) to find the probability of a specific event of the form $E = \bigcap \{X_v \in S_v\}$, as the conditioning procedure in the join tree on E without normalization finds $\Pr\{\cdot, E\}$, and this can be summed over any clique to find $\Pr\{E\}$. (This is done in the example which follows.)

An example

We give an entirely fictitious example from medical diagnosis with eight vertices presented in Figure 8.7 and Table 8.2. The specification of the distribution via the conditional probability tables is

$$p(A)\,p(B)\,p(C)\,p(D|A,B)\,p(E|A)\,p(F|C)\,p(G|D,E)\,p(H|D,F).$$

To form the moral graph (Figure 8.8) we have to join a to b, d to e and d to f; this is already a triangulated graph. The cliques are abd, ade, cf, deg and dfh. One ordering by maximum cardinality search, starting from a, for the vertices is $abdegfhc$ and for the cliques is $C_1 = abd, C_2 = ade, C_3 = deg, C_4 = dfh, C_5 = cf$. The separators are then $S_2 = ad, S_3 = de, S_4 = d, S_5 = f$ and so the marginal representation is

$$\frac{p(A,B,D)\,p(A,D,E)\,p(D,E,G)\,p(D,F,H)\,p(C,F)}{p(A,D)\,p(D,E)\,p(D)\,p(F)}.$$

In forming the join tree we have considerable freedom, as C_4 can be linked to any of its predecessors. One choice (joining C_4 to C_3) would make the join tree into a chain, but we chose the tree shown in Figure 8.9.

The marginal probabilities may be found from $p(C_1) = p(D\,|\,A,B)p(A)p(B)$, $p(C_2) = p(E\,|\,A)p(S_2)$, $p(C_3) = p(G\,|\,D,E)p(S_3)$, $p(C_5) = p(F\,|\,C)p(C)$ and $p(C_4) = p(H\,|\,F,D)p(S_5)p(S_4)$. We find, writing tables in lexicographic order (false before true, last index varies fastest).

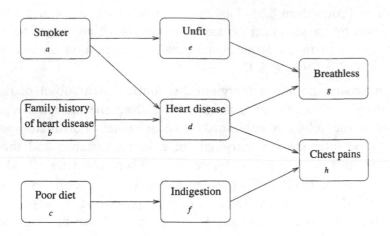

Figure 8.7: The DAG for an artificial medical diagnosis problem.

$\Pr\{A\} = 0.5 \qquad \Pr\{B\} = 0.2 \qquad \Pr\{C\} = 0.3$
$\Pr\{D\,|\,F,F\} = 0.05 \quad \Pr\{D\,|\,F,T\} = 0.3 \quad \Pr\{D\,|\,T,F\} = 0.2 \quad \Pr\{D\,|\,T,T\} = 0.5$
$\Pr\{E\,|\,F\} = 0.3 \qquad \Pr\{E\,|\,T\} = 0.5$
$\Pr\{F\,|\,F\} = 0.1 \qquad \Pr\{F\,|\,T\} = 0.4$
$\Pr\{G\,|\,F,F\} = 0.01 \quad \Pr\{G\,|\,F,T\} = 0.5 \quad \Pr\{G\,|\,D = T\} = 1$
$\Pr\{H\,|\,F,F\} = 0 \qquad \Pr\{H\,|\,F,T\} = 1 \quad \Pr\{H\,|\,T,F\} = 0.5 \quad \Pr\{H\,|\,T,T\} = 1$

Table 8.2: The conditional probability tables for the DAG of Figure 8.7. For each vertex the condition is on its parents in alphabetical order.

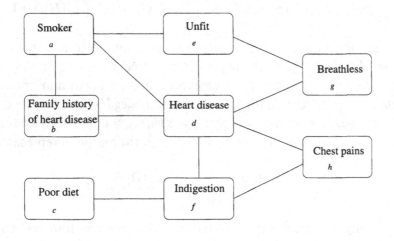

Figure 8.8: The moral graph associated with Figure 8.7.

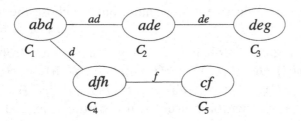

Figure 8.9: The clique tree associated with Figure 8.8. The separators S_i are marked on the edges.

ABD	0.38	0.02	0.07	0.03	0.32	0.08	0.05	0.05
AD	0.45	0.05	0.37	0.13				
ADE	0.315	0.135	0.035	0.015	0.185	0.185	0.065	0.065
DE	0.50	0.32	0.10	0.08				
DEG	0.495	0.005	0.16	0.16	0	0.1	0	0.08
D	0.82	0.18						
CF	0.63	0.07	0.18	0.12				
F	0.81	0.19						
DFH	0.6642	0	0	0.1558	0.0729	0.0729	0	0.0342

Note the treatment of clique $C_5 = cf$; we need the marginal for f, and this demands that we process C_5 before C_4. It is natural to think of C_4 depending on C_5, but to produce a tree we have to label the edge in the opposite direction.

We can also find the initial marginal probabilities via a potential representation and message-passing. Suppose we take initial potentials as $p(A)p(B)p(D\,|\,A,B)$, $p(E\,|\,A)$, $p(G\,|\,D,E)$, $p(H\,|\,F,D)$ and $p(F\,|\,C)p(C)$. Message-passing then multiplies ϕ_{C_2} by $p(A,D)$, ϕ_{C_3} by $p(D,E)$ and ϕ_{C_4} by $p(D)$ and by $p(F)$ to form the marginal representation.

Suppose a patient presents symptoms of breathlessness and chest pains, and is a smoker. What is the probability of heart disease? We condition on the evidence $A = G = H = $ T. We illustrate the message-passing approach by sending messages to $C_1 = abd$ at the root of the tree. We enter $G = H = $ T at cliques C_3 and C_4. For $A = $ T we have a choice of cliques, and choose C_1. We start at C_4 with a message over $S_4 = d$ to C_1, the new-to-old ratio of the separator distributions.

DFH	0	0	0	0.1558	0	0.0729	0	0.0342
D	0.1558			0.1071				
msg	0.1558/0.82 = 0.19		0.1071/0.18 = 0.595					

Clique C_3 sends a message over $S_3 = de$ to C_2:

DEG	0	0.005		0	0.16		0	0.1	0	0.08
DE	0.005			0.16			0.1		0.08	
msg	0.005/0.5 = 0.01		0.16/0.32 = 0.5		1		1			

This is then incorporated into clique C_2 and a message sent over $S_2 = ad$ to C_1:

ADE	0.00315	0.0675	0.035	0.015	0.00185	0.0925	0.065	0.065
AD	0.07065		0.05		0.09435		0.13	
msg	0.157		1		0.255		1	

Clique C_1 then incorporates two messages and its own constraint. We need only give the results for $A = \text{T}$:

BD	$0.32 \cdot 0.19 \cdot 0.255$	$0.08 \cdot 0.595 \cdot 1$	$0.05 \cdot 0.19 \cdot 0.255$	$0.05 \cdot 0.595 \cdot 1$
=	0.015504	0.0476	0.002423	0.02975
D	0.017927	0.07735		

so $\Pr\{\text{Evidence}\} = 0.095277$ and $\Pr\{D \mid \text{Evidence}\} = 0.812$.

As C_1 is the root, we should then pass messages back down the tree, but our question has already been answered from the marginal in C_1. It will be convenient at this stage to normalize, that is to divide the marginal distribution C_1 by $\Pr\{\text{Evidence}\}$. The messages sent to C_2 and C_4 are then approximately

AD	0	0	0.188/0.09435	0.812/0.13
D	0.188/0.1558	0.812/0.1071		

This modifies the clique marginals to (approximately)

ADE	0	0	0	0	0.0037	0.1843	0.406	0.406
DFH	0	0	0	0.188	0	0.553	0	0.259

These send messages to C_3 and C_5 of

DE	0.00037/0.005	0.1843/0.16	0.406/0.1	0.406/0.08
F	0.553/0.81	0.447/0.19		

and those cliques can be updated to

DEG	0	0.00037	0	0.1843	0	0.406	0	0.406
CF	0.430	0.165	0.123	0.282				

After these calculations it is often easy to answer further questions. Suppose we discover that the patient's family has a history of heart disease. This amounts to new evidence that $B = \text{T}$. Rather than go through the full procedure of propagating messages, we can just examine the marginal distribution of C_1 to see that the conditional probability of heart disease is now $0.02975/(0.002423 + 0.02975) \approx 0.925$. Similar calculations will often allow us to evaluate the value of 'buying' various items of new evidence, and so decide which to obtain (Lauritzen & Spiegelhalter, 1988, §5.5; Jensen & Liang, 1995).

Mixed models

Lauritzen (1989, 1992) and Olesen (1993) consider analogues of these procedures for conditional Gaussian distributions (page 41). The distinction between discrete and continuous variables leads to consideration of marked graphs (with the marks differentiating discrete and continuous random variables) and stronger concepts of decomposability and moralization.

There is a considerable literature about multivariate normal belief nets, but for pattern recognition we need at least one discrete variable, the true class.

Simulation-based calculations

The calculations of the previous subsection work well for sparsely-connected graphs with discrete random variables which take a small number of values. The approach can be extended to conditional Gaussian distributions (page 41) when the evidence on continuous variables is a precise value (Gammerman *et al.*, 1995), but most attempted extensions run into overwhelming computational complexity.

Henrion (1988) (and Henrion *et al.*, 1991) used a stochastic simulation method he called *(probabilistic) logic sampling*. This uses (8.12) for *unconditional* simulation of the whole collection of random variables. This is easy; we move through the DAG in vertex order, at each stage sampling conditionally on the values at the parents of the current vertex (and these values must be already known). To condition, run many simulations and only keep those which are consistent with the conditions, using these to compute frequencies of any events of interest. Provided the evidence has positive probability this will work, although if the probability of the evidence is low, only a small proportion of the runs will be retained, and so the process may be slow. If we need to condition on point values of continuous variables, this approach will be impossible (or if we replace the point value with a small interval, possible but impractically slow).

The connection to the Gibbs sampler was made by Hrycej (1990), Chavez & Cooper (1990), Shachter & Peot (1990), and, more elegantly, by York (1992).

This too has been considered in genetics: Ott (1989), Ploughman & Boehnke (1990) and Kong (1991).

Pearl (1987) suggested the use of iterative simulation methods such as the Gibbs sampler (see Section A.3). This approach had previously been used in image analysis and has since become popular in mainstream Bayesian statistics. These methods are now often called *MCMC* for 'Markov chain Monte Carlo'. We have a collection of random variables on a graph; as in the join tree each random variable might itself be a collection. For the Gibbs sampler we pick a vertex, and sample from X_v conditional on all the random variables at all the other vertices. For a local Markov distribution (to which we restrict attention) the conditional distribution depends only on the values of the random variables at neighbouring vertices. (We have already seen that the neighbours in the moral graph consist of the parents, children, and other parents of those children.) If the graph is sparsely connected, this may be a small enough set for the conditional simulation to be performed quite easily. To actually run the Gibbs sampler, vertices are picked sequentially, either at random or in some pre-assigned order.

Conditioning is easy for the type of evidence we are considering; each random variable is simulated conditionally on the event(s) concerning it.

This seems a very attractive method. If we want some aspect of the distribution of a subset of random variables, all we need to do is to simulate the whole set, and compute the frequency of the desired event(s). By taking a large enough sample we can compute the desired probabilities to any accuracy required. The difficulties are

- we have to run the Gibbs sampler long enough to ensure that we have a close enough approximation to the asymptotic distribution, and

- we need independent or close-to-independent samples from the distribution, which we can achieve by taking samples sufficiently far apart in the run of the Gibbs sampler.

These can be achieved, but the convergence can be extremely slow; Ripley & Kirkland (1990) give a dramatic example in which equilibrium has not been approached after each random variable in the system has been sampled 10,000 times. In fact the second point is unnecessary in our application if we just count to obtain the frequency of events, since we are estimating an expectation, and expectations are additive. Thus

$$\frac{1}{N} \sum_{t=1}^{N} I[X_V^{(t)} \in E]$$

will be an unbiased estimator of $\Pr\{E\}$ provided the process is started in the equilibrium state (by a sample from the correct distribution). However, it remains true that the estimator will be very variable unless the process has been run many times longer than the interval between approximately independent samples, and that we will have to discard an initial part of the run to combat the first point.

For this problem it is known that there are better MCMC samplers than Gibbs (Peskun, 1973), but the simplicity of Gibbs sampling may prevail.

The Gibbs sampler step applied to the moral graph obtained from a recursive model needs a method to sample from the distribution at a vertex conditional on the parents, children and 'mates' of the vertex, plus any evidence restriction on the random variable. It is easy to see that the conditional density (ignoring any evidence) is of the form

$$p(X_v \mid X_{\partial\{v\}}) \propto p(X_v \mid X_a, a \text{ a parent of } v)$$

$$\times \prod_{\text{children } c} p(X_c \mid X_a, a \text{ a parent of } c). \quad (8.13)$$

(Pearl, 1988, p. 218, gives a formal proof, but this is immediate.) This can be an awkward distribution to sample by standard rejection sampling (Ripley, 1987), but it is easy to sample from the first term, then use rejection sampling on each of the child terms separately. (Gammerman *et al.*, 1995, derive this by a very indirect route involving *auxiliary variables*; Besag & Green, 1993.) If evidence is involved at the node this can be incorporated by conditioning in the simulation given the parents.

An attraction of the Gibbs sampler to computer scientists is its intrinsic parallelism. We can update the sample at vertices which are not neighbours simultaneously; samples can also be updated by different processors asynchronously, at least if we ensure that messages are received from neighbours before processing begins. There are message-passing policies which prevent simultaneous updating of neighbours without needing a central controller; some are sketched in Pearl (1988, pp. 219–222).

Simulated annealing (Ripley, 1987; Aarts & Korst, 1989) can be used to find the most plausible explanation (the most probable combination of other variables given those observed) but is likely to be very slow to do so. (This method was suggested by Hrycej, 1992, Theorem 11.2.3.)

There is no reason why Gibbs sampling should be applied to a single variable at a time, and *blocked* variants have been explored in image analysis, in which a group of variables is sampled given its neighbours; the idea being that although sampling may be more difficult, the process may traverse the sample space more rapidly and so converge to equilibrium faster. Choosing suitably sized groups of variables is a black art at present.

8.3   Learning the network structure

So far we have assumed that the graph or DAG was given by the experts. Is there any hope of inducing the network from examples? Several approaches have been pursued.

From the traditional statistical viewpoint an undirected graphical model represents a restriction of a full dependence, for example (and most commonly) a log-linear model for a contingency table. The model-building strategy is to move around the space of possible models, and eventually to select a small number of simplified models which are complex enough to explain the patterns in the examples, yet have comprehensible interpretations. (It is this last point which often leads to a consideration of only subsets of graphical models; Edwards & Havárnek, 1985; Upton, 1991.) The strategy can involve either a series of hypothesis tests or searching to minimize a penalized fit criterion such as AIC. (Lauritzen *et al.*, 1994, give a case study of both for the CHILD network of Spiegelhalter *et al.*, 1993.) This approach usually involves a great deal of user control, although it is beginning to be partially automated. Since interpretability is paramount, a domain-expert will always be needed to monitor the process.

A variant on the testing approach is to use a series of tests for conditional independence to find the Markov boundary of each vertex (those vertices on which $\Pr\{X_v \mid X_{V \setminus \{v\}}\}$ depends), and then construct a graph with respect to which the distribution is local Markov (Fung & Crawford, 1990). There will be very many tests, some of which will give the wrong answer by chance. One way to combat this problem of multiple comparisons (suggested by Fung & Crawford, 1990) is to use the significance level α of the tests as a parameter of the procedure, and select it by cross-validating a relevant measure of performance. Many fewer tests would be needed to establish a graph with respect to which the distribution is pairwise Markov, since we need only test for each pair of vertices. Each test will be a test of conditional independence in a many-way contingency table, so there will difficulties with sparse tables unless the training set is large. The Fung–Crawford approach has the advantage of allowing small sets of neighbours to be considered first.

There are proposals to use these methods in insurance, where a database of millions of cases is available.

Another tradition, from the social sciences, for inferring a causal network from data (often in a regression setting) is represented by Glymour *et al.* (1987) and Spirtes *et al.* (1993). Finding a DAG makes the problem considerably harder, although it does allow the specification of smaller conditional probability tables, and the deduction of causal relationships. An edge between a and b will be absent in the DAG if there is a set S such that $X_a \perp\!\!\!\perp X_b \mid X_S$, so a search over subsets is needed. Since we know that the undirected version of the DAG is a subgraph of the moral graph, finding a graph with respect to which the distribution is local Markov provides a good starting point. The main result of Spirtes *et al.* (1993), their Theorem 3.4, applies only to a subclass of possible distributions (those with some DAG as a perfect map) and for the example on page 249 their procedures induce a graph with no edges. (They make no attempt to check their condition in their examples, but it would be straightforward to check that the data distribution is recursive on the induced DAG.)

In the regression setting this is called *path analysis*.

If a follows b in the order then we can take S to be the parents of a.

Markov trees

Chow & Liu (1968) (see also Pearl, 1988, Chapter 8) extend the contingency-table approach by seeking the belief *tree* that best approximates (in Kullback–Leibler divergence) a general multi-way discrete distribution. They show that:

- given the tree topology, the conditional distribution of each node given its parent in the best approximation is that in the original distribution, and
- the best tree is any minimal spanning tree with distance between a and b given by the mutual information of (X_a, X_b),

$$I(X_a, X_b) = E \log \frac{p(X_a, X_b)}{p(X_a)p(X_b)}.$$

Since there are efficient algorithms for finding minimum spanning trees (Cormen *et al.*, 1990, Chapter 24; Sedgewick, 1990, Chapter 31) this allows a best approximating Markov tree to be constructed.

The proof of these two properties is easy. We assume for simplicity that we have discrete random variables. Given a tree, let $p(i)$ denote the parent of vertex i (empty for the root). Any recursive model P' can be expressed as

$$P'\{X\} = \prod_i P'\{X_i \mid X_{p(i)}\}$$

from (8.12) on specializing to a tree. Now consider the Kullback–Leibler divergence between the true distribution P and a recursive model P':

$$
\begin{aligned}
d(P, P') &= \sum_{x_V} P\{x_V\} \log \left[P\{x_V\}/P'\{x_V\} \right] \\
&= \sum_{x_V} P\{x_V\} \log P\{x_V\} - \sum_{x_V} P\{x_V\} \sum_i \log P'\{x_i \mid x_{p(i)}\} \\
&= \sum_{x_V} P\{x_V\} \log P\{x_V\} - \sum_i \sum_{x_i,\, x_{p(i)}} P\{x_i, x_{p(i)}\} \log P'\{x_i \mid x_{p(i)}\} \\
&= \sum_{x_V} P\{x_V\} \log P\{x_V\} \\
&\quad - \sum_i \sum_{x_{p(i)}} P\{x_{p(i)}\} \left[\sum_{x_i} P\{x_i \mid x_{p(i)}\} \log P'\{x_i \mid x_{p(i)}\} \right].
\end{aligned}
$$

This can be minimized by maximizing each term in square brackets; the maximum occurs at $P'\{x_i \mid x_{p(i)}\} = P\{x_i \mid x_{p(i)}\}$ and the minimized divergence is

$$d(P, P') = \sum_{x_V} P\{x_V\} \log P\{x_V\} - \sum_{x_V} P\{x_V\} \sum_i \log P\{x_i \mid x_{p(i)}\}$$

$$= \sum_i I(X_i, X_{p(i)}) + \sum_{x_V} P\{x_V\} \log P\{x_V\} - \sum_i \sum_{x_i} P\{x_i\} \log P\{x_i\}.$$

A tree can always be turned into a causal graph by choosing a root and directing edges away from the root. This can be convenient in specifying the conditional probabilities, especially if the root is chosen judiciously. However, if we are inducing structure from data, the conditional probabilities will normally be estimated from the same dataset. In that case the unknown true distribution is replaced by frequencies in the training set, and the simplest way to specify the Markov tree is via a marginal representation, with $\Pr\{X_a, X_b\}$ for adjacent vertices estimated by the frequency in the training set. (The consistency conditions are automatically satisfied.) This can also be seen as fitting an unrestricted Markov tree to the training set by maximum likelihood.

Priors for belief networks

The full Bayesian procedure is to put a prior on the topology of belief networks, and then average the results over belief networks weighted by their posterior probabilities. This principle has been pursued by Cooper & Herskovits (1992), in tandem with learning the conditional probability tables from data. There is an enormous number of different networks; for example Cooper & Herskovits (1992, p. 319) quote approximately 4.2×10^{18} for ten vertices. Cooper & Herskovits make some very strong assumptions on the prior on topologies (such as a uniform distribution) to simplify computation. All such assumptions are unrealistic, as considering that set of vertices implies a belief in a sparsely connected network. (However, the prior may be swamped by the data and so be practically irrelevant.) My own prior might often be approximated by one on the number of edges.

The combinatorics of DAGs are considered by Robinson (1977).

The full Bayesian approach is normally computationally impossible as we cannot average over all topologies. Full averaging can be replaced by averaging over the few most plausible topologies, maybe even just the most plausible. This can be considered as the approach of traditional model selection and of Chow & Liu (1968), with slightly different measures of plausibility. (Cooper & Herskovits suggest how to calculate a topology with close to highest posterior probability, given an ordering on the vertices and asking that parents precede children.)

To make use of Bayesian methods we need prior distributions over the parameters (in the conditional probability tables for recursive models or clique marginals for decomposable models) as well as efficient means to integrate out those parameters to find the posterior probabilities of models. This is possible for Dirichlet priors for conditional probability tables and the *hyper-Dirichlet* priors introduced by Dawid & Lauritzen (1993) for clique marginals (which have a consistency condition). Indeed, with these priors the posterior distribution of the random variables is Markov after integrating out the current distribution of the parameters.

Madigan & Raftery (1994) confine their model averaging, as we saw in Chapter 2. They employ a stepwise search procedure through the space of models, adding or deleting an edge at a time (and, for undirected graphs, staying within the class of decomposable models by removing an edge only if it is a member of just one clique, and only adding an edge if it does not create a chordless cycle).

It is possible to use simulation methods, but they will also only average over a small subset of the topologies, so the method will need to be carefully constructed to give useful results. Madigan & York (1995) illustrate the use of MCMC methods to traverse the model space.

Hidden variables

A Markov or belief network can have one or more vertices representing unobserved latent variables. This device is widely used in medical applications, for example to represent the true (rather than reported) test result (as in Figure 8.2). They will cause observations to have missing values, and so complicate the learning of conditional probability tables. Unsurprisingly, the EM algorithm and variations (see Section A.2) have been used (Spiegelhalter *et al.*, 1993, Section 7).

Hidden variables can also be allowed within topologies inferred from data, in which case their interpretation is not specified in advance. Pearl (1988, Section 8.3) considers hidden vertices in trees, and other methods have been developed (Liu *et al.*, 1991; Verma & Pearl, 1991).

8.4 Boltzmann machines

One very specific case of our networks, the Boltzmann machine (Hinton & Sejnowski, 1983; Ackley *et al.*, 1985; Rumelhart & McClelland, 1986, Chapter 7), has a place in the history of neural networks. A Boltzmann machine has binary random variables at a finite set of vertices V which are completely connected, and the conditional distribution at each vertex given all the other random variables is Bernoulli with 'success' parameter θ_v given by a logistic regression on the other vertices, so $\text{logit}(\theta_i) = w_{i0} + \sum_{j \neq i} w_{ij} X_j$. As the network is completely connected, the graph properties are trivial; the expressive power comes from the restriction on the conditional distributions. The 'connection weights' w_{ij} are restricted to be symmetric ($w_{ij} = w_{ji}$) and without loss of generality we can take $w_{ii} = 0$. The joint distribution is then

$$\Pr\{X_V\} = \frac{1}{Z} \exp \sum_{i<j} w_{ij} X_i X_j \qquad \text{where} \quad Z = \sum_{x_V} \exp \sum_{i<j} w_{ij} x_i x_j.$$

Boltzmann machines are used to 'learn' an input–output distribution, that is the joint distribution of a set of binary random variables, some of which are designated inputs I and some outputs O. There will normally also be further units (designated 'hidden', H). Let $S = I \cup O$, the variables which are 'visible'. Once the parameters (all the w_{ij}) are given, we have a joint distribution over X_V, which gives a joint distribution over X_S, and hence the conditional distribution $X_O | X_I$. Thus a Boltzmann machine models the full joint distribution of inputs and outputs (the latter indicating classes).

So far the weights have been unspecified; the issue is to choose them to best approximate a given joint distribution of inputs and outputs, specifically the empirical distribution of a training set $\mathcal{T}$. This is done by maximum likelihood fitting of the parameters. However, since the joint distribution is unknown as a function of the weights, the necessary quantities are estimated by simulation, by Gibbs sampling.

The precise procedure used is gradient ascent, which only entails estimating the derivative of the log-likelihood with respect to the weights. To avoid handling w_{i0} separately, we assume a vertex 0 and implicitly condition on $X_0 \equiv 1$. The log-likelihood is

$$L = \sum_{\mathcal{T}} \log \Pr_{\mathbf{w}}\{X_S = x_S\} = \sum_{\mathcal{T}} \left[\log \sum_{x_H} \exp \sum_{i<j} w_{ij} x_i x_j - \log Z \right].$$

Consider just one summand L_1 of the log-likelihood. We have

$$\frac{\partial \log Z}{\partial w_{ij}} = \frac{1}{Z} \sum_{x_V} x_i x_j \exp \sum_{i<j} w_{ij} x_i x_j = \mathsf{E}[X_i X_j] = \Pr\{X_i = X_j = 1\}$$

and

$$\frac{\partial L_1}{\partial w_{ij}} = \frac{\sum_{x_H} x_i x_j \exp \sum_{i<j} w_{ij} x_i x_j}{\sum_{x_H} \exp \sum_{i<j} w_{ij} x_i x_j} - \frac{\partial \log Z}{\partial w_{ij}}$$

$$= \frac{E[X_i X_j I(X_S = x_S)]}{\Pr\{X_S = x_S\}} - \Pr\{X_i = X_j = 1\}$$

$$= \Pr\{X_i = X_j = 1 \mid X_S = x_S\} - \Pr\{X_i = X_j = 1\}.$$

Thus

$$\frac{\partial L}{\partial w_{ij}} = \sum_{x_S \in \mathcal{T}} \left[\Pr\{X_i = X_j = 1 \mid X_S = x_S\} - \Pr\{X_i = X_j = 1\} \right]. \quad (8.14)$$

Each summand in (8.14) is estimated by simulation. Two runs of the Gibbs sampler are needed, one for the unconditioned network, and one conditioned on X_S. (In the terminology of the field, the inputs and outputs are 'clamped'.) A small step is made uphill, the gradient re-estimated and so on. Once an approximation to a (local) maximum of the likelihood is found, future cases can be 'presented' by conditioning on X_I and running the Gibbs sampler to find the conditional distribution of X_O. Note that missing or partially observed values are easily accommodated by not (fully) conditioning, both during training and during prediction.

Unfortunately, the convergence of the Gibbs sampler has proved to be problematic even in toy problems, since at every step of steepest ascent the Gibbs sampler has to run to convergence for each example in the training set. For example, Kohonen *et al.* (1988) found that Boltzmann machines out-performed feed-forward neural networks on a toy problem, but were too slow to use on their real example.

The *mean field approximation* (Peterson & Anderson, 1987; Haykin, 1994, §8.13) avoids the simulation in performing (8.14) by approximating the probabilities. Specifically, the random variables X_i are replaced by their means, so $\Pr\{X_i = X_j = 1\}$ is replaced by the products of the means of X_i and X_j which, since they are binary, is $\theta_i \theta_j$. (This is suggested by a saddle-point approximation given in the references.) This reduces the problem to calculating (θ_v). We also replace the actual input variables by their means, so $\text{logit}(\theta_i) = w_{i0} + \sum_{j \neq i} w_{ij} \theta_j$ in this approximation, and (θ_v) is the solution to this non-linear system of equations. If we want the conditional distribution, we know X_S and thus solve the mean-field equations with the constraint $\theta_S = X_S$. As all simulation is avoided, mean-field Boltzmann learning is much faster than using Gibbs sampling. It appears to work well even for small systems (Peterson & Anderson, 1987; Hinton, 1989b).

Attempts have been made to improve the performance of the Boltzmann machine by abandoning its symmetry; it has every unit connected to every other, and each can influence the other. If we consider a DAG there is no immediate feedback of influences. Logistic units have been considered in analogues of Boltzmann machines by Apolloni & de Falco (1991) and Neal (1992a, b). Assume that the vertices of the DAG are labelled so that parents precede children. Then we assume that

$$\Pr\{X_V = x_V\} = \prod_i \Pr\{X_i \mid X_j, j < i\} = \prod \frac{\exp X_i \sum_{j<i} w_{ij} X_j}{1 + \exp \sum_{j<i} w_{ij} X_j} \quad (8.15)$$

and $w_{ij} = 0$ unless j is a parent of i. As we saw for Henrion's logic sampler, unconditional simulation of a recursive model is very easy, but as the joint distribution is known explicitly unconditional simulation is not needed. From (8.15) we have

$$\frac{\partial}{\partial w_{ij}} \log \Pr\{X_V = x_V\} = X_i X_j - \frac{\partial}{\partial w_{ij}} \log\left[1 + \exp \sum_{j<i} w_{ij} X_j\right]$$

$$= X_i X_j - X_j \frac{\exp \sum_{j<i} w_{ij} X_j}{1 + \exp \sum_{j<i} w_{ij} X_j} = X_i X_j - X_j \theta_i$$

where θ_i is the 'success' probability for X_i conditional on its parents. Hence

$$\frac{\partial}{\partial w_{ij}} \Pr\{X_S = x_S\} = \sum_{x_H} \Pr\{X_V = x_V\} \frac{\partial}{\partial w_{ij}} \log \Pr\{X_V = x_V\}$$

$$= \mathsf{E}\left[(X_i - \theta_i) X_j I(X_S = X_s)\right]$$

$$\frac{\partial}{\partial w_{ij}} \log \Pr\{X_S = x_S\} = \mathsf{E}\left[(X_i - \theta_i) X_j \,\middle|\, X_S = x_S\right]$$

so the gradient of the log-likelihood is

$$\frac{\partial L}{\partial w_{ij}} = \sum_{x_S \in \mathscr{F}} \mathsf{E}\left[(X_i - \theta_i) X_j \,\middle|\, X_S = x_S\right] \quad (8.16)$$

where θ_i depends on the parents of X_i. This is evaluated by Gibbs sampling (and we have already seen the form of the Gibbs sampler for a belief net).

A special case of this belief-net Boltzmann machine was considered earlier by Yair & Gersho (1990a, b), under the name of a *Boltzmann perceptron network*. They have general inputs X_I, binary hidden units X_V and binary outputs X_I. The hidden units depend on the output units via individual logistic regressions, and the output units depend on the inputs and hidden units via a multiple logistic regression (so the outputs are mutually exclusive). The architecture is then very similar to a single-hidden-layer neural network, except that the hidden units are randomly on or off with the probability that their output value would be in the corresponding neural network. This makes the posterior probabilities encoded by a set of weights slightly different, for although the average output of a hidden unit is its probability θ_h, its effect enters non-linearly into the output probabilities *via* the softmax output stage. (This difference disappears in the mean-field approximation; Hopfield, 1987.) However, we can average correctly by replacing $\beta_{hk}\theta_h$ by $\log(1 + \exp\beta_{hk}\theta_h)$ in the output multiple logistic regression. This gives a slightly different output layer for the neural network, but back-propagation can still be applied to find $\partial L/\partial w_{ij}$ without simulation.

> Note that $\log(1 + \exp\beta_{hk}\theta_h) \approx \beta_{hk}\theta_h$ when $|\beta_{hk}\theta_h|$ is small.

Other variants of Boltzmann machines have been proposed. The *radial basis Boltzmann machines* of Kappen (1995) have binary or continuous input and output units and continuous hidden units with fixed inhibiting connections between the hidden units. A stochastic diffusion (Langevin) simulation system is used. The inhibition between hidden units essentially allows only one of them to be on at a time, and so restricts the solution space. This variant appears to be able to solve realistically sized problems.

8.5 Hierarchical mixtures of experts

Hierarchical 'mixtures of experts' (HMEs; Jordan & Jacobs, 1994) are a way to specify the conditional distribution of class c given features $\mathbf{x}$ that has connections to several topics, and has already been touched on in Chapter 2. It is most closely related to belief networks. The mixture of experts idea was introduced by Jacobs *et al.* (1991), and hierarchies of experts by Jordan & Jacobs (1992).

The idea is that there are a number of classifiers C_i, each of which produces for an input $\mathbf{x}$ a posterior distribution over classes. Each of these can be thought of as appropriate for a particular subpopulation S of cases, and another 'gating' classifier G tells us the proportions of those subpopulations at $\mathbf{x}$. Since the subpopulation is unknown, the posterior probabilities over classes is

$$p(c \mid \mathbf{x}) = \sum_s p(c \mid \mathbf{x}, S = s)\Pr\{S = s \mid \mathbf{x}\}. \qquad (8.17)$$

N

Note that this is a model-based approach to stacked generalization.

In the hierarchical form, (8.17) is used to define a classifier, and the process repeated to combine classifiers of this form. Normally only a few levels of the hierarchy are used, but Waterhouse & Robinson (1994) use up to ten levels, combining a pair of networks at each stage. The terminology is perhaps a little imprecise: there is a single layer of 'experts' but the mixture is defined hierarchically. Thus (8.17) still holds for an HME, but $\Pr\{S = s \,|\, \mathbf{x}\}$ is parametrized hierarchically.

The classifiers used in HMEs could be quite general, but in all the examples presented in the references they are logistic discriminants.

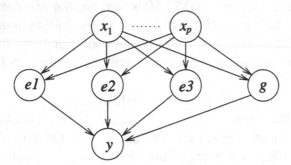

Figure 8.10: The belief network for a 'mixture of experts'.

We can view a 'mixture of experts' as a belief network. The features $\mathbf{x}$ provide a set of vertices I which are numbered first and connected to all expert and gating vertices. Each 'expert' is represented by a vertex which has as inputs all the feature vectors X_I and has as its state variable a class. The 'gating' classifier has as inputs the features, and state variable one of the experts. The output vertex O has state the actual class, and inputs the states of the experts and the gating classifier. This is shown in Figure 8.10. The extension to a hierarchy is immediate: at each stage we combine two or more subnets by adding a gating classifier with state the label of a subnet connected to the feature vectors, and an output node connected to the gate and the outputs of the subnets. Despite this representation, the marginal distribution of X_I is never specified, as we always work conditionally on the feature vectors.

We can then use the methods of belief nets to find the posterior probabilities $\Pr\{S \,|\, \mathbf{x}, y\}$ given an example and its true class. So far we have implicitly assumed that all the classifiers are fully specified, but in fact they are logistic discriminants with a vector of parameters. These parameters can be chosen by maximum likelihood by a variety of algorithms. If parameters θ specify the 'experts' and ϕ the gating process, the log-likelihood is

$$L(\theta, \phi; \mathscr{T}) = \sum_{(\mathbf{x},y)\in\mathscr{T}} \log \sum_{s} p_s(y \,|\, \mathbf{x}; \theta_s)\Pr\{S = s \,|\, \mathbf{x}; \phi\}.$$

In many problems this is simple enough to maximize directly. Alternatively we can use the EM algorithm (Section A.2), by viewing S, the true subpopulation, as a missing value. The log-likelihood for the complete data is

$$L(\theta, \phi; \{(\mathbf{x}, y, s)\}) = \sum_{(\mathbf{x}, y, s)} \log p_s(y \mid \mathbf{x}; \theta_s) + \log \Pr\{S = s \mid \mathbf{x}; \phi\}$$

and at step i we have to maximize the expectation of this over S evaluated at the current parameter estimates,

$$Q((\theta, \phi), (\theta, \phi)^{(i)}) =$$
$$\sum_{\mathcal{T}} \sum_s \Pr\{S = s \mid y, \mathbf{x}, \phi^{(i)}\} \Big[\log p_s(y \mid \mathbf{x}; \theta_s) + \log \Pr\{S = s \mid \mathbf{x}; \phi\} \Big].$$

The maximization over Q then splits into separate maximizations over the parameters θ_s in each expert and over ϕ. For θ_s we have a maximum likelihood estimate with case weights $\Pr\{S = s \mid y, \mathbf{x}; \phi^{(i)}\}$. For ϕ we have to maximize the mutual information between $\Pr\{S = s \mid y, \mathbf{x}; \phi^{(i)}\}$ and $\Pr\{S = s \mid \mathbf{x}; \phi\}$. For a hierarchically specified gate this further divides into maximal mutual information problems at each level.

Both Jordan & Jacobs (1994) and Waterhouse & Robinson (1994) use variants of this EM algorithm, for example not maximizing fully at the M step, and finding 'on-line' versions. Waterhouse & Robinson (1994) found a number of difficulties with their version, which appeared quite prone to reach a local maximum of the log-likelihood. They suggest alleviating this by having a large pool of experts, some of which are then effectively ignored.

Figure 8.11: The belief network for Bayesian inference on a 'mixture of experts'.

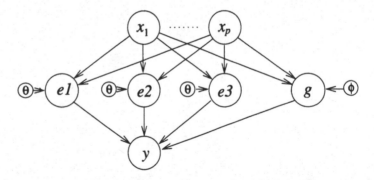

An alternative (Peng *et al.*, 1994) is to consider Bayesian inference. As in Spiegelhalter & Lauritzen (1990), we can add a parent to each classifier which contains its parameter vector. Thus Figure 8.10 becomes Figure 8.11. Gibbs sampling can then be used to find the posterior distribution of the parameters given the training set $\mathcal{T}$, and so to integrate out the posterior distribution of the parameters to find the predictive posterior distribution. Estimating the parameters by maximum likelihood is essentially the 'plug-in' version, since the maximum likelihood parameter values will give the posterior mode if flat priors are used for the parameters (as Peng *et al.*, 1994, did).

9

Unsupervised Methods

Unsupervised methods are used when no classes are defined *a priori*, or when they are but the data are to be used to confirm that these are suitable classes. Examples of the latter type are quite common in biology, where species are often defined by physical characteristics, and datasets of biochemical measurements become available. The interesting question is then whether the physical and biochemical measurements define the same classification. A variant of this occurs with our *Leptograpsus* crabs data. There the division into species was based on colour, and the interesting question is whether this is supported by morphological differences. Our analyses hitherto have been to find *supporting* morphological differences, but this begs the question of whether there might be even more striking differences unrelated to colour.

There are differences related to sex rather than colour, for example, but sex is also recorded.

Unsupervised methods are generally designed for *visualization*, either to show views of the data which indicate groups, or to show affinities between the examples by displaying similar examples close together. *Dendrograms* are a one-dimensional display of similarity, with the height of the join indicating (dis)similarity. For example, Figure 9.1 shows a dendrogram of the Cushing's syndrome data. Each pair is joined in the tree, and the height at which they are joined is an indication of their dissimilarity. This plot shows clearly that one point (labelled u) is very different from the rest, and does tend to group the diseases together, imperfectly. However, this is two-dimensional data, and the data can be plotted as in Figure 1.2 on page 11. This shows that we too would have difficulty separating the groups.

Groupings found by unsupervised methods are usually referred to as *clusters*. They are usually taken to be disjoint (we do not allow an animal to belong to two species) but sometimes it is helpful to allow some overlap (botanical populations may contain hybrids). Finding clusters is one of the uses of the word 'classification', and the book

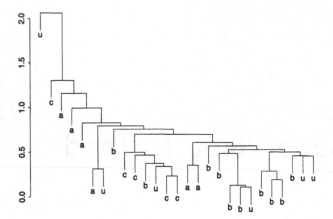

Figure 9.1: Dendrogram of the Cushing's syndrome data by the 'single-link' method.

by Gordon (1981) entitled *Classification* is entirely concerned with unsupervised methods, mainly clustering.

Unsupervised methods are sometimes used to classify. For example, we could use Figure 9.1 to select the closest grouping, and take a majority vote amongst its true classes. This has been advocated (Fuzzy ARTMAP: Carpenter *et al.*, 1992; Carpenter & Grossberg, 1994), but is dangerous as the unsupervised groupings may reflect a completely different classification of the data (colour *vs* sex for our *Leptograpsus* crabs, or, as in fact occurs, overall size).

Our exposition will move from visualization towards finding structure in the data. We start with methods to show linear or smooth non-linear transformations of the dataset which will reveal interesting structure in low-dimensional plots (usually two-dimensional scatterplots). We then consider the class of methods sometimes known as *multidimensional scaling* which produce low-dimensional plots (again, usually in one or two dimensions) in which similar data points are plotted close together.

The last two sections are concerned with clustering. The first covers methods to produce a few large clusters, and to produce taxonomic hierarchies such as Figure 9.1. The last section concerns methods which produce many clusters, but link them in a one- or two-dimensional layout wherein nearby clusters are more similar than distant clusters.

9.1   Projection methods

Projection methods choose one or more linear combinations of the original features to maximize some measure of 'interestingness'. Equivalently, the space of features is rotated in $\mathbb{R}^p$, and the first few dimensions of the rotated space are retained.

Principal components

Principal components occur in a number of problems and by different names: they are same thing as the Karhunen–Lòeve expansion of Watanabe (1969) and Devijver & Kittler (1982), for example. Jolliffe (1986) and Jackson (1991) devote whole books to this topic.

Suppose the data are $n \geqslant p$ vectors $\mathbf{x}_i \in \mathbb{R}^p$ forming the rows of an $n \times p$ data matrix X. We will assume that the column means are zero, that is that each feature has mean zero in the given sample.

A is a $p \times q$ matrix, each column of which gives the coefficients of a linear combination.

The idea is to take $q < p$ linear combination $XA \in \mathbb{R}^q$ which in one of a number of senses best represent the original data. This is done by taking the *singular value decomposition* of the data matrix X (Golub & Van Loan, 1989) $X = U\Lambda V^T$, where Λ is a diagonal matrix with decreasing non-negative entries (λ_i), U is an $n \times p$ matrix with orthonormal columns, and V is a $p \times p$ orthogonal matrix. Then the principal components are the columns of XV. Since X and Λ must have the same rank, at most p of the singular values (the diagonal elements of Λ) will be non-zero. Then we have the following properties:

Proofs are given at the end of this section.

This is the usual definition of principal components.

1 The first singular value (the first column of XV) is the linear combination $a^T\mathbf{x}$ for a of unit length with the largest variance, the second is the combination of largest variance which is uncorrelated with the first, and so on.

2 The first $q < p$ columns of XV are the linear projection of X into q dimensions with the largest variance. (The covariance matrix of a q-dimensional projection is a $q \times q$ matrix, and this one is largest as measured by the trace and also by the determinant.)

3 Let $\widetilde{X} = U\Lambda_q V^T$ be the matrix formed by setting all but the first q singular values (diagonal elements of Λ) to zero. Then $\widetilde{X}$ is the best possible rank-q approximation to X in several senses, including the Frobenius norm, the square root of the sum of squares of the elements.

4 Another way to express this is that if we project onto the first q principal components we have the most accurate rank-q reconstruction of the original data points.

5 Yet another way to express this is to say that the points of the q-dimensional projection onto the first q principal components lie in a q-dimensional space, and this is the best-fitting q-dimensional space as measured by the sum of the squares of the distances from the data points to their projections into the space.

In summary, if you want a reduction to $q < p$ dimensions by linear combinations of the features, the principal components have many optimality properties. Note that the first two properties show that the measure of 'interestingness' that is maximized is the variance.

The emphasis on variance reveals the Achilles' heel of principal components: they depend on the units in which the features are measured. In a biological problem in which we might have lengths, volumes and weights, the principal components will depend critically on the units used. Even when all the measurements are lengths, do we want to regard variation in the length of a small part as equivalent to variation in the length of the whole organism? Usually not; in biological problems the first principal component will normally be a measure of overall size, and be of little interest. So unless we have good *a priori* reasons to regard the *variances* of the features to be comparable, we would normally make them equal by rescaling all the features to have variance one.

There is an older approach to principal components which is better known but numerically less stable. This is to form the covariance matrix S of the observations, and take its eigenvalue decomposition CDC^T. As S is a covariance matrix, hence non-negative definite, the eigenvalues will be real and non-negative. Now (for centred data) $(n-1)S = X^TX = V\Lambda U^T U\Lambda V^T = V\Lambda^2 V^T$, so $D = \Lambda^2/(n-1)$ and $C = V$. Thus the principal components may be found from the eigendecomposition of S. It is customary to advocate using the eigendecomposition of the correlation matrix rather than the covariance matrix, which is the same procedure as rescaling the features to unit variance before calculating the covariance matrix.

Viruses example

We consider the *Tobamovirus* group of the viruses example, which has $n = 38$ examples with $p = 18$ features. Figure 9.2 shows plots of the first two principal components with 'raw' and scaled variables. As the data here are counts, there are arguments for both, but principally for scaling as the counts vary in range quite considerably between variables. Virus 11 (Sammon's opuntia virus) stands out on both plots: this is the one virus with a much lower total (122 rather than 157–161). Both plots suggest three subgroupings.

In both cases the first two principal components have about equal variances, and together contain about 69% and 52% of the variance in the raw and scaled plots respectively.

Figure 9.2: Principal
component (top row)
and Sammon mapping
(bottom row) plots of
the *Tobamovirus* group
of the viruses example.
The plots in the left
column are of the raw
data, those in the right
column with variables
rescaled to have unit
variance. The points are
labelled by the index
number of the virus.

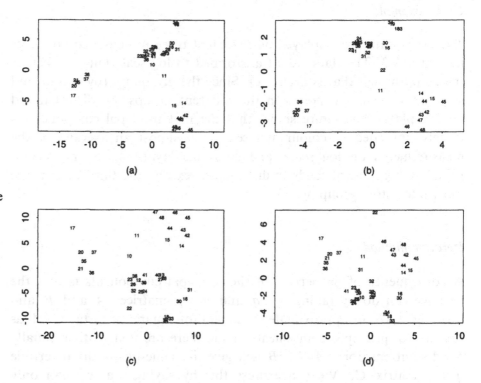

Figure 9.3: Pairwise
scatterplots of the first
three principal
components of the
Leptograpsus crabs
data. Males are coded
as capitals, females as
lower case, colours as
the initial letter of blue
or orange.

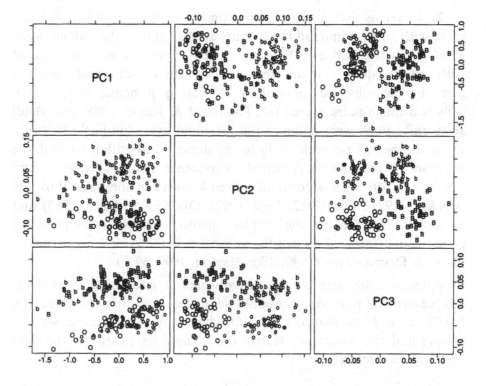

Crabs example

The crabs data are displayed on their first three principal components in Figure 9.3. The data were transformed to log scale but not further scaled (although that is arguable). Since the groupings (by colour and sex) are known, the points in the different groups are distinguished on the plots. The figure shows that the first principal component is largely unrelated to colour and sex: it is almost an average of the measurements on log scale, and so is displaying size. The second principal component tends to distinguish sex and the third colour, the most interesting grouping.

Iterative methods

A consequence of property 4 of the principal components is that the best we can do by taking $p \times q$ and $q \times p$ matrices A and B and forming XAB to approximate X in sum-of-squares is to take XA as the first q principal components. Now there are many other equally good solutions, for $XACC^{-1}B$ will give the same fit for any invertible $q \times q$ matrix C. We can express this by saying that we can only optimize over the subspace spanned by the q linear combinations.

It is obvious that XAB is the outcome of a feed-forward neural network with no bias unit, all linear units and q units in the hidden layer. Thus the best possible fit by least squares of such a network trained with output equal to input (an 'auto-encoder' or 'auto-associator') is given by the subspace spanned by the first q principal components. This is a much-rediscovered fact (Bourlard & Kamp, 1988, and Baldi & Hornik, 1989, were amongst the first) and every 'on-line' algorithm to fit the neural network leads to an iterative algorithm to find the subspace spanned by the principal components. This has led to around 100 papers on such algorithms. Well-known versions are those of Brockett (1991), Oja (1982, 1989, 1992), Oja & Karhunen (1985) and Sanger (1989). One of the simplest methods that extracts principal components (rather than just the subspace) is the APEX algorithm of Kung & Diamantaras (1990). (See Haykin, 1994, §9.8.)

Although the idea of these algorithms is interesting, they seem unnecessary in practice. Singular value decomposition routines can handle quite large matrices X, and when they cannot cope, we can always find the covariance matrix on a single pass through the data and find its eigenvalues.

Robustness

The extraction of principal components is based on variances, and so is sensitive to the presence of outliers. Outliers in high-dimensional data are notoriously difficult to find, although they often emerge as a side-effect from some projection pursuit methods.

It is highly desirable to use a method of extracting principal components which is less sensitive to outliers. This can be achieved by taking an eigendecomposition of a robustly estimated covariance or correlation matrix. There is a slight catch, in that it is essential to robustly estimate the means, and that this must be done by estimating the means of all features simultaneously (Rousseeuw & Leroy, 1987, p. 250). So the real task is to estimate the vector of means and the covariance matrix. There have been many attempts to do this; see for example Devlin *et al.* (1981). Modern approaches are discussed in Section 2.5.

An alternative approach is to find a projection maximizing a robust measure of variance in q dimensions. This would have to be done iteratively, as for the projection pursuit methods described below.

Proofs

As our results depend on various properties of principal components, these are proved here for those who are interested in the details.

Proposition 9.1 *Consider an $n \times p$ matrix X with singular value decomposition $X = U \Lambda V^T$. The best approximation in Frobenius norm to X by a matrix of rank $k \leqslant \min(n, p)$ is given by*

$$\widetilde{X} = U \mathrm{diag}(\lambda_1, \ldots, \lambda_k, \ldots, 0) V^T.$$

This is also the best approximation by a projection onto a subspace of dimension at most k, the projection onto the space spanned by the first k columns of U, and maximizes the Frobenius norm of a projection of X onto a subspace of dimension at most k.

Proof: We have $\|X - \widetilde{X}\|^2 = \|\Lambda - \Lambda_k\|^2 = \sum_{k+1}^{\min(n,p)} \lambda_i^2$. Now $\widetilde{X}$ corresponds to a projection onto the space spanned by the first k columns of U, say U_k, since that projection gives

$$U_k(U_k^T U_k)^{-1} U_k^T X = U_k U_k^T U \Lambda V^T = U_k [I \ 0] \Lambda V^T = U \Lambda_k V^T.$$

Consider any approximation Y of rank at most k. This can be written as $Y = AB$ where A is $n \times k$ and B is $k \times p$ (for example, via the SVD of Y). Now consider the best approximation of the form AC for any $k \times p$ matrix C. Since the squared Frobenius norm is the sum of the squared lengths of the columns, this is solved by regressing each column of X in turn on A; the optimal choice is $\widehat{C} = (A^TA)^{-1}A^TX$ and

$$\|X - Y\|^2 \geqslant \|X - A\widehat{C}\|^2 = \|[I - P_A]X\|^2 = \|X\|^2 - \|P_AX\|^2$$

where $P_A = A(A^TA)^{-1}A^T$ is the projection matrix onto span(A). Now we choose P_A to maximize $\|P_AX\|^2$:

$$\|P_AX\|^2 = \|P_AU\Lambda\|^2 = \sum_1^{\min(n,p)} \lambda_j^2\|P_Au_j\|^2 = \sum_1^{\min(n,p)} \lambda_j^2 p_j^2$$

and $p_j \leqslant 1$ (it is the length of a projection of a unit-length vector), $\sum p_j^2 = \|P_AU\|^2 = \|P_A\|^2 = k$. It is then obvious that the maximum is attained if and only if the first k p_j's are one, the rest zero, so

$$\|X - Y\|^2 \geqslant \|X\|^2 - \|P_AX\|^2 \geqslant \|X\|^2 - \sum_1^k \lambda_i^2 = \sum_{k+1}^{\min(n,p)} \lambda_i^2 = \|X - \widetilde{X}\|^2.$$

Any projection of X onto a subspace of k dimensions has rank at most k.   □

Proposition 9.2 *Consider n observations of p features forming a matrix X. Then the projection of Proposition 9.1*

(a) *minimizes the sum of squared lengths from points to their projections onto any subspace of dimension at most k,*

(b) *maximizes the trace of the covariance matrix of the projected variables onto any subspace of dimension at most k, and*

(c) *maximizes the sum of squared inter-point distances of the projections onto any subspace of dimension at most k.*

Proof: Without loss of generality we can centre the observations, so each variable has mean zero. Part (a) follows from the squared Frobenius norm of $X - P_AX$ being the sum of squared lengths of its rows. For (b) the squared Frobenius norm of P_AX is the sum of squares of the projected variables, that is $n - 1$ times the sum of the variances of the variables, which is the trace of the covariance matrix (and is invariant to the choice of a basis for that subspace). For (c) consider any projection P_AX. Let d_{rs} be the distance between observations r

and s, and $\tilde{d}_{rs}$ the distance under projection (which is smaller, as it is a projection). Let $\mathbf{y}_r$ be the rth projected observation. Then, using $\sum \mathbf{y}_r = 0$ (since the projections are still centred),

$$\sum_{r,s} \tilde{d}_{rs}^2 = \sum_{r,s} \|\mathbf{y}_r - \mathbf{y}_s\|^2 = \sum_{r,s} \|\mathbf{y}_r\|^2 + \|\mathbf{y}_s\|^2 - 2\mathbf{y}_r^T \mathbf{y}_s$$

$$= 2n \sum_r \|\mathbf{y}_r\|^2 = 2n \|P_A X\|^2$$

which is maximized according to Proposition 9.1. □

Proposition 9.3 *The principal components defined by property 1 are given, in order, by columns of V. The first k principal components span a subspace with the properties of Proposition 9.2.*

Proof: Consider a linear combination $y = a^T \mathbf{x}$ with $\|a\| = 1$. Then

$$\text{Var}(y) = a^T \text{Var}(\mathbf{x})a = \frac{1}{n-1} a^T X^T X a = \frac{1}{n-1} a^T V \Lambda V^T a$$

$$= \frac{1}{n-1} \sum \lambda_i^2 a_i'^2$$

where $a' = V^T a$ also has unit length (and this corresponds to rotating to a new basis for the feature variables). It is clear that the maximum occurs when a' is the first coordinate vector, or a the first column of V. Now consider the second principal component $b^T \mathbf{x}$. It must be uncorrelated with the first, so

$$0 = [Xa]^T [Xb] = [U\Lambda a']^T [U\Lambda b'] = \lambda_1^2 b_1'$$

and it is obvious that the maximum variance under this constraint is given by taking b' as the second coordinate vector. An inductive argument gives the remaining principal components.

Using the principal component variables we have $X = U\Lambda$, so it clear that the subspace spanned by the first k columns is the approximation of Propositions 9.1 and 9.2. □

Proposition 9.4 *Consider an orthogonal change XB to k new variables. Amongst such transformations the first k principal components have maximal variance, both in the sense of the trace and of the determinant of the covariance matrix.*

Proof: The trace statement is Proposition 9.2(b), but we will give an alternative proof. Consider the SVD of XB, and let its singular values be $\mu_1, \ldots, \mu_k$. We will show $\mu_j \leqslant \lambda_j, j = 1, \ldots, k$, which suffices as the trace of the variance matrix is proportional to the sum of the squared singular values, and the determinant is proportional to their product.

Consider a variable $y = \mathbf{x}^T a$ which is a unit-length linear combination of the first j principal components of the B set, but is orthogonal to the first $j-1$ original principal components. (A dimension argument shows that such a variable exists. Since B is orthogonal y is also a unit-length combination of the original variables and of their principal components.) Thus y has variance at least μ_j^2 and at most λ_j^2, hence $\mu_j \leqslant \lambda_j$. □

Projection pursuit methods

Projection pursuit methods seek a q-dimensional projection of the data that maximizes some measure of 'interestingness', usually for $q = 1$ or 2 so that it can be visualized. This measure would not be the variance, and would normally be scale-free. Indeed, most proposals are also affine invariant, so they do not depend on the correlations in the data either.

The methodology was named by Friedman & Tukey (1974), who sought a measure which would reveal groupings in the data. Later reviews (Huber, 1985; Friedman, 1987; Jones & Sibson, 1987) have used the result of Diaconis & Freedman (1984) that a randomly selected projection of a high-dimensional dataset will appear similar to a sample from a multivariate normal distribution to stress that 'interestingness' has to mean departures from multivariate normality. Another argument is that the multivariate normal distribution is elliptically symmetrical, and cannot show clustering or non-linear structure, so all elliptically symmetrical distributions should be uninteresting.

The simplest way to achieve affine invariance is to 'sphere' the data before computing the index of 'interestingness'. Since a spherically symmetric point distribution has covariance matrix proportional to the identity, we transform the data to have identity covariance matrix. This can be done by transforming to principal components, discarding any components of zero variance (hence constant) and then rescaling each component to unit variance. As principal components are uncorrelated, the data are sphered. Of course, this process is susceptible to outliers and it may be wise to use a robust version of principal components.

The *idea* goes back to Kruskal (1969, 1972). Kruskal (1969) needed a snappier title!

The discussion of Jones & Sibson (1987) included several powerful arguments against sphering, but as in principal component analysis something of this sort is needed unless a particular common scale for the features can be justified.

Specific examples of projection pursuit indices are given below. Once an index is chosen, a projection is chosen by numerically maximizing the index over the choice of projection. A q-dimensional projection is determined by a $p \times q$ orthogonal matrix and q will be small, so this may seem like a simple optimization task. One difficulty is that the index is often very sensitive to the projection directions, and good views may occur within sharp and well-separated peaks in the optimization space. Another is that the index may be very sensitive to small changes in the data configuration and so have very many local maxima. Rather than use a method which optimizes locally (such as quasi-Newton methods) it will be better to use a method which is designed to search for isolated peaks and so makes large steps. In the discussion of Jones & Sibson (1987), Friedman says

'It has been my experience that finding the substantive minima of a projection index is a difficult problem, and that simple gradient-guided methods (such as steepest ascent) are generally inadequate. The power of a projection pursuit procedure depends crucially on the reliability and thoroughness of the numerical optimizer.'

and our experience supports Friedman's wisdom. It will normally be necessary to try many different starting points, some of which may reveal projections with large values of the projection index. Posse (1990, 1995b) considers an almost random search which Posse (1995a) finds to be superior to his implementation of the optimization methods of Jones & Sibson and of Friedman.

Once an interesting projection is found, it is important to remove the structure it reveals to allow other interesting views to be found more easily. If clusters (or outliers) are revealed, these can be removed, and both the clusters and the remainder investigated for further structure. If non-linear structures are found, Friedman (1987) suggests non-linearly transforming the current view towards joint normality, but leaving the orthogonal subspace unchanged. This is easy for $q = 1$; any random variable with cumulative distribution function F can be transformed to a normal distribution by $\Phi^{-1}(F(X))$. For $q = 2$ Friedman suggests doing this for randomly selected directions until the two-dimensional projection index is small.

Here Φ is the cumulative distribution function of a standard normal.

Projection indices

A very wide variety of indices have been proposed, as might be expected from the many ways a distribution can look non-normal. A projection index will be called repeatedly, so needs to be fast to compute. Recent attention has shifted towards indices which are rather crude approximations to desirable ones, but very fast to compute (being based on moments).

For simplicity, most of our discussion will be for one-dimensional projections; we return to two-dimensional versions at the end. Thus we seek a measure of the non-normality of a univariate random variable X. Our discussion will be in terms of the density f even though the index will have to be estimated from a finite sample. (This can be done by replacing population moments by sample moments or using some density estimate for f such as the kernel methods of Section 6.1.)

The original Friedman–Tukey index had two parts, a 'spread' term and a 'local density' term. Once a scale has been established for X (including protecting against outliers), the local density term can be seen as a kernel estimator of $\int f^2(x)\,dx$. The choice of bandwidth is crucial in any kernel estimation problem; as Friedman & Tukey were looking for compact non-linear features (cross-sections of 'rods'—see Tukey's contribution to the discussion of Jones & Sibson, 1987) they chose a small bandwidth. Even with efficient approximate methods to compute kernel estimates, this index remains one of the slowest to compute.

Jones & Sibson (1987) introduced an entropy index $\int f \log f$ (which is also very slow to compute) and indices based on moments such as $[\kappa_3^2 + 1/4\kappa_4^2]/12$, where the κ's are cumulants, the skewness and kurtosis here. These are fast to compute but sensitive to outliers (Best & Rayner, 1988).

Friedman (1987) motivated an index by first transforming normality to uniformity on $[-1, 1]$ by $Y = 2\Phi(X) - 1$ and using a moment measure of non-uniformity, specifically $\int (f_Y - 1/2)^2$. This can be transformed back to the original scale to give the index

$$I^L = \int \frac{[f(x) - \phi(x)]^2}{2\phi(x)}\,dx.$$

Here ϕ is the standard normal density.

This has to be estimated from a sample, and lends itself naturally to an orthogonal series expansion, the Legendre series for the transformed density.

The index I^L has the unfortunate effect of giving large weight to fluctuations in the density f in its tails (where ϕ is small), and so will display sensitivity to outliers and the precise scaling used for the density. This motivated P. Hall (1989) to propose the index

$$I^H = \int \left[f(x) - \phi(x)\right]^2 \mathrm{d}x$$

and Cook *et al.* (1993) to propose

$$I^N = \int \left[f(x) - \phi(x)\right]^2 \phi(x) \, \mathrm{d}x.$$

Both of these are naturally computed via an orthogonal series estimator of f using Hermite polynomials (Thisted, 1988, §5.3.2). Note that all three indices reduce to $\sum_0^\infty w_i(a_i - b_i)^2$, where a_i are the coefficients in the orthogonal series estimator, and b_i are constants arising from the expansion for a standard normal distribution.

To make use of these indices, the series expansions have to be truncated, and possibly tapered as well (see Section 6.1). Cook *et al.* (1993) make the much more extreme suggestion of keeping only a very few terms, maybe the first one or two. These still give indices which are zero for the normal distribution, but which are much more attuned to large-scale departures from normality. For example, I_0^N is formed by keeping the first term of the expansion of I^N, $(a_0 - 1/2\sqrt{\pi})^2$ where $a_0 = \int \phi(x)f(x)\,\mathrm{d}x = \mathrm{E}\phi(X)$, and this is maximized when a_0 is maximal. In this case the most 'interesting' distribution has all its mass at 0. The minimal value of a_0 gives a local maximum, attained by giving equal weight to ± 1. Now of course a point mass at the origin will not meet our scaling conditions, but this indicates that I_0^N is likely to respond to distributions with a central clump or a central hole. To distinguish between them we can maximize a_0 (for a clump) or $-a_0$ (for a central hole). These heuristics are borne out by experiment.

In principle the extension of these indices to two dimensions is simple. Those indices based on density estimation just need a two-dimensional density estimate and integration (and so are likely to be even slower to compute). Those based on moments use bivariate moments. For example, the index I^N becomes

$$I^N = \iint \left[f(x, y) - \phi(x)\phi(y)\right]^2 \phi(x)\phi(y)\,\mathrm{d}x\,\mathrm{d}y$$

and bivariate Hermite polynomials are used. To maintain rotational invariance in the index, the truncation has to include all terms up to a given degree of polynomial.

All the indices described so far are implemented for $q = 2$ in XGobi, a freely available data visualization tool from Bellcore (Swayne *et al.*, 1991).

Eslava & Marriott (1994) defined two indices for $q = 2$ specifically designed to display all clusters; conventional indices have a tendency to superimpose clusters in the projection. Suppose the projected points (or those not very near the origin) have ordered polar angles θ_i. The *polar nearest neighbour* index (to be minimized) is the average of $\min(|\theta_i - \theta_{i-1}|, |\theta_{i+1} - \theta_i|)$, the angular separation from the remaining points. Their other criterion maximizes the mean radial distance, or equivalently for sphered data, minimizes the variance of the radial distance. Posse (1995a, b) has another two-dimensional index also based on ideas of radial symmetry, using a chi-squared index of departure from normality averaged over univariate projections.

One of the very few examples of a method which is both biologically motivated *and* practically useful is the projection index of Intrator (1990, 1992) and Intrator & Cooper (1992) based on the BCM model of neuron selectivity put forward by Bienenstock *et al.* (1982). The BCM model is based on a one-dimensional projection $c = a^T x$ of a signal x, which is chosen to maximize

The BCM model appears to be very well supported by experiment.

$$c^3 - c^2 \mathsf{E}c^2 \qquad (9.1)$$

where the 'threshold' $\Theta = \mathsf{E}c^2$ is adjusted according to the distribution of the population of examples (in practice the training sample). Notice that (9.1) is not scale-free, and will be negative for large c and hence a. Thus there is no need to normalize a to unit length. There is a natural on-line algorithm to minimize (9.1), namely

The details, especially the constants, differ from paper to paper. In the original BCM paper $\Theta = (\mathsf{E}c)^2$.

$$a \leftarrow a + \mu(a^T x)[(a^T x) - \tfrac{4}{3}\Theta]x$$

where Θ will also be updated from time to time. (Intrator & Cooper, 1992, discuss the stability of the differential equation limit of this update.) We can also consider several BCM neurons with lateral inhibition, in which case c is replaced by $c_k - \eta \sum_{j \neq k} c_j$ for neuron k.

The BCM neuron is itself a projection index, but as it is based on moments it will be sensitive to outliers. Intrator replaces $c = a^T x$ by $\ell(c)$ for the usual logistic function ℓ; this effectively transforms to $[0,1]$ by the inverse of the logistic cumulative distribution function before computing the index. Not only does this give a one-dimensional projection index but the lateral inhibition BCM network may be used to project onto $q > 1$ dimensions. Applications are shown by Intrator (1991, 1992) and Intrator & Gold (1993).

There is no unanimity over the merits of these indices (except the moment index, which seems universally poor). Some workers have reported that the Legendre index is very sensitive to outliers, and this is our experience. Yet Posse (1995a) found it to work well, in a study that appears to contain no outliers. The natural Hermite index is particularly sensitive to a central hole and hence clustering. The best advice is to try a variety of indices.

Viruses example

Figure 9.4 shows six views of the main group of the viruses dataset obtained by (locally) optimizing various projection indices; this is a small subset of hundreds of views obtained in interactive experimentation in XGobi. With only 38 points in 18 dimensions, there is a lot of scope for finding a view in which an arbitrarily selected point appears as an outlier, and there is no clear sense in which this dataset contains outliers (except point 11, whose total residue is very much less than the others). When viewing rotating views of multidimensional datasets (a *grand tour* in the terminology of Asimov, 1985) true outliers are sometimes revealed by the differences in speed and direction which they display—certainly point 11 stands out in this dataset.

Not many views showed clear groupings rather than isolated points. The Friedman–Tukey index was most successful in this example. Eslava-Gómez (1989) studied all three groups (which violates the principle of removing known structure).

This example illustrates a point made by Huber (1985, §21); we need a *very* large number of points in 18 dimensions to be sure that we are not finding quirks of the sample but real features of the generating process. Thus projection pursuit may be used for hypothesis formation, but we will need independent evidence of the validity of the structure suggested by the plots.

Crabs example

Various projections of the crabs data are shown in Figure 9.5. This is a rather different example, with 200 examples on only five variables, and with four groups suspected in advance. The first term of the natural Hermite expansion does find other local maxima, but the view shown in Figure 9.5(b) is the most commonly found. The other indices are less successful. View (b) is close to a local maximum for Friedman's index, but more often just the colour forms are separated as shown in view (d). The Friedman–Tukey index does not recognize these clusters, and instead finds views such as (c) which seem to have no interpretation.

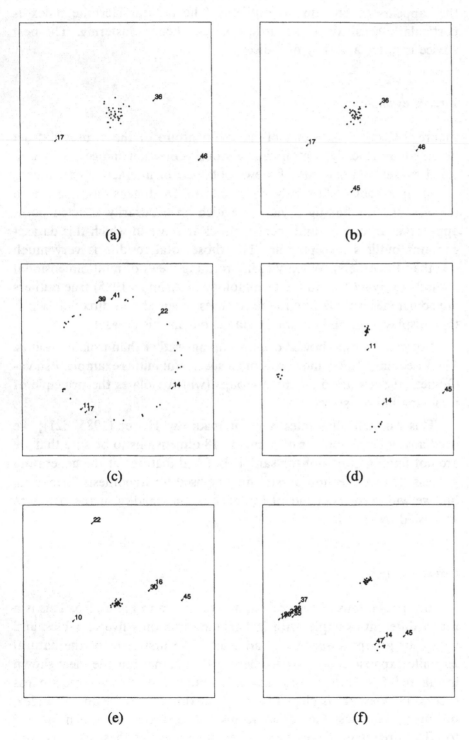

Figure 9.4: Projections of the *Tobamovirus* group of the viruses data found by projection pursuit. Views (a) and (b) were found using the natural Hermite index, view (c) by minimizing a_0, and views (d, e, f) were found by the Friedman–Tukey index $\int f^2$ with different choices of bandwidth for the kernel density estimator.

Figure 9.5: Projections of the *Leptograpsus* crabs data found by projection pursuit. View (a) is a random projection. View (b) was found using the natural Hermite index, view (c) by the Friedman–Tukey index and view (d) by the Friedman (1987) index.

This figure is available in colour for download from www.cambridge. org/9780521717700

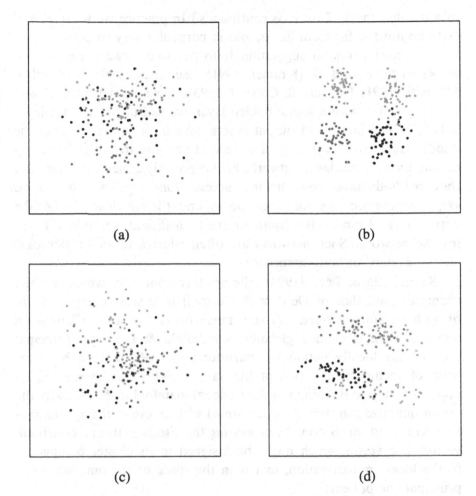

(a)   (b)

(c)   (d)

It is worth noting that the successful projections have ignored the most variable direction, size, in favour of a view with more structure, unlike all the other methods we illustrate on this example.

Non-linear feature extraction

One of the characterizations of the principal components was that if we took a linear map $F: \mathbb{R}^p \to \mathbb{R}^q$ and another linear map $G: \mathbb{R}^q \to \mathbb{R}^p$, the most accurate reconstruction $G(F(\mathbf{x}))$ in the sense of least squares is given by using the first q columns of V to form the first q principal components and for the reconstruction map G.

Can we do better with non-linear mappings F and G? The answer must be yes, for the diagonal Cantor construction can map invertibly $\mathbb{R}^p$ into $\mathbb{R}$! (Write each of the components $x_1, \dots, x_p$ in a binary expansion and interleave the expansions to obtain the binary expansion of a number in $\mathbb{R}$. This F is continuous.) In practice we want F and G to be not too far from linear, and in particular very smooth.

The most common suggestion is to use feed-forward neural networks to fit F and G (Kramer, 1991; Usui *et al.*, 1991; Cottrell & Metcalfe, 1991; DeMers & Cottrell, 1993). If we model both F and G by networks with a single hidden layer, we end up with a five-layer network. The input and output layers have linear units, as does the middle layer (with q units). The second and fourth layers have sigmoidal units. Multi-layer networks are notoriously difficult to train, and these methods have shown limited success. None appears to have used skip-layer connections nor weight decay, and it is not clear whether the current lack of success is intrinsic or due to inefficient methods of training the network. Such networks are often referred to as 'bottlenecks', 'auto-encoders' or 'auto-associators'.

Kambhatla & Leen (1994) take another approach, which in their examples (and that of DeMers & Cottrell) is at least competitive in fit with an auto-encoder, but very much faster to train. Whereas an auto-encoding is trained globally, Kambhatla & Leen use principal components locally within the partitioning of $\mathbb{R}^p$ defined by some form of vector quantization of the dataset. This can be seen as an approximation to defining a q-dimensional manifold in $\mathbb{R}^p$. Clearly the vector quantization should be performed with an eye to the approximation error, and this is done by measuring the distances from a codebook vector to a vector which might be assigned to its cluster orthogonal to the local approximation, that is in the space of the omitted $p - q$ principal components.

Principal curves and surfaces

Principal curves are defined by Hastie & Stuetzle (1989) as a mapping of a dataset in $\mathbb{R}^p$ to a one-dimensional manifold in $\mathbb{R}^p$. Let $\mathbf{f}(\lambda)$ be a smooth curve in $\mathbb{R}^p$ parametrized by $\lambda \in \mathbb{R}$. Then for any data point $\mathbf{x} \in \mathbb{R}^p$ we seek the nearest point $\lambda(\mathbf{x})$ on the curve in Euclidean distance. The curve is called a *principal curve* for a distribution on $\mathbb{R}^p$ if $\mathsf{E}[X \mid \lambda(\mathbf{x}) = \lambda] = \mathbf{f}(\lambda)$, that is the mean of those points that project to a point on the curve is that point.

There are many possible parametrizations of a one-dimensional curve; the most natural is in terms of arc length λ from a fixed point on the curve.

We have defined a principal curve for a distribution, and the natural way to find such a curve is to project a distribution onto a candidate curve $\mathbf{f}(\lambda)$, and to take as the next iteration the conditional expectation

There could be more than one nearest point, but this will be exceptional. For definiteness, we choose the nearest point with largest λ.

$E\{X \mid \lambda(\mathbf{x}) = \lambda\}$ and re-parametrize this in terms of arc length. There are no known guarantees that this algorithm will converge. For a set of data, the points only project to a discrete set of values of λ, and the conditional expectation must be replaced by a smoothing operation. We have a set of values $(\lambda_i, \mathbf{x}_i)$. Whereas we considered scatterplot smoothers in Chapter 4 for univariate $\mathbf{x}$, these methods extend readily to multidimensional $\mathbf{x}$. In most methods we smooth each coordinate of $\mathbf{x}$ separately.

We can think of principal curves as an algorithm to map $\lambda = F(\mathbf{x})$ by projecting to the nearest point and then projecting back by $\mathbf{x} = \mathbf{f}(\lambda)$. As such it is very similar to the method of Kambhatla & Leen (1994), but handles the projection step in a smoother way. Tibshirani (1992) proposed a variant on the original principal curves idea.

In principle this technique can be extended to manifolds of $q > 1$ dimensions, called *principal surfaces*, although q-dimensional smoothing is much harder unless data are abundant, even for $q = 2$.

As with all projection techniques, principal curves and surfaces depend critically on the scaling of the features; current algorithms also depend on choosing well the degree of smoothing.

9.2   Multidimensional scaling

In multidimensional scaling we are given the distances d_{rs} between every pair of observations. These could be genuine distances in some high-dimensional space, or they could be surrogates for the Euclidean distances. For example, some favourite examples use 'official' road distances between major towns and the scheduled flight times between cities. The latter need not even be symmetric, but we will confine attention to symmetric distances. Thus we suppose we are given non-negative symmetric numbers d_{rs} which we will call *dissimilarities* to indicate that they need not be genuine distances. In particular, they need *not* satisfy the triangle inequality

These are the distances allowed for expense claims in bureaucracies.

$$d_{rt} \leqslant d_{rs} + d_{st}$$

satisfied if they were produced by a metric. (Gower & Legendre, 1986, explore when dissimilarities are metric.)

Most of the work on multidimensional scaling has been developed in the psychological literature, but has also been discussed in ecology under the name of *ordination*. The recent short book by Cox & Cox (1994) has considerable detail on the various methods and their history.

Dissimilarities

If we are given an $n \times p$ matrix X of data to be considered as n
p-variate continuous observations, there are several ways to measure
the distance between the pairs of observations. If these observations
are categorical, there are even more ways. Since many of them produce
measures of distance which do not satisfy the triangle inequality, the
more general term dissimilarity is used. A dissimilarity is just a non-
negative symmetric function on pairs of objects; we will usually assume
that the self-dissimilarities are zero. Kaufman & Rousseeuw (1990, §1.2)
review many definitions of dissimilarities.

Exactly the same choices of a distance measure occur when k-
nearest neighbour methods are used in supervised classification.

For continuous data, the most obvious dissimilarity is Euclidean
distance computed from $d^2 = XX^T$. This does however depend critically on the scales in which the features are measured. One way out
we saw for principal component analysis is to rescale the features to
unit variance, and in projection pursuit we saw the idea of 'sphering'
the data. In this context sphering implies using Mahalanobis distances
with respect to a covariance matrix Σ, which could be the covariance
matrix of these observations if $n > p$. Another idea is the *Manhattan*
or L_1 distance, that sums the absolute differences in features.

For categorical data, the most commonly used dissimilarity is based
on the *simple matching coefficient*, that is the proportion c_{rs} of features
which are common to the two observations r and s. As this is
between zero and one, the dissimilarity is found by $d_{rs} = 1 - c_{rs}$. For
binary features, it might be thought that having a feature present in
both observations should be considered a more important indication
of similarity than having it absent in both. (Think of types of pottery
found in neolithic graves.) The *Jaccard coefficient* c_{rs} considers the
proportion of features which are present in one or other observation
which are found in both. Once again, $d_{rs} = 1 - c_{rs}$. Coefficients between
zero and one which are high for similar observations are quite common,
and called *similarity coefficients*.

For ordinal data, the most appropriate treatment seems to be to
use the ranks as if they were continuous data, probably after rescaling
to the range $[0, 1]$ so that every ordinal feature is given equal weight.
There then arises the question of how to handle mixtures of continuous,
categorical and ordinal features. The definition of Gower (1971) has
been widely adopted. For each feature f we define a dissimilarity d_{rs}^f,
and an indicator I_{rs}^f which is one only if feature f is recorded for both
observations. Further, $I_{rs}^f = 0$ if we have a categorical feature and an

absence–absence match. Then

$$d_{rs} = \frac{\sum_f I_{rs}^f d_{rs}^f}{\sum_f I_{rs}^f}.$$

(9.2)

The classical or metric method

In the classical or metric method of multidimensional scaling, often known as *principal coordinate analysis* (Gower, 1966) but going back to Schoenberg (1935), Young & Householder (1938) and Torgerson (1952, 1958), we assume that the dissimilarities were derived as Euclidean distances between n points in p dimensions, for unknown p. Given the distances, we obviously cannot recover the observations themselves, since the distances are invariant to rigid motions (translations, rotation and reflections) of $\mathbb{R}^p$. It transpires that this is the only freedom allowed.

Proposition 9.5 *For any symmetric matrix T, define the matrix*

$$T' = -\frac{1}{2}\left[T - \frac{(T1)1^T}{n} - \frac{1(T1)^T}{n} + \frac{1^T T1}{n^2}\right]$$

by subtracting row and column means and adding back the overall mean, or, equivalently, by removing row means then column means.

(a) *Given any configuration X of n points in $\mathbb{R}^p$, the matrix $T = (d_{rs}^2 = \|x_r - x_s\|^2)$ gives a non-negative definite $T' = XX^T$. Such a set of distances is called* Euclidean.

(b) *Given a symmetric $n \times n$ matrix T with non-negative definite T', we can find a configuration of points in $\mathbb{R}^{(n-1)}$ such that $T = (d_{rs}^2)$.*

(c) *A necessary and sufficient condition for an $n \times n$ matrix T to be a squared distance matrix is that $w^T T w \leqslant 0$ for all w with $w^T 1 = 0$.*

(d) *Any two configurations of n points with the same (d_{rs}^2) differ only by a shift and a rigid motion of $\mathbb{R}^p$, so lie in (shifted) subspaces of the same minimal dimension, the rank of T'.*

Proof: (a) Without loss of generality, centre the data so every column of X has zero mean. Then $T = (\|x_r - x_s\|^2) = (\|x_r\|^2 + \|x_s\|^2 - 2x_r^T x_s) = E1^T + 1E^T - 2XX^T$ where $E = (\|x_r\|^2)$. Let $e = E^T 1$ so $T1 = nE + e1$ and $1^T E1 = 2ne$. Thus

$$-2T' = E1^T + 1E^T - 2XX^T - E1^T - e11^T/n$$
$$- 1E^T - e11^T/n + 2ne11^T/n^2$$
$$= -2XX^T$$

which is non-positive definite.

(b) Let $T' = CD^2C^T$ be the eigendecomposition of T', noting that the eigenvalues are non-negative, and by construction T' has zero column sums and so has rank r at most $(n-1)$. Take X as the first r columns of CD, so $T' = XX^T$. This configuration is centred, since $\|X1\|^2 = 1^T T'1 = 0$. Note that $(\|x_r\|^2) = \text{diag}(XX^T) = \text{diag}(T')$, so T' determines $T = (d_{rs}^2)$ and (under zero means) this gives the same T' by result (a).

(c) Note that $[(I - 11^T/n)\mathbf{w}]^T T [(I - 11^T/n)\mathbf{w}] = -2\mathbf{w}^T T'\mathbf{w}$ which is negative if T' is non-negative definite.

(d) The procedure of (b) constructs a canonical configuration which is obtained by a shift (to zero mean) and a rigid motion from either configuration. $\square$

Given a Euclidean dissimilarity on n points, this proposition produces a data matrix in $r \leqslant n - 1$ dimensions with distances equal to the dissimilarity, and part (d) shows that this is the minimal number of dimensions needed. If we want a lower-dimensional view, Proposition 9.2 tells us to take the first q principal components of X, and this corresponds to taking only the q largest eigenvalues of T' and the first q columns of CD. This is the optimal approximation in the sense of minimizing the sum of squared dissimilarities minus squared distances over projections, and hence gives most weight to representing large dissimilarities accurately.

If the set of dissimilarities is not Euclidean, we can seek an approximation by a Euclidean set in $\mathbb{R}^k$ for small k. We know that $T' = CDC^T$ cannot be non-negative definite, but we can set all the negative elements and the small positive elements of D to zero and use the columns of CD corresponding to the large positive eigenvalues. If the dissimilarities are close approximations to Euclidean distances in a small number of dimensions, we expect to find a small number of large positive eigenvalues, the rest being near zero. If this is not the case, one of the other techniques may be preferable (but is likely to be much more computationally intensive).

One common mistake with classical scaling is to supply squared distances: these are not likely to be simply representable by distances.

Sammon mapping

Sammon mapping is a multidimensional scaling technique introduced by Sammon (1969) and widely known even where other methods of multidimensional scaling are unheard of. Given a dissimilarity d on n points it constructs a k-dimensional configuration with distances $\tilde{d}$ to (locally) minimize

$$E(d, \tilde{d}) = \sum_{i \neq j} \frac{(d_{ij} - \tilde{d}_{ij})^2}{d_{ij}}.$$

Note that this is undefined if there are pairs with zero dissimilarity. In contrast to principal coordinate analysis, this gives weight to representing small dissimilarities accurately, which may be desirable if the plot is being used to detect clusters.

Sammon used a diagonal Newton method to locally optimize E; this is a Newton method in which the off-diagonal part of the Hessian is ignored, and the step length reduced by a 'magic' factor of 0.3–0.4. Details of the algorithm are given in his paper. We have found that it is quite often necessary to use a smaller step-length factor (or even to use a crude search over step length) to avoid divergence. Most implementations seem to use a random starting point, but starting from a classical solution can save much CPU time, *if* it is a good approximation.

The Sammon mapping for the main group of the viruses example is shown in Figure 9.2 on page 291. This shows much less compact groupings than the principal components plots. As Sammon mapping is a more accurate representation of small distances, this should caution against over-interpretation of those groups.

In the viruses example the Sammon algorithm does not converge at all unless the 'magic' factor is reduced to around 10^{-3}. This is not uncommon behaviour when some points have to move very close to each other in the optimization run. Using a random starting configuration (as is common practice) produced very much worse local minima, with E around 0.3–0.5 rather than 0.07 for the configuration shown. Virus 22 (sunn-hemp mosaic virus) is clearly separated in the mapping of the scaled data.

Ordinal methods

What are known as non-metric or ordinal methods of scaling do not attempt to match the dissimilarity by a distance, but to choose a configuration whose distances have similar order properties, that is that points which have larger dissimilarity from a given point should be farther away. For such a method it is immaterial whether we supply (approximate) distances or squared distances, and the fit will be invariant to overall scale of the dataset, as well as to rigid motions.

A configuration X gives Euclidean distances δ_{rs} between pairs of points. We choose an increasing function θ so that $\theta(d_{rs})$ is close to δ_{rs}. The sum of squares of the differences is used, for then θ can be found by isotonic regression, for which there are simple algorithms (Barlow *et al.*, 1972). This is then minimized over the configuration (standardized to have unit sum of squares from the origin) by a gradient descent algorithm. Equivalently we minimize

$$STRESS^2 = \frac{\sum_{r,s}[\theta(d_{rs}) - \delta_{rs}]^2}{\sum_{r,s}\delta_{rs}^2}$$

Isotonic regression is the name for fitting an increasing or decreasing function by least squares. The solution is piecewise constant.

over θ and the configuration of points giving rise to distances (δ_{rs}). This is differentiable with respect to the configuration points (Kruskal, 1971; de Leeuw, 1984).

One detail in the implementation of ordinal methods is the treatment of tied dissimilarities. Clearly if $d_{ij} < d_{kl}$ we want $\theta(d_{ij}) \leqslant \theta(d_{kl})$, but how should we consider $d_{ij} \leqslant d_{kl}$? If we insist that $\theta(d_{ij}) \leqslant \theta(d_{kl})$, we are attempting to preserve the equality of tied dissimilarities, which can be a considerable constraint on the solution. It is normal practice to allow such ties to be broken.

The idea of ordinal methods is due to Shepard (1962a, b) and was developed into an objective method by Kruskal (1964a, b).

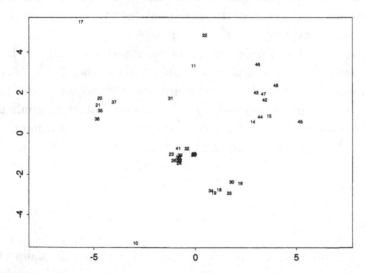

Figure 9.6: Non-metric multidimensional scaling plot of the *Tobamovirus* group of the viruses example. The variables were scaled before Euclidean distance was used. The points are labelled by the index number of the virus.

Figure 9.6 shows a local minimum for ordinal multidimensional scaling for the scaled viruses data. This fit is similar to that by Sammon mapping in Figure 9.2, but the subgroups are more clearly separated, and viruses 10 (frangipani mosaic virus), 17 (cucumber green mottle mosaic virus) and 31 (pepper mild mottle virus) have been clearly

The fit is poor, with $STRESS \approx 17\%$, and we found several local minima differing in where the outliers were placed.

separated. Figure 9.7 shows the distortions of the distances produced by the Sammon and ordinal scaling methods. Both show a tendency to increase large distances relative to short ones for this dataset, and both have considerable scatter.

Figure 9.7: Distortion plots of Sammon mapping and non-metric multidimensional scaling for the viruses data. For the right-hand plot the fitted isotonic regression is shown as a step function.

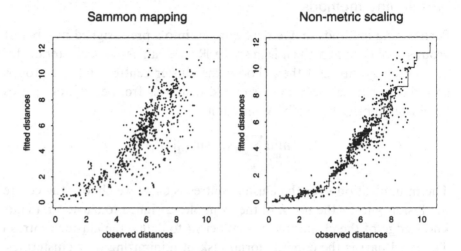

Figure 9.6 shows some interpretable groupings. That on the upper left is the cucumber green mottle virus, the upper right group is the ribgrass mosaic virus and two others, and a group at bottom centre-right (16, 18, 19, 30, 33, 34) are the tobacco mild green mosaic and odontoglossum ringspot viruses.

9.3 Clustering algorithms

Clustering algorithms are methods to divide a set of n observations into g groups so that members of the same group are more alike than members of different groups. If this is successful, the groups are called clusters. The number of groups g may be pre-assigned, or it may be decided by the algorithm. Formally, a cluster algorithm produces a mapping $c: \{1, \ldots, n\} \rightarrow \{1, \ldots, g\}$ associating a group with every example. Some (but not all) clustering algorithms work by representing each group by a representative point (not necessarily an example), and these have close links with vector quantization (Section 6.3).

All of these methods are just algorithms: even those which aim to optimize a criterion are not guaranteed to find the global optimum. Like all unsupervised methods they are judged by their results; a successful clustering produces groups which can be interpreted by domain experts.

Of the many books on clustering, Kaufman & Rousseeuw (1990) is one of the most practically oriented and has example FORTRAN programs which can be obtained from file servers. Older and more comprehensive references are Anderberg (1973), Hartigan (1975), Späth (1985) and Jain & Dubes (1988).

Partitioning methods

Partitioning methods divide the examples into a pre-assigned number of groups. For data in a Euclidean space $\mathbb{R}^p$ we can assign a cluster centre **m** to each group, and then choose the cluster centres and the groups so as to minimize the sum of squared distances from each example to its cluster centre. Formally, we minimize

$$\min_{c,\mathbf{m}_j} \sum_i \|\mathbf{x}_i - \mathbf{m}_{c(i)}\|^2.$$

The minimization over the cluster centres is easy; we choose the centre of cluster j to be the mean of the examples assigned to cluster j. (Thus knowledge of the clustering c is sufficient to define the cluster centres.) The hard part is the combinatorial task of minimizing over clusterings.

This method is sometimes called k-means or c-means, although those terms are also used to refer to specific algorithms. Early references are Forgy (1965), Jancey (1966) and MacQueen (1967), but the ISODATA algorithm of Ball (1965) and Hall & Ball (1965) (and Hall & Khanna, 1977) is closely related). All algorithms start with some division of the examples into k groups or a set of k cluster centres. In Forgy's algorithm all examples are re-assigned simultaneously to their nearest cluster centre, each cluster centre moved to the group's mean and this process repeated. A group can become empty in this algorithm, so it may choose less than k groups. MacQueen's algorithm differs in that each example is considered in turn, and the cluster centres are updated whenever an example is assigned to a group. Both variants always reduce the sum of squared distances, and so must converge. The ISODATA algorithm is a variant of Forgy's in which groups are split or merged (so k changes dynamically); MacQueen also considered splitting and merging.

> This algorithm goes back to Lloyd (1957), which was unpublished until 1982.
> Jancey's algorithm doubles the size of the moves.

A specific algorithm for k-means is given (including FORTRAN code) by Hartigan & Wong (1979). This is based on transferring observations from one group to another; other algorithms also allow the exchange of observations between clusters. Koontz *et al.* (1975) give a branch-and-bound algorithm to find the global minimum of the k-means criterion, which is feasible for small sets of examples. There are also random algorithms based on the idea of simulated annealing (Flanagan *et al.*, 1989; Zeger *et al.*, 1992).

> See the glossary.

See the glossary.

Note that *k*-means can assign any future example to one of the
k clusters, since it defines a partition of the whole feature space by
the Dirichlet tessellation of the cluster representatives. Most cluster
methods do not have this predictive aspect.

Figure 9.8 shows the clusters for 6-means for the virus data. The
iterative process has to be started somewhere, and in this case was
initialized from a hierarchical clustering discussed below. The choice of
6 clusters was by inspection of the visualization plots discussed above
and the dendrograms shown in Figure 9.9 (on page 320).

Figure 9.8: The clusters
suggested by *k*-means
for *k* = 6 for the virus
data displayed on the
ordinal
multidimensional
scaling plot.

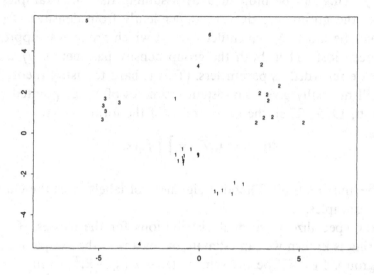

k-medoids

The *k*-means algorithm must choose centres in $\mathbb{R}^p$, and so can only be
used when the dataset is available and consists of continuous features.
We can overcome these restrictions by insisting that the cluster centres
be examples. Then we seek a clustering *c* and cluster centres $\mathbf{x}_{m_j}$ which
minimize

$$\sum_i d^2_{im_{c(i)}}$$

since this squared dissimilarity is the squared Euclidean distance for
(scaled) continuous measurements. This may be dominated by outliers,
so it is usual to use dissimilarities without squaring. This is known
as *k*-median or *k*-medoid clustering (Vinod, 1969). Once again there
is a local minimization algorithm which reduces the criterion and so
must converge (Kaufman & Rousseeuw, 1990, §2.4). This first selects *k*
centres, then considers swapping a centre with an example which is not
a centre and selects the most advantageous such swap. The process is

repeated until convergence. In general this finds a local minimum, but for $k = 2$ it finds the global minimum. For $k > 2$ Massart *et al.* (1983) give a branch-and-bound algorithm to find the global minimum which is only practicable for small sets of examples.

Clusters of different size and shape

A different extension of k-means is to allow the distance measure to vary between clusters, that is to allow the size and/or shape of clusters to vary. This can be motivated by assuming that the examples from each of the groups are drawn independently from densities $f_j(\mathbf{x}; \theta_j)$, but that the labels S_i which determined which group was appropriate have been lost. Then both the group density parameters θ_j *and* the labels are regarded as parameters. (This is hard to justify theoretically, and will normally give inconsistent estimates of θ_j, as pointed out by Marriott, 1975.) Then the likelihood is of the form

$$\ell((\theta_j), (s_i); \mathcal{T}) = \prod_i f_{s_i}(\mathbf{x}_i; \theta_{s_i})$$

and the 'maximum likelihood' assignment of labels gives the clustering of the examples.

Now specialize to normal distributions for the classes. Once the clustering is known we can estimate the means as the sample means for each group. Let W_k be the sum of $(\mathbf{x}_i - \overline{\mathbf{x}})(\mathbf{x}_i - \overline{\mathbf{x}})^T$ within group k. Then the profile log-likelihood becomes

$$L((s_i); \mathcal{T}) = \text{const} - \sum_j \text{trace}(W_j \Sigma_j^{-1}) + n_j \log |\Sigma_j|$$

where n_j is the number of observations assigned to group j.

Next we make some assumptions about the covariance matrices Σ_j. If these are assumed equal to the identity (or to a common multiple of the identity) we recover the k-means criterion (since trace W_k is the sum of squares to the cluster centre for group j). If we assume that the variances are equal but otherwise unknown, we find $\widehat{\Sigma} = \sum W_j/n$ and the clustering is chosen to minimize $|\widehat{\Sigma}|$ or equivalently $|W|$ for $W = \sum W_j$. (Up to a scale factor, W is the within-group covariance matrix we used for linear discriminant analysis in Chapter 3.) This criterion was proposed by H. P. Friedman & Rubin (1967). It can be thought of as applying k-means while allowing the 'sphering' of the data to be adjusted. Thus its view of clusters is as ellipsoids of the same size, shape and orientation. Any of these restrictions can be relaxed. Scott & Symons (1971) relaxed all, and Banfield & Raftery (1993) discuss intermediate cases. The latter also allow a 'uniform' cluster to pick up outliers.

These criteria can be optimized by algorithms of the types used for k-means and k-medoids.

Adaptive resonance theory

The adaptive resonance theory of Carpenter & Grossberg (1987a, b, 1990) and Carpenter *et al.* (1991a, b, 1992) (see also Moore, 1989; Georgiopoulos *et al.*, 1990, 1991; Huang *et al.*, 1995) is closely related to adaptive versions of k-means such as ISODATA and MacQueen's algorithm, but was expressed in a pseudo-biological language that clouds its simplicity. There are a variety of on-line algorithms that group the input examples in up to a pre-specified number k of clusters.

The first algorithm, ART 1, works with binary inputs. Let $\|\mathbf{x}\| = \sum |x_i|$ be the L_1 norm, for binary vectors the number of non-zero elements. For each of k groups there is a prototype $\mathbf{w}_j$ which is initially set to the vector of all ones (and is called 'uncommitted'). When an example $\mathbf{x}$ is presented, it is compared in turn with each $\mathbf{w}_j$ in order of decreasing $\mathbf{w}_j^T \mathbf{x}/(\epsilon + \|\mathbf{w}_j\|)$ until a prototype is found with $\mathbf{w}_j^T \mathbf{x} > \rho\|\mathbf{x}\|$. If such a $\mathbf{w}_j$ is found, the example is assigned to cluster j and $\mathbf{w}_j$ is updated by the bitwise operation

$$\mathbf{w}_j \leftarrow \mathbf{w}_j \text{ AND } \mathbf{x}.$$

This algorithm has two parameters. The tolerance ϵ is infinitesimal, serving to break ties in favour of prototypes with more positive elements. The parameter $\rho < 1$ is called the *vigilance*, and controls the diffuseness of the clusters. Note that the first uncommitted prototype to be considered will be selected. Once all prototypes are committed it is possible that none will be selected and the input is then rejected.

ART 1 is restricted to binary inputs and is highly sensitive to noise, since $\mathbf{w}_j$ can only be made smaller during updating. We can extend the process to inputs in $[0, 1]$ by replacing the AND operation by a bitwise minimum. Adding 'momentum' (Moore, 1989) changes the update rule to

$$\mathbf{w}_j \leftarrow (1 - \beta)\min(\mathbf{w}_j, \mathbf{x}) + \beta\mathbf{w}_j$$

for $\beta \in [0, 1)$. *Complement coding* includes both a feature y and its 'complement' $1 - y$, so that $\|\mathbf{x}\| = p$ for all examples. With all these changes we have 'fuzzy ART'. It is often assumed that $\beta = 1$ if the prototype is uncommitted: this is called 'fast-commit slow-recode' learning.

Adaptive resonance theory provides a large family of algorithms, but only a little analysis has been performed on their properties. It is unclear if they have any advantages over the earlier adaptive k-means algorithms. It is clear that they have a major disadvantage originally pointed out by Moore (1989), of sensitivity to noise. The update rule can only reduce the coordinates of the prototypes, so if a large number of examples are presented, each having added noise, the prototypes will shrink towards the zero vector. Prototypes which are close to the zero vector will fail the vigilance test, since $\mathbf{w}_j^T \mathbf{x} = \sum w_{ji}x_i \leqslant \max[w_{ji}]\|\mathbf{x}\|$. Thus for large enough training sets true clusters will be divided repeatedly into groups which depend on the order of presentation of the examples.

We assume that the noise allows each coordinate to take values smaller than ρ.

Methods based on mixtures

Suppose we believe that the examples come from a mixture of sources, and each has a parametrized density $f_i(\mathbf{x}; \theta_i)$. The proportions w_i of the mixtures are also unknown. The fitting of such mixture densities to data is discussed in Section 6.4. Once the mixture density has been fitted, we can ask for any future observation $\mathbf{x}$ what is the posterior probability that it belongs to component i; this is

$$p(i \mid \mathbf{x}) = \frac{\widehat{w}_i f_i(\mathbf{x}; \widehat{\theta}_i)}{\sum \widehat{w}_j f_j(\mathbf{x}; \widehat{\theta}_j)}.$$

Now if we view this as a classification problem, we would assign the observation to the component with highest posterior probability. This can be applied to the training examples to produce a clustering method, which partitions the data into a group (possibly empty) for each component (Wolfe, 1970).

This method is often confused with the likelihood-based partitioning method. Both employ models which are mixtures of components. However, the maximum likelihood method estimates the parameters in the components from classified data, then optimizes over the classifications, whereas the mixture method fits the parameters in the components and the mixing proportions from unclassified data.

Fuzzy clustering

In partitioning methods, each example is definitely assigned to one cluster. Fuzzy logic allows degrees of membership of sets, so would allow us to divide the membership of example i into proportions u_{iv} for group v. These membership proportions must be non-negative and sum to one. From the perspective of probability theory, u_{iv} can be interpreted as a posterior probability of having been generated by component v of a mixture, although in the interpretation below, perhaps u_{iv}^2 is closer to a posterior probability.

The earliest and best-known fuzzy clustering technique is the *fuzzy k-means* method of Dunn (1974) and Bezdek (1974). This minimizes

$$\min_{(u_{iv})} \sum_i \sum_j u_{iv}^2 \| \mathbf{x}_i - \mathbf{m}_v \|^2.$$

Here the cluster centre is found as the weighted mean of the whole set of examples with weights u_{iv}^2.

This method has the same disadvantages as k-means of being restricted to continuous data with X available. Kaufman & Rousseeuw (1990, §4.4) construct a fuzzy equivalent of k-medoids, to minimize

$$\min_{(u_{iv})} \sum_v \frac{\sum_{i,j} u_{iv}^2 u_{jv}^2 \, d_{ij}}{2 \sum_i u_{iv}^2}.$$

It can be shown that the variant of this with squared dissimilarities reduces to fuzzy k-means.

AutoClass

AutoClass (Cheeseman *et al.*, 1988a, b) is a widely-used 'Bayesian classification system' which is based on mixtures. There are J unknown classes. The major simplifying assumption made is that called idiot's Bayes in Chapter 8, that within each class the features are independent. (For a multivariate normal distribution this corresponds to assuming a diagonal covariance matrix.) A normal distribution is used for continuous features, and a general discrete distribution for discrete features. Conjugate priors are used for the parameters in the component models.

Under this assumption it claims a full Bayesian solution, including a random number J of classes. In practice the integration over the parameters for each class density is too difficult, and the usual approximations (expansions about MAP estimators) are used. The value of J is set to a large quantity by trial-and-error, and classes with negligible posterior estimates of proportions are omitted.

Mode separation

Earlier methods of partitioning were based on the idea of separating the modes of a multimodal density, implicitly assumed to be a mixture. For example, Henrichon & Fu (1968) considered projecting onto the first principal component, forming a density estimate (by a kernel estimator or just a histogram), and splitting at each local minimum of the density estimate. Such a procedure is highly sensitive to the precise density estimate used, much more so than would have been realized in 1968. Further procedures are described by Devijver & Kittler (1982, Chapter 11), but all suffer from the need to estimate densities in a high-dimensional space.

Hierarchical clustering

Biologists are used to taxonomic hierarchies: species are grouped into genera which are grouped into families and so on. Thus we can think of clusters of clusters. Hierarchical methods of clustering produce a tree, usually known as a *dendrogram*, such as Figure 9.1. This can be read in two directions. From the bottom up, we start with n clusters and the clustering changes at each level as two existing clusters are joined. (This is the *agglomerative* view.) In the *divisive* view, we start with one cluster and successively split clusters into two parts until this is no longer possible. These two views represent different families of algorithms. It is not necessary to split into two parts or to combine just two clusters, but this is easier to compute and so normally done.

Hierarchical methods avoid specifying how many clusters are appropriate by providing the user with many different partitions by cutting the tree at some level (and normally this will achieve a partition into any specified number of clusters). Sometimes this can help to choose an appropriate number, but users should be warned that none of these partitions may be particularly good, even under the criterion used in the hierarchical algorithm.

The *levels* on Figure 9.1 represent a dissimilarity between examples; we can define the tree-dissimilarity $\tilde{d}_{rs}$ as the minimum height in the tree at which examples r and s belong to the same cluster. Such dissimilarities obey not just the triangle inequality but the stronger *ultrametric* property

$$d_{rt} \leqslant \min(d_{rs}, d_{st}).$$

Thus we can think of hierarchical clustering as approximating a given dissimilarity by an ultrametric dissimilarity.

Agglomerative algorithms

The essence of an agglomerative algorithm is very simple: pick the two clusters with smallest dissimilarity and merge them. Starting is easy (use each example as a cluster), but we are then faced with defining the dissimilarity between our merged cluster and all other clusters. There are many methods to do so, and no consensus as to which is best. Two simple ideas are to define the dissimilarity between two clusters to be the minimum and the maximum dissimilarity between pairs, one from each cluster, and these give rise to clustering algorithms known as single-link and complete-link clustering respectively.

In *single-link clustering*, two examples will be joined at level λ if and only if we can find a chain of links of pairs of examples with dissimilarity less than λ. Thus the tree-dissimilarity $\tilde{d}_{rs} \leqslant d_{rs}$, and it can be shown that single-link gives the largest ultrametric dissimilarity with this property. It will tend to produce long and loosely connected clusters, since only a single link is required.

In contrast, *complete-link clustering* joins two clusters if and only if all members of one cluster are close to the other cluster, and so tends to produce 'compact' clusters, and relatively similar objects can remain separated up to quite high levels in the tree.

There are many other rules for combining clusters. The only other one that is widely considered is *group-average clustering*, in which the combined dissimilarity of two groups is the average of all dissimilarities between members of each group. Unlike single- and complete-link, this depends on the scale of the dissimilarities; the other two are equivariant to increasing transformations (such as the square) of the dissimilarities. We also note that using the increase in the k-means criterion on merging the clusters is often attributed to Ward (1963). By standard analysis of variance computations, this attributes a squared dissimilarity of

$$\frac{2n_A n_B}{n_A + n_B} \|\mathbf{m}_A - \mathbf{m}_B\|^2$$

to clusters A and B.

Figure 9.9 shows dendrograms produced by single-link, complete-link and group-average clustering for the viruses data. All identify viruses 10, 11, 17, 22 and 31 as loosely connected to the rest, and single-link also highlights virus 46. (We note that 10, 11, 17, 31, 46 and 48 are called 'miscellaneous' in the original source.) Nevertheless, each graph gives the impression of three or four major groupings of viruses.

Divisive algorithms

Divisive algorithms are much less known (and so much less used). They do have the advantage that if most interest is on the upper levels of the dendrogram (for example to produce a partition into k clusters for small k) they are much more likely to produce rational clusterings.

single-link   complete-link   group average

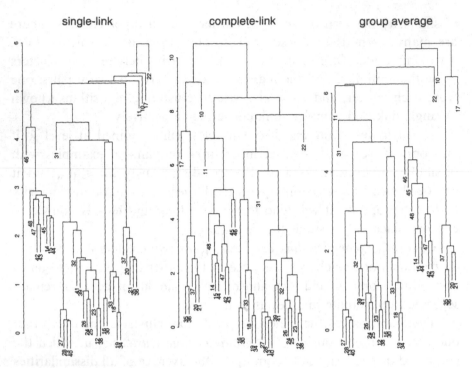

Figure 9.9: Dendrograms from three common hierarchical clustering techniques applied to the scaled viruses data. Such plots show the dissimilarity at which clusters are merged on the vertical scale and so show the construction process from bottom to top.

At the first step, a divisive method has to consider the $2^{n-1}-1$ partitions of n examples into two non-empty sets. This is computationally infeasible, so of course only a small proportion of those partitions are actually considered. By analogy to agglomerative methods, we might seek a division into two clusters A and B such that the dissimilarity between A and B is maximized. This is infeasible, but we can attempt to approximate it by an iterative method. We could, for example, use any of the partitioning methods into $k = 2$ clusters that we have discussed earlier. For example, Ward's method could be used divisively by applying 2-means recursively. (This seems not to appear in the clustering literature, but is known as an algorithm for *tree-structured vector quantization*; Gersho & Gray, 1992, §12.4.)

Macnaughton-Smith *et al.* (1964) proposed a method for general dissimilarities which is discussed in detail by Kaufman & Rousseeuw (1990, Chapter 5). We first select a single example whose average dissimilarity to the remaining examples is greatest, and transfer that example to cluster B. For all remaining examples of cluster A we compare the average dissimilarity to B with that to the remainder of A. If any examples in A are on average nearer to B, we transfer to B that for which the difference in average dissimilarity is greatest. If there are no such examples the process stops. This process splits a single cluster. We can then split each of the clusters that are created (unless one is a singleton or all its members have zero dissimilarity from each other) and repeat the process as far as is required. The splits are not uniquely ordered; Kaufman & Rousseeuw suggest splitting first the cluster with the largest diameter (maximum dissimilarity between members), which will be biased towards clusters with many members.

Divisive hierarchical clustering is reminiscent of the classification tree methods discussed in Chapter 7. As there, we can restrict the combinatorial explosion by confining attention to splits which involve just one of the features; such methods are called *monothetic*. In Chapter 7 the value of a split was computed from the distributions of the class variable in the two daughters; here it must be expressed by the difference in the clusters on the feature variables themselves. The obvious idea is to use the dissimilarity between the two daughters, calculated for example by group-averaging.

For *binary* variables we can interpret monothetic methods a little further. A split on a binary variable will generate clusters that differ only on the remaining variables, and we want these clusters to be as different as possible. Thus we seek one variable whose difference most accurately reflects the difference of all. This is the aim of *association analysis* (Williams & Lambert, 1959).

Examples

We will apply clustering methods to the crabs example. Since we saw in Figure 9.3 that the variation was dominated by crab size, the data were adjusted to crabs of common size, effectively by dividing each measurement by the geometric mean of all five measurements on that crab. Figure 9.10 shows some partitions into four clusters, which we know in advance to be the correct number. The k-means algorithm does rather well (but the clusters are near to spherical here). It is not surprising that the hierarchical clustering does badly; it has merged 200 examples and past groupings will tend to dominate at the last stages of agglomeration. The 'maximum likelihood' clustering with ellipsoidal clusters of the same size and shape should do well but does not, probably because the optimizer used seems less effective.

Using mixtures of four normals with either a common covariance (which in this problem is close to the truth) or separate covariances did slightly better than k-means, but took considerably longer, the EM algorithm converging in about 10 iterations when started from the centres of the k-means solution but about 50–100 iterations from a random start.

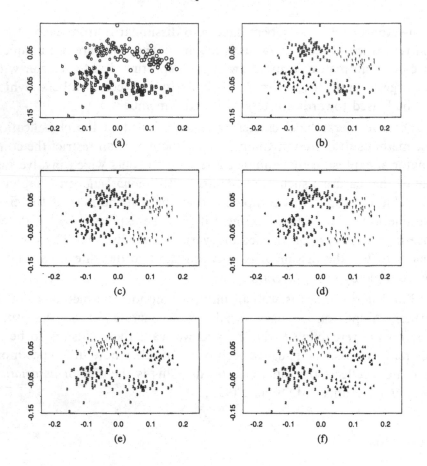

Figure 9.10: Sammon mapping plots of the *Leptograpsus* crabs data adjusted for overall size of the example. Plot (a) shows the true classification. The other five plots show a division into four clusters. Plot (b) shows *k*-means, initialized by the (c), complete-link hierarchical clustering. Plot (d) shows 'maximum likelihood' clustering with ellipsoidal clusters. Plot (e) shows the classification by Macnaughton-Smith *et al.*'s divisive method, and (f) shows the 'hardened' classification from fuzzy clustering.

We also tried all the applicable programs of Kaufman & Rousseeuw (1990). The results for divisive clustering and fuzzy clustering are shown in Figure 9.10. In this example *k*-medoids does well, as well as *k*-means. On the other hand, fuzzy clustering shows little discrimination, allocating most objects around 40–50% to one cluster, with appreciable proportions to at least two others. The clustering shown is 'hardened' by taking the cluster with the largest membership coefficient. Macnaughton-Smith *et al.*'s divisive method started with a 117/83 division, approximately by sex, then split the sexes by colour form. Group-average clustering had all the females in one group, apart from a small group of 5 outliers on the far left of the plot.

Fuzzy clustering was slow, at least 10 times slower than any other method considered.

9.4   Self-organizing maps

The *self organizing map* is an algorithm developed by Teuvo Kohonen (1982a, b, 1989, 1990a, 1995). This is usually described in the language of neural networks (involving 'weights') and had a biological motivation discussed in these references, Kohonen (1993) and Ritter *et al.* (1992). It is, however, just a specific type of clustering algorithm.

In the k-means method we saw that an example was assigned to the cluster whose representative m_j is nearest to the example. This is precisely what happens in SOM, but the training algorithm attempts to assign some structure to the representatives m_j. A large number of clusters are chosen, and arranged on a regular grid in one or two dimensions. (Both square and hexagonal grids have been used.) The idea is that the representatives (called 'weights' by Kohonen) are spatially correlated, so that representatives at nearby points on the grid are more similar that those which are widely separated.

This process is conceptually similar to multidimensional scaling. That maps similar examples to nearby points in a q-dimensional space. If we were to discretize the q-dimensional space, for example by dividing it into a grid of square bins, we would have a mapping from the space of possible examples into a discrete space that provides a clustering. Further, we could average the examples which are mapped to each bin to provide a representative for each non-empty bin, and the representatives in nearby bins would be similar. This is precisely the spirit of SOM, and it is often used to provide a crude version of multidimensional scaling. Indeed Kohonen says

> 'I just wanted an algorithm that would effectively map similar patterns (pattern vectors close to each other in the input signal space) onto contiguous locations in the output space.' (Kohonen, 1995, p. VI.)

We have a spatial smoothness property of the cluster representatives which Kohonen refers to as *topological ordering*. Cherkassky & Mulier (1994) draw analogies with principal curves, but those with multidimensional scaling seem closer.

Kohonen defined an 'on-line' algorithm, so examples are presented in some order (possibly random) until convergence. The cluster representatives are initially assigned at random in some suitable distribution. Whenever an example x is presented, the closest representative m_j is found. Then

$$m_i \leftarrow m_i + \alpha[x - m_i] \qquad \text{for all neighbours } i \qquad (9.3)$$

for all representatives i which are neighbours of j on the grid. Both the constant α and the definition of 'neighbour' are allowed to change with time. A typical specification is that α might decline linearly from 1.0 to 0.04 over 1000 examples, then linearly to zero over the second thousand, while the definition of a 'neighbour' is a grid point i within distance r of j, where r declines linearly from 6 to 1 over the first 1000 examples.

This defines just an algorithm, and the result will depend on the random initialization, the order of presentation of the examples and the tuning of the constants. Clearly it will be necessary to start with fairly large neighbourhoods, or no global order will emerge. Rather than update all clusters within the neighbourhood equally, it is natural to have a distance-weighted factor within the update, so

$$\mathbf{m}_i \leftarrow \mathbf{m}_i + h_{ij}(t)[\mathbf{x} - \mathbf{m}_i] \qquad \text{for all } i \qquad (9.4)$$

where h_{ij} depends on the proximity of i to j, for example $h_{ij}(t) = \alpha_t \exp -[d(i,j)/\sigma_t]^2$. It is possible that some representatives may never get updated unless the initial neighbourhoods are very large. On the other hand, if the neighbourhoods are large, the representatives get updated in blocks, and it is wasteful to have so fine a grid. It is clearly better to refine the grid rather than shrink the neighbourhoods, an idea Haykin (1994) attributes to Luttrell (*cf.* Luttrell, 1989).

It is helpful to note what happens if we take neighbourhoods so small that they only contain one point. Then there will never be any connection between representative points, and we might expect SOM to reduce to k-means clustering. It does. Although the algorithm appears to update only the representative for the cluster that $\mathbf{x}$ joins and not the one it leaves, the latter is achieved by the continual presentation of examples and the 'forgetting' property for $\alpha > 0$. Thus $\mathbf{m}_j$ is an exponentially-weighted average of all examples which have ever been assigned to cluster j, and will eventually become the average of a stable set. This suggests that we can regard SOM as a spatially smooth version of k-means, and assess the degree of fit of a particular solution by the quantization error, the sum of squared differences between examples and the corresponding cluster centres.

Analysis of the algorithm has been hampered by the lack of an 'energy' function that the algorithm can be considered to minimize, and Erwin *et al.* (1992) showed that in general no such function exists. However, if we restrict attention to randomly sampling from a training set and take fixed neighbourhoods, clearly a suitable energy function is

$$V = \mathsf{E} \sum_j h_{Ij} \|\mathbf{X} - \mathbf{m}_I\|^2$$

where I is the cluster to which the randomly chosen input $\mathbf{X}$ is assigned (Růžička, 1993). For randomly sampled inputs from a population, few results are known except for the special case of just one feature on $[0, 1]$

Figure 9.11: SOM mapping of the crabs data to a 6 × 6 grid. The labels of those examples mapped to each cluster are distributed randomly within the circle representing the cluster. As before the coding is upper case for males, lower case for females, 'B' for blue and 'O' for orange.

and a linear grid of representations (Cottrell & Fort, 1987; Bouton & Pagès, 1993, 1994; Fort & Pagès, 1993).

A batch version of SOM has been proposed much more recently (Kohonen, 1995, §3.14). This is a simple adaptation of Forgy's algorithm for *k*-means; simultaneously for all clusters the representative is updated to the (weighted) means of examples which are mapped to a neighbour of the cluster. This step is iterated, slowly decreasing the size of the neighbourhoods.

The results of SOM mapping of the crabs data to a 6 × 6 grid are shown in Figure 9.11. (The grid size was chosen to allow a reasonable number of the 200 examples to be mapped to each representative.) This figure illustrates the difficulty of displaying an SOM map. We have five-dimensional data, so cannot show the representatives directly, neither on the grid nor as points in the feature space. What we can do is map each example to its nearest representative (its cluster centre) and display the clustering, as we show in the figure.

Contiguity-constrained clustering

Kohonen's SOM produces a grid of clusters. Often we wish to group those clusters into super-clusters, preserving the spatial smoothing. (This is pertinent for Figure 9.11.) One way to do so is to use segmentation methods from image analysis. Many of these reduce to agglomerative hierarchical clustering methods with contiguity constraints. Suppose we consider the group-average method of clustering, but only allow clusters to be merged if they are neighbours (that is that there are members $a \in A$ and $b \in B$ which are neighbours on the grid). Then by construction clusters will always be connected subgraphs of the grid. Such algorithms are discussed by Gordon (1981), Murtagh (1985, 1995a) and Beaulieu & Goldberg (1989) and applied to a grid of SOM clusters by Murtagh (1995b).

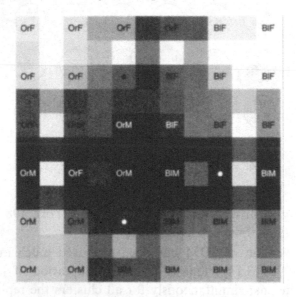

Figure 9.12: Ultsch representation of the SOM representatives for Figure 9.11. The rows and columns between the units represent the magnitude of the gradient (black being high); the greylevel at each unit represents the median of the surrounding gradients. The label is the most common class mapped to that representative.

Other visualization strategies for SOMs are given by Ultsch (1993a, b), which display the similarity of the representatives $\mathbf{m}_j$ by showing the magnitude of the gradient, viewing the representatives as a vector field (Figure 9.12).

10

Finding Good Pattern Features

In this chapter we consider the problem of what features should be included when designing our classifier. We should make clear at the outset that this is an impossible problem; there may be no substitute for trying them all and seeing how well the resulting classifier works. However, this may be computationally impracticable, and unless a large test set is available it may be impossible to avoid selection effects, of choosing the best of a large class of classifiers on that particular test set and not for the population.

To illustrate the difficulty, consider a battery of diagnostic tests $T_1, \ldots, T_m$ for a fairly rare disease, which perhaps around 5% of all patients tested actually have. Suppose test T_1 correctly picks up 99% of the real cases and has a very low false positive rate. However, there is a rare special form of the disease that T_1 cannot detect, but T_2 can, yet T_2 is inaccurate on the normal disease form. If we test the diagnostic tests one at a time, we will never even think of including T_2, yet T_1 and T_2 together may give a nearly perfect classifier by declaring a patient diseased if T_1 is positive or T_1 is negative and T_2 is positive. This illustrates that considering features one at a time may not be sufficient.

Our aim in this chapter is to indicate single features which are likely to have good discriminatory power (*feature selection*) or linear combinations of features with the same aim (*feature extraction*). Unfortunately the methods described can be quite effective with conventional statistical methods (linear and quadratic classifiers) but rather ineffective with modern non-linear classifiers. One reason that this is the last chapter of the book is that its methods are being supplanted by the model selection methods discussed in earlier chapters.

Throughout this chapter we will work with the conventional 0–1 loss although some of the ideas can be extended to situations with genuine costs for erroneous classifications. Thus here the Bayes risk is the error rate of the Bayes rule. We will also concentrate on $K = 2$ classes, as this suffices to illustrate the principles involved.

10.1   Bounds for the Bayes error

The 'gold standard' for a classifier was seen in Chapter 2 to be the Bayes error, the risk of the Bayes classifier. The Bayes classifier does however depend on the information available, and the Bayes error will be higher if only some of the features are measured. Thus it is of interest to estimate the Bayes error as a function of the variables included in the classifier design. Exact calculations are impossible (except in trivial problems) but we can obtain reasonable upper and lower bounds.

Thinking in terms of the Bayes error tells us immediately which features we ideally need, $p(c \mid \mathbf{x})$ for $K - 1$ of the classes. Of course this is unattainable in practice, although it is one view of the derivation of linear and quadratic discriminants for normal class distributions.

In a two-class problem the Bayes error is

$$E^* = \int \min[p(1 \mid \mathbf{x}), p(2 \mid \mathbf{x})]\, p(\mathbf{x})\, d\mathbf{x} = \int \min[\pi_1 p_1(\mathbf{x}), \pi_2 p_2(\mathbf{x})]\, d\mathbf{x}$$

and since $\min(a, b) \leqslant a^s b^{1-s}$ for any $s \in [0, 1]$, we have

$$E^* \leqslant \int [\pi_1 p_1(\mathbf{x})]^s [\pi_2 p_2(\mathbf{x})]^{1-s}\, d\mathbf{x} = \pi_1^s \pi_2^{1-s} \exp -J_C \qquad (10.1)$$

say, where

$$J_C = -\log \int p_1(\mathbf{x})^s p_2(\mathbf{x})^{1-s}\, d\mathbf{x}. \qquad (10.2)$$

This is known as the Chernoff bound on the Bayes error (Chernoff, 1952, 1973). The special case of $s = 1/2$ was derived earlier by Bhattacharyya (1943) and is therefore known as the Bhattacharyya bound. Because of its greater simplicity it is much more widely used.

We will evaluate the Bhattacharyya bound for two normal distributions. It becomes

$$E^* \leqslant \sqrt{\pi_1 \pi_2} \exp -J_B$$
$$J_B = \tfrac{1}{8}(\boldsymbol{\mu}_1 - \boldsymbol{\mu}_2)^T [\tfrac{1}{2}(\Sigma_1 + \Sigma_2)]^{-1}(\boldsymbol{\mu}_1 - \boldsymbol{\mu}_2)$$
$$+ \tfrac{1}{2}\log \frac{|\tfrac{1}{2}(\Sigma_1 + \Sigma_2)|}{\sqrt{|\Sigma_1||\Sigma_2|}} \qquad (10.3)$$

The second term of J_B disappears if the two covariance matrices are equal; in that case the Chernoff bound is tightest for $s = 1/2$.

Devijver & Kittler (1982, p. 58) point out that we can also obtain a lower bound on the Bayes risk in terms of the Bhattacharyya coefficient. The 1-nn rule has asymptotic risk

$$E_1 = \int 2p(1\,|\,\mathbf{x})p(2\,|\,\mathbf{x})\,p(\mathbf{x})\,\mathrm{d}\mathbf{x} \leqslant \int \sqrt{p(1\,|\,\mathbf{x})p(2\,|\,\mathbf{x})}\,p(\mathbf{x})\,\mathrm{d}\mathbf{x}$$

since $p(1\,|\,\mathbf{x})p(2\,|\,\mathbf{x}) \leqslant 1/4$. Thus $E^* \leqslant E_1 \leqslant \sqrt{\pi_1\pi_2}\,\exp -J_B$. However, Proposition 6.1 gives $E_1 \leqslant 2E^*(1-E^*)$ which we can invert to give

$$\tfrac{1}{2}\left[1 - \sqrt{1 - 4\pi_1\pi_2\exp -2J_B}\right] \leqslant \tfrac{1}{2}\left[1 - \sqrt{1 - 2E_1}\right] \leqslant E^* \qquad (10.4)$$

$$1 - 2E_1 \geqslant$$
$$1 - 4E^*(1-E^*)$$
$$= (1 - 2E^*)^2, \text{ and}$$
$$E_1, E^* \leqslant 1/2, \text{ so}$$
$$\sqrt{1 - 2E_1} \leqslant 1 - 2E^*.$$

Of course, these bounds are only of any use if we know (or can estimate accurately) the Bhattacharyya coefficient.

The Chernoff and Bhattacharyya coefficients are only two of a large class of *separation measures* which indicate how dissimilar two probability distributions are, in our case applied to the two class densities. Other measures are the divergence

$$J_D = \int \left\{\pi_1 p_1(\mathbf{x}) - \pi_2 p_2(\mathbf{x})\right\} \log \frac{p_1(\mathbf{x})}{p_2(\mathbf{x})}\,\mathrm{d}\mathbf{x} \qquad (10.5)$$

and the Patrick–Fisher coefficient (Patrick & Fisher, 1969)

$$J_P = \left[\int \left\{\pi_1 p_1(\mathbf{x}) - \pi_2 p_(2(\mathbf{x})\right\}^2\,\mathrm{d}\mathbf{x}\right]^{1/2}. \qquad (10.6)$$

The idea is to use one of the J coefficients to indicate how good a set of features is likely to be; large values of the coefficient indicate that it is likely that a classifier with low error rate can be found (although this is only guaranteed for J_C and J_B). The divergence J_D is signed, so we would look for large absolute value.

10.2   Normal class distributions

In practice the class probability densities $p_i(\mathbf{x})$ are unknown, but progress can be made if we assume that they are normal distributions. We have already seen at (10.3) the expression for J_B. There is a similar expression for J_C, and we have

$$J_D = \tfrac{1}{2}(\mu_1 - \mu_2)^T[\Sigma_1^{-1} + \Sigma_2^{-1}](\mu_1 - \mu_2)$$
$$\qquad + \tfrac{1}{2}\,\mathrm{trace}\,[\Sigma_1^{-1}\Sigma_2 + \Sigma_2^{-1}\Sigma_1 - 2I] \qquad (10.7)$$

$$J_P = \frac{1}{\sqrt{(2\pi)^p|2\Sigma_1|}} + \frac{1}{\sqrt{(2\pi)^p|2\Sigma_2|}} - \frac{2}{\sqrt{(2\pi)^p|\Sigma_1 + \Sigma_2|}} \times$$
$$\qquad \exp -\tfrac{1}{2}(\mu_1 - \mu_2)^T[\Sigma_1 + \Sigma_2]^{-1}(\mu_1 - \mu_2). \qquad (10.8)$$

Note that for equal covariance matrices ($\Sigma_1 = \Sigma_2 = \Sigma$), $J_D = 8J_B = (\mu_1 - \mu_2)^T \Sigma^{-1}(\mu_1 - \mu_2)$ is the Mahalanobis distance between the class means. We saw on page 22 that the expected error rate for the linear classifier depended only on the Mahalanobis distance, so maximizing J_B or J_D is equivalent to minimizing the expected error rate of the linear classifier.

Other class separation measures commonly used are trace($W^{-1}B$) and $|B|/|T|$ for the between-class B, within-class W and total T covariance matrices defined in Section 3.1. These too are obviously closely related to linear discrimination, and for $K = 2$ the trace measure reduces to the Mahalanobis separation of the means.

Why should we use these measures rather than fit a linear or quadratic classifier and measure its performance directly? If computation permits, there is no reason not to assess performance directly, especially if the performance can be assessed on the actual distributions rather than normal distributions. Even for assumed normal distributions we can compute the expected error rate, using numerical integration for $K > 2$ classes or for quadratic classifiers. So separation measures are best seen as a computational short cut for suboptimal feature selection. (Feature extraction in linear classifiers is simple: use the linear discriminants.)

> For example, using a test set or cross-validation.

Most feature selection methods such as forwards and backwards selection and branch-and-bound (discussed in the next section) change the set of features under consideration by adding or deleting a single feature at a time. The various measures depend on the means μ_i and variance matrices Σ_i. These can readily be found by taking the appropriate subsets of the mean vector and covariance matrix for all features, but it is worth noting that updating formulae for the inverses, determinants and traces that occur in the separation measures are available (for example in Devijver & Kittler, 1982, pp. 266–267).

We have acted as if the class means and variances are known. In practice they are estimated from data, and we may bias-correct the formulae for separation measure by similar ideas to those used in Section 2.4. The correction for J_B is given in Hjort (1986, §10.3).

10.3   Branch-and-bound techniques

The simplest feature selection strategies are stepwise ones. Suppose we wish to choose that combination of $k < p$ features which maximizes some measure J of class separation or classifier performance. We will assume that J is monotone, so that adding features is guaranteed not to decrease *J*. *Forwards selection* adds a feature at a time, at each stage choosing the addition that most increases *J*. *Backward elimination* starts with all *p* features and at each step drops the feature whose presence least increases *J*. The backward and forward procedures are optimal at each stage, but are unable to anticipate interactions between features of the sort we considered at the beginning of the chapter. Exchange strategies would start with a subset of size *k*, perhaps found by forward or backward methods, then try exchanging a feature in the set with any outside it.

These strategies are all heuristics to avoid considering all of the very large number of subsets of size *k* of *p* variables for even moderate *p*. We can often find a subset which is guaranteed to be best of size *k* without considering all subsets by the technique of *branch-and-bound* which is well known in combinatorial optimization and artificial intelligence (Winston, 1992, Chapter 5), and was considered in this context by Narendra & Fukunaga (1977). In choosing subsets of a regression, the procedure is best known from the algorithm of Furnival & Wilson (1974).

Branch-and-bound allows us to eliminate subsets *A* from consideration if we know that a larger subset *A'* has a value of *J* which is below that of our current best estimate α of the maximum value of $J(A)$ over subsets of size *k*, for by monotonicity, necessarily $J(A) < \alpha$. An initial estimate of α is found by one of the heuristic searches (or it is set to $-\infty$). We start by considering the set *A* of all the features, and search the tree of subsets of size at least *k* found by dropping one feature at a time. Whenever we find a subset *A* with $J(A) < \alpha$ we prune the tree at that point, to ensure that we do not consider any subsets of *A* (which can also occur elsewhere on the tree with the variables in a different order, and should also be pruned). Whenever we find a subset of size *k* with $J(A) > \alpha$, we remember the subset and increase α to this $J(A)$.

There are a number of strategies for the actual search of the tree. We would like to consider 'good' subsets first, and we need to reach the leaves (the subsets of size *k*) to be able to increase α. So the search needs to be depth-first in 'good' subsets, and is aided by having a good initial estimate of α.

10.4 Feature extraction

Feature extraction is generally used to mean the construction of linear combinations $a^T \mathbf{x}$ of continuous features which have good discriminatory power between classes. It is naturally part of finding linear classifiers, and it is also often used as a data reduction technique, to reduce the number of features to be input to a non-linear classifier.

The simplest (and by far the most commonly used) method of feature extraction is to take the principal components of $\mathbf{x}$. This was done, for example, by Candela & Chellappa (1993) in studies of fingerprint images, and by Grother & Candela (1993) in studies of hand-written zip codes. Apparently principal components have nothing to do with discriminatory power (they are an unsupervised technique) and it is easy to envisage (and find) examples where they have little discriminatory power. In problems where the features have been carefully scaled and are highly correlated (like images), large variance of a linear combination may imply that it varies across classes.

Blue *et al.* (1994) is the journal version of these reports.

It is possible, at least in principle, to maximize a measure of class separation over one or a few linear combinations of the features. This can be seen as a supervised version of projection pursuit (Section 9.1), in which the measure of 'interestingness' of the projection is related to how well it separates the known classes. Of course, we have to know the class-conditional densities on the projection, but they can be estimated by the methods of Chapter 6, especially by kernel methods. Devijver & Kittler (1982, §8.2.2) suggest that the Patrick–Fisher measure is most suitable for feature extraction with Gaussian kernel estimation since its derivative with respect to the projection direction a can be found analytically.

A

Statistical Sidelines

This appendix explains more of the background of some statistical ideas which are used at several points in the main text, but may not be well-known even to statistical readers.

A.1 Maximum likelihood and MAP estimation

A prototypical statistical problem is to estimate the value of some parameter θ from a finite set $\mathscr{T} = (X_i)$ of data. (In the parlance of pattern recognition, we will refer to this as the *training set*.) Since θ is described as a parameter, this implies the existence of a family of probability densities $p(x; \theta)$ for $\theta \in \Theta$, and we will assume that the observations x_i are independent samples from an unknown density p_0, which might be $p(\cdot; \theta)$ for some θ, but need not be.

Two technical asides. The assumption of independence is easily circumvented by taking all the observations as X_1. Readers not used to measure-theoretic treatments of probability theory will associate densities with continuous distributions and probability mass functions with discrete ones. As probability mass functions *are* densities in the rigorous theory (with respect to counting measure) it is permissible to call both 'densities' and we do so.

The *likelihood* is a function of θ defined by

$$\ell(\theta; \mathscr{T}) = p(\mathscr{T}; \theta) = \prod_i p(x_i; \theta).$$

Although it is another expression of the joint density, the notation reflects the change in emphasis to fixed data and varying parameter. The *maximum likelihood estimator* (MLE) then associates with each training set a value of θ which maximizes $\ell(\theta; \mathscr{T})$, or

$$\widehat{\theta}(\mathscr{T}) = \underset{\theta}{\operatorname{argmax}} \, \ell(\theta; \mathscr{T}).$$

We will almost always drop the dependence on $\mathcal{T}$ and regard $\widehat{\theta}$ as a random variable.

For sufficiently regular problems (for which the likelihood is differentiable and the maximum occurs in the interior of Θ) the maximum likelihood will occur at a stationary point of the log-likelihood, and this is the most common way to find $\widehat{\theta}$. Beware though that the MLE need not occur at a stationary point, even a local maximum, but could occur at the boundary of Θ.

In the Bayesian paradigm, the parameter vector θ is random, and so itself has a distribution. The posterior density of θ can be found by Bayes' formula as

$$p(\theta \mid \mathcal{T}) \propto p(\mathcal{T} \mid \theta)\, p(\theta) = \ell(\theta; \mathcal{T})\, p(\theta).$$

A MAP (maximum *a posteriori*) estimator of θ maximizes $p(\theta \mid \mathcal{T})$ or, equivalently, $\ell(\theta; \mathcal{T})\, p(\theta)$. Thus the maximum likelihood estimator is a MAP estimator for the 'flat' prior over Θ, the possibly improper distribution with uniform density. This highlights the problem with a MAP estimator for a continuous parameter; it finds the mode of a *density*. Densities are with respect to an underlying measure, and the MAP will depend on that measure. This implies that it will not transform in a sensible way. Suppose θ is a parameter expressing a variance. Do we want the MAP of the variance. θ, the standard deviation $\sqrt{\theta}$, the precision $\kappa = 1/\theta$ or the log-variance $\log \theta$? The maximum likelihood estimator will transform in the way you would expect (we say it is *equivariant* to 1–1 transformations) but a MAP estimator will not. Only if the posterior density is highly concentrated about its mode and we allow only smooth transformations is the MAP estimator approximately equivariant. Thus MAP estimators are most useful as a simple summary of a highly concentrated posterior distribution.

A.2   The EM algorithm

The Expectation–Maximization (EM) algorithm is a device to help find maximum likelihood estimators in a problem with unobserved data. Suppose we have data X which have been observed and data Y which have not, and a vector of parameters θ. The goal is to find the maximum likelihood estimator of θ given the observed data X in a situation in which the joint density $p(x, y; \theta)$ is known explicitly, but the marginal density of X, $p(x; \theta)$, can only be found by numerical summation or integration from the joint density.

One application is to problems where we have a series of pairs (X_i, Y_i) of which only X_i is observed (and Y_i is often the label for a component of a mixture). Then the log-likelihood is

$$L(\theta; (X_i)) = \sum_i \log \int p(x_i, y; \theta) \, dy$$

and the presence of the logarithm inhibits any simplification. Another application is to missing observations of a few of the features. The idea has many precursors, including the Baum–Welch algorithm in speech recognition (see Baum *et al.*, 1970), but was developed in some generality by Dempster *et al.* (1977).

Let $Q(\theta, \theta') = \mathsf{E}[\log p(X, Y; \theta) \mid X; \theta']$ which is also a function of the observed data X; the conditional expectation is over the values of Y, and is evaluated as if θ' were the true parameter. The EM algorithm starts at some value $\theta^{(0)}$ and alternates two steps:

E   Find $Q(\theta, \theta^{(i-1)}) = \mathsf{E}[\log p(X, Y; \theta) \mid X; \theta^{(i-1)}]$.

M   Choose $\theta^{(i)}$ to maximize $Q(\theta, \theta^{(i-1)})$.

Each iteration increases the log-likelihood $L(\theta; X) = \log p(X; \theta)$. Write

$$L(\theta; X) = \log p(X, Y; \theta) - \log p(Y \mid X; \theta)$$

and take expectations using the density $p(Y \mid X; \theta')$ to obtain

$$L(\theta; X) = Q(\theta, \theta') - \mathsf{E}[\log p(Y \mid X; \theta) \mid X; \theta'].$$

Now consider expectations E' with respect to $p(Y \mid X; \theta')$ and let $h(Y) = p(Y \mid X; \theta)/p(Y \mid X; \theta')$. Then $\mathsf{E}' \log h(Y) \leqslant \mathsf{E}' h(Y) - 1 = 0$ from $\log x \leqslant x - 1$. Thus

$$\mathsf{E}' \log p(Y \mid X; \theta) \leqslant \mathsf{E}' \log p(Y \mid X; \theta').$$

Now suppose $Q(\theta, \theta') > Q(\theta', \theta')$, so

$$\begin{aligned} L(\theta; X) &= Q(\theta, \theta') - \mathsf{E}' \log p(Y \mid X; \theta) \\ &> Q(\theta', \theta') - \mathsf{E}' \log p(Y \mid X; \theta') \\ &= L(\theta'; X). \end{aligned}$$

The increase in likelihood at an EM iteration will be positive provided $Q(\theta, \theta^{(i-1)}) > Q(\theta^{(i-1)}, \theta^{(i-1)})$, and this is so unless $\theta^{(i-1)}$ is already the maximizer. These arguments still apply to what is often called a GEM (generalized EM) algorithm in which $Q(\theta, \theta^{(i-1)})$ is not fully maximized, but $\theta^{(i)}$ is chosen to increase its value (except at a global maximum, and perhaps at a local maximum).

The convergence properties of (G)EM algorithms are often stated rather loosely. At each step the log-likelihood is increased. In a problem which does have a finite maximum to the likelihood (and by no means all mixture problems do) the sequence $L(\theta^{(i)}; X)$ is bounded above, and so has a limit. That limit need not be a local maximum, but it will be under mild regularity conditions. (Convergence properties are discussed by Boyles, 1983, and Wu, 1983; the proofs in Dempster *et al.*, 1977, are flawed.) Under further regularity conditions we can show the less important condition that the sequence $(\theta^{(i)})$ itself converges to a local maximizer of the likelihood.

It is often useful to know the Hessian at the (local) maximum likelihood solution, for example to find asymptotic standard errors. Louis (1982) gives an algorithm to do so, based on the complete-data likelihood.

There is also a Bayesian view of the EM algorithm as a way to find posterior modes for a subset of the parameters. Write $\theta = (\phi, \psi)$ which in the Bayesian paradigm is a random vector, and suppose we wish to find a mode of the posterior density $p(\phi \mid X)$. Now take ψ as the unobserved data. Since

$$\log p(\phi \mid X) = \log p(\phi, \psi \mid X) - \log p(\psi \mid \phi, X)$$

the same arguments apply to $Q(\phi, \phi') = \mathsf{E}\big[\log p(\phi, \psi \mid X) \mid \phi', X\big]$, so the EM algorithm can be used to help find a MAP estimator of ϕ.

There are 'on-line' versions of the EM algorithm, given for example by Titterington (1984), Celeux & Diebolt (1992) and Jordan & Jacobs (1994).

Mixture distributions

Most applications of the EM algorithm are either to missing data or mixture distributions. The latter are often particularly simple, and were discussed by Dempster *et al.* (1977). Suppose we have a density of the form

$$p(\mathbf{x}) = \sum_i w_i f_i(\mathbf{x}; \phi_i)$$

where the parameters ϕ of the densities may have common components (for example a common covariance matrix in a Gaussian mixture), and the mixing weights (w_i) are unrestricted (apart from forming a discrete distribution for the component I). The parameter vector θ

encompasses the weight distribution (w_i) and all ϕ_i. We regard the component I as the missing data. Then the E step gives

$$Q(\theta, \theta') = \sum_{\mathbf{x} \in \mathscr{T}} \sum_i \pi_i(\mathbf{x}) \left[\log f_i(\mathbf{x}; \phi_i) + \log w_i \right]$$

where

$$\pi_i(\mathbf{x}) = \mathsf{E}\left[I(I = i) \mid X = \mathbf{x}; \theta' \right] = \frac{w_i f_i(\mathbf{x}; \phi_i)}{\sum_j w_j f_j(\mathbf{x}; \phi_j)}.$$

To maximize this over (w_i) we need only consider the second term of $Q(\theta, \theta')$, so we are maximizing $\sum \pi_i \log w_i$ where (π_i) is the average of the $\pi_i(\mathbf{x})$ over the training set. We have

$$\sum \pi_i \log w_i / \pi_i \leqslant \sum \pi_i w_i / \pi_i - 1 = 0,$$

so $\sum \pi_i \log w_i \leqslant \sum \pi_i \log \pi_i$ and $\widehat{w}_i = \pi_i$. For the parameters ϕ_i we maximize the weighted log-likelihood

$$\sum_i \sum_{\mathbf{x} \in \mathscr{T}} \pi_i(\mathbf{x}) \log f_i(\mathbf{x}; \phi_i).$$

If there are no common parameters, each ϕ can be found separately.

A.3 Markov chain Monte Carlo

Markov chain Monte Carlo methods are iterative methods to simulate from distributions that are not easily simulated by more direct methods. They have been used to simulate stochastic processes for many years (Metropolis *et al.*, 1953; Ripley, 1977, 1979; Geman & Geman, 1984; Ripley, 1987), but have recently become popular in mainstream Bayesian statistics following their espousal by Gelfand & Smith (1990) and Gelfand *et al.* (1990). (See Geyer, 1992; Smith & Roberts, 1993; Besag & Green, 1993; Tierney, 1994; Besag *et al.*, 1995; and Gelman *et al.*, 1995, for recent reviews.)

We will consider a finite collection $X_v, v \in V$ for random variables, and use the notation of Chapter 8, that X_A denotes the collection $X_a, a \in A \subset V$. We are interested in sampling from the whole collection X_V or from the conditional distribution of X_A given X_{A^c}, often with the aim of finding aspects of the marginal distribution of some subset A. We can in principle sample successively from the marginal distribution of X_1, then from $X_2 \mid X_1$, $X_3 \mid X_1, X_2$ and so on, but these distributions may not be known sufficiently explicitly to sample from. Suppose we do know how to sample from the conditional distribution of X_v given

$X_{V\setminus\{v\}}$ for each random variable (as is required at the last step of successive sampling). Then starting with some set of values for X_V, we can pick a variable, and sample it conditionally on the rest. This is repeated for all the variables in some order, and is what is known as the *Gibbs sampler*. This term is due to Geman & Geman (1984), who showed that for discrete random variables with a finite state space and no zero-probability configurations the joint distribution of X_V after n samples converges to the required joint distribution provided only that each random variable is visited infinitely often. Thus we can visit the variables in random or systematic order, and visit some more often than others.

The restriction to a strictly positive joint distribution is an essential one, as without it the Gibbs sampler may fail. Consider three binary random variables A, B, C such that B and C are independent given $A = 0$, but $A = 1$ implies $B = C = 1$. The Gibbs sampler will eventually reach the state $A = B = C = 1$ and be unable to escape. Assuming *irreducibility* (that we can move with positive probability from any configuration with positive probability to any other in a finite number of steps) saves convergence (Ripley, 1987, §4.7). Sometimes irreducibility can be retrieved by grouping vertices or by taking zero probabilities to be the limit of very small probabilities (Sheehan & Thomas, 1993), but there are practical limits to the value of these 'tricks'. It is widely assumed that this result still holds for continuous random variables, although the theory is much more complicated and further mild conditions are required (Chan, 1993; Smith & Roberts, 1993, Appendix A; Tierney, 1994).

In the discrete case there are many other MCMC schemes. We confine attention to configurations with positive probability (without any loss of generality). Suppose we have a transition kernel $q(x_V, x'_V)$ which gives the conditional probability of moving from configuration x_V to configuration x'_V. A *Metropolis* algorithm generates a move according to this conditional probability, and accepts it with probability $\min\{1, p(x'_V)/p(x_V)\}$. This converges to the joint distribution provided

(a) the kernel q is symmetric,

(b) the process is irreducible, and

(c) the process is *aperiodic*. (This says that the feasible return times to a state have period one, and precludes only returning with positive probability at even times, for example.)

A *Hastings* algorithm (Hastings, 1970) allows an asymmetric kernel q, and accepts the move with probability

$$\min\left\{1, \frac{p(x'_V)q(x'_V, x_V)}{p(x_V)q(x_V, x'_V)}\right\}.$$

This process converges if it is irreducible and aperiodic. Suppose in a Metropolis–Hastings algorithm we consider moves which change only one random variable. If this is X_v we have

$$\frac{p(x'_V)}{p(x_V)} = \frac{p(x'_v \mid X_{V\setminus\{v\}} = x_{V\setminus\{v\}})}{p(x_v \mid X_{V\setminus\{v\}} = x_{V\setminus\{v\}})}$$

which needs only the conditional distributions as for the Gibbs sampler. Unlike the Gibbs sampler it needs only ratios of distributions, so these need not be normalized. There are a few continuous analogues of the Metropolis–Hastings method.

'Blocked' Gibbs sampler methods use the Gibbs sampler on blocks of random variables rather than on single variables. Sometimes this is a device to ensure irreducibility or to encourage faster convergence. On the other hand, it can be quite natural. Consider a general mixture distribution as at the end of Section A.2. We can treat the set of indicators (I_i) of the unknown components for each member of the training set as a block, and sample these simultaneously given the real parameters.

Much of the theoretical work when using an MCMC method is proving irreducibility, and in practical examples this can be lengthy (Grenander *et al.*, 1991). In practice the difficulty is knowing when the process has reached equilibrium, and how long to wait between sampling x_A if (approximately) independent samples are required. There is much discussion of empirical methods of assessing convergence in the surveys cited at the beginning of this section, but none would recognize the pseudo-equilibrium behaviour reported by Ripley & Kirkland (1990), who give an example in which equilibrium has not been approached after each random variable in the system has been sampled 10,000 times, although the process has appeared stable for 9,500 passes. So great care is needed!

A.4   Axioms for conditional independence

We will use the notation $X \perp\!\!\!\perp Y \mid Z$ of Dawid (1979, 1980) to say that random variables X and Y are independent given Z. For discrete random variables it is clear what this means:

$$\Pr\{X = x, Y = y \mid Z = z\} = \Pr\{X = x \mid Z = z\}\Pr\{Y = y \mid Z = z\}$$
$$\text{whenever } \Pr\{Z = z\} > 0.$$

If just Z is discrete we can ask that the joint conditional density factorizes. In general we may ensure that the conditional densities exist and require

$$p(x, y \mid z) = p(x \mid z) \, p(y \mid z)$$

except for z belonging to a set visited by Z with probability zero. Alternatively, we will have $p(x \mid y, z) = p(x \mid z)$, and conditional independence will hold whenever there is a version of $p(x \mid y, z)$ which is a function of x and z alone (since then $p(x \mid z) = \mathsf{E}[p(x \mid Y, z) \mid Z = z] = p(x \mid y, z)$ for any y). Alternatively (and only for those thoroughly familiar with probability theory) we can use regular conditional probabilities (Freedman, 1971, §10.10) which define a conditional probability for each value of Z, and use the usual definition of independence, the factorization of $\mathrm{Pr}\{X \in A, Y \in B \mid Z = z\}$.

We will regard a group of random variables as still a random variable, so in the following X, Y, Z and W may be collections of random variables.

The following properties are easily derived from the definitions.

$$X \perp\!\!\!\perp Y \mid Z \iff Y \perp\!\!\!\perp X \mid Z \qquad \text{(A.1)}$$

$$X \perp\!\!\!\perp Y, W \mid Z \implies X \perp\!\!\!\perp Y \mid Z \qquad \text{(A.2)}$$

$$X \perp\!\!\!\perp Y, W \mid Z \implies X \perp\!\!\!\perp Y \mid Z, W \qquad \text{(A.3)}$$

$$X \perp\!\!\!\perp W \mid Z, Y \ \text{and} \ X \perp\!\!\!\perp Y \mid Z \implies X \perp\!\!\!\perp Y, W \mid Z. \qquad \text{(A.4)}$$

Note that on interchanging the roles of W and Y in (A.3) we may replace (A.2–A.4) by

$$X \perp\!\!\!\perp W \mid Z, Y \ \text{and} \ X \perp\!\!\!\perp Y \mid Z \iff X \perp\!\!\!\perp Y, W \mid Z. \qquad \text{(A.5)}$$

For strictly positive densities (only) we have also

$$X \perp\!\!\!\perp W \mid Z, Y \ \text{and} \ X \perp\!\!\!\perp Y \mid Z, W \implies X \perp\!\!\!\perp Y, W \mid Z. \qquad \text{(A.6)}$$

Properties (A.4) and (A.6) are different, since in general neither of $X \perp\!\!\!\perp Y \mid Z$ and $X \perp\!\!\!\perp Y \mid Z, W$ implies the other.

The graphical interpretations of conditional independence discussed in Chapter 8 are all deducible from these axioms. This has led Pearl and his co-workers to term concepts which respect (A.1–A.4) *graphoids* and graphoids which also obey (A.6) are called *positive graphoids*. This is not pure axiomatization; there are other concepts which obey these axioms (such as embedded multi-valued dependencies of attributes in databases; Fagin, 1977). If we allow non-disjoint collections of variables we also need to know

$$X \perp\!\!\!\perp Y \mid Y, W. \qquad \text{(A.7)}$$

J. Q. Smith (1989) uses (A.7), (A.1) and (A.5) as his axiom system for a discussion of conditional independence on DAGs (less powerful than that described in Chapter 8). From these axioms we deduce

$$X, Z \perp\!\!\!\perp Y, Z \mid Z \iff X \perp\!\!\!\perp Y \mid Z$$

so we can confine attention to disjoint collections of random variables.

The graphical representations of conditional independence discussed in Chapter 8 have stronger properties. Write $A \perp B \mid C$ if set C separates A from B. Separation on an undirected graph satisfies

$$X \perp Y \mid Z \iff Y \perp X \mid Z \tag{A.8}$$

$$X \perp Y \cup W \mid Z \implies X \perp Y \mid Z \tag{A.9}$$

$$X \perp Y \mid Z \implies X \perp Y \mid Z \cup W \tag{A.10}$$

$$X \perp W \mid Z \cup Y \text{ and } X \perp Y \mid Z \cup W \implies X \perp Y \cup W \mid Z. \tag{A.11}$$

Note that (A.10) is stronger than (A.3) (and we have already said is not true for conditional independence), and that we do have the intersection condition (A.11). The first three conditions are immediate. Condition (A.11) is easily proved by contradiction. (Suppose there is a path from X to $Y \cup W$ avoiding Z. Then this path can be truncated if necessary to have no interior vertices in $Y \cup W$. If it ends in Y, it avoids $Z \cup W$, and if it ends in W it avoids $Z \cup Y$.) Graph separation is also transitive:

$$X \perp Y \mid Z \implies X \perp \{v\} \mid Z \text{ or } Y \perp \{v\} \mid Z \text{ if } v \notin X \cup Y \cup Z. \tag{A.12}$$

On the other hand, d-separation on a DAG satisfies (Pearl, 1988, p. 128)

$$X \perp Y \mid Z \iff Y \perp X \mid Z \tag{A.13}$$

$$X \perp Y \cup W \mid Z \implies X \perp Y \mid Z \tag{A.14}$$

$$X \perp Y \cup W \mid Z \implies X \perp Y \mid Z \cup W \tag{A.15}$$

$$X \perp W \mid Z \cup Y \text{ and } X \perp Y \mid Z \implies X \perp Y \cup W \mid Z \tag{A.16}$$

$$X \perp W \mid Z \cup Y \text{ and } X \perp Y \mid Z \cup W \implies X \perp Y \cup W \mid Z \tag{A.17}$$

$$X \perp Y \mid Z \text{ and } X \perp W \mid Z \iff X \perp Y \cup W \mid Z. \tag{A.18}$$

The first four conditions map to properties of conditional independence, and the fifth is valid for conditional independence under strict positivity. We also have

$$X \perp Y \mid Z \text{ and } X \perp Y \mid Z \cup \{v\}$$
$$\implies X \perp \{v\} \mid Z \text{ or } Y \perp \{v\} \mid Z \tag{A.19}$$

$$\{a\} \perp \{b\} \mid \{c, d\} \text{ and } \{c\} \perp \{d\} \mid \{a, b\}$$
$$\implies \{a\} \perp \{b\} \mid \{d\} \text{ or } \{a\} \perp \{b\} \mid \{d\} \tag{A.20}$$

none of which are necessary for probabilistic conditional independence.

Shenoy & Shafer (1990) also axiomatize marginalization, which enables them to give an abstract version of the computations of marginal probabilities on join trees; Dempster & Kong (1988) had already shown the version of these computations for Dempster–Shafer belief functions. The local computations are about taking a potential representation of a distribution (the density is taken to be a product of potential functions on cliques) and rearranging the terms of the product while keeping the product constant so that the terms are related to marginals. To pass a message from C_i to C_j we need to be able to

1   form the marginal on $S = C_i \cap C_j$ and

2   combine this new marginal with the existing marginal on C_j.

For this to be valid, marginals have to be combined in a way that is commutative, associative, marginalization has to be consistent (marginalizing from C to $S \subset C$ must be the same as first marginalizing to T then S if $C \supset T \supset S$), and marginalization must be distributive over combination. (This means that combining marginals on C_i and C_j and then marginalizing to C_j is the same as marginalizing from C_i to $S = C_i \cup C_j$ and then combining marginals on S and C_j.) These axioms are true for other systems of combination and marginalization that arise in belief function theory and database theory.

A.5   Optimization

Optimization, while not strictly statistical, is used in many statistical procedures. In statistical applications it is only necessary to find a parameter estimate to within a small fraction of its standard error, so for our applications it is more important that the optimization algorithm is quick to reach an approximately right answer than that its convergence (to machine precision) is fast.

We will concentrate on good methods for estimating many parameters. More detailed expositions are given by Gill *et al.* (1981), Dennis & Schnabel (1983) and Fletcher (1987). These methods are iterative. A generic minimization algorithm is of the form:

1   Choose an initial point $\mathbf{x}_0$.

2   Select a search direction $\mathbf{p}$.

3   Select a step length α, set $\mathbf{s} = \alpha\mathbf{p}$ and $\mathbf{x} \leftarrow \mathbf{x} + \mathbf{s}$. We normally ensure that $f(\mathbf{x})$ is decreased.

4   Return to step 2 unless the convergence criteria are met.

The convergence criteria will be problem-specific.

Methods based on Taylor expansions

Suppose we have a function $f : \mathbb{R}^m \to \mathbb{R}$ which we wish to minimize. We assume that f is differentiable, and let $g = \nabla f$ denote the gradient. A first-order Taylor expansion about any point $\mathbf{x}_0$ gives

$$f(\mathbf{x}) \approx f(\mathbf{x}_0) + g(\mathbf{x}_0)^T (\mathbf{x} - \mathbf{x}_0)$$

and so the value of the function will be reduced by moving to $\mathbf{x}_0 - \eta g(\mathbf{x}_0)$ for small enough η, provided the derivative is not zero at $\mathbf{x}_0$. This is known as the method of *steepest descent*, since amongst all unit-length vectors $\mathbf{a}$, $\mathbf{a}^T g(\mathbf{x}_0)$ is smallest when $\mathbf{a} \propto -g(\mathbf{x}_0)$.

The Taylor expansion provides no idea how to choose η, and there are two main strategies. One is to perform a line search for the minimum of $f(\mathbf{x}_0 + \eta \mathbf{a})$. This entails that the next step will be orthogonal to this one, and the method tends to move in a series of zig-zag steps. The other strategy is to increase η as far as possible while $f(\mathbf{x}_0) - \eta \| g(\mathbf{x}_0) \|^2$ remains an adequate approximation to $f(\mathbf{x} - \eta g(\mathbf{x}_0))$.

The curvature of the surface $f(\mathbf{x})$ can give us information about the step length, so we now assume that f is continuously twice differentiable and has Hessian matrix $H(\mathbf{x})$. A second-order Taylor expansion gives

$$f(\mathbf{x}) \approx f(\mathbf{x}_0) + g(\mathbf{x}_0)^T (\mathbf{x} - \mathbf{x}_0) + \tfrac{1}{2}(\mathbf{x} - \mathbf{x}_0)^T H(\mathbf{x}_0)(\mathbf{x} - \mathbf{x}_0), \quad \text{(A.21)}$$

and the minimum of the right-hand side occurs at $\mathbf{x}_0 - H(\mathbf{x}_0)^{-1} g(\mathbf{x}_0)$ provided that $H(\mathbf{x})$ is positive-definite; otherwise the right-hand side does not have a unique minimum. Algorithms based on this expansion are called *Newton methods*. It is not always obvious that $H(\mathbf{x})$ is positive-definite, but careful algorithms will check this, and adjust it to be so if it seems likely that it fails to be positive-definite only through rounding errors.

Note that Newton methods require no line search, but do not ensure that $f(\mathbf{x})$ is reduced at each step. They can diverge dramatically. Thus practical algorithms will reduce the step length to ensure that the step reduces $f(\mathbf{x})$ and perhaps that (A.21) is a reasonable approximation. Eventually the full step length will always be used, and the convergence then is second-order, that is $\| \mathbf{x}_{t+1} - \mathbf{x}_t \| = O(\| \mathbf{x}_t - \mathbf{x}_{t-1} \|^2)$.

The convergence properties of Newton methods are unsurpassed, but they are not necessarily so well behaved away from the minimum. The Hessian $H(\mathbf{x})$ measures the curvature at $\mathbf{x}$, but this may not be useful except very close to $\mathbf{x}$. The methods described next build up a quadratic approximation similar to (A.21) which is valid at about the length scale of a current step of the algorithm. The other drawback of

Newton methods is that $H(\mathbf{x})$ may not be known or be very expensive to compute, much more expensive than function values and derivatives.

Methods based on quadratic approximations

Suppose our function f is actually quadratic with Hessian G (which does not depend on $\mathbf{x}$), so (A.21) is an equality everywhere. We can garner information about G from the gradients at $\mathbf{x}$ and $\mathbf{x}+\mathbf{s}$ at the beginning and end of a step, since

$$g(\mathbf{x}+\mathbf{s}) = g(\mathbf{x}) + G\mathbf{s}.$$

A *quasi-Newton* method uses this to build up an approximation B to G. Initially B is set to the best possible guess (the identity matrix if no further information is available). Each step is along $-B^{-1}g(\mathbf{x})$, and after each step B is updated by $B \leftarrow B + U$, where the correction U is chosen so that after correction $B\mathbf{s} = g(\mathbf{x}+\mathbf{s}) - g(\mathbf{x})$. If the function is quadratic, we will learn G exactly after m linearly independent steps, and for a general function we will build up a local quadratic approximation to f.

> An older name for quasi-Newton methods is *variable metric* methods.

There are many ways to find a correction term U. It is widely believed that the most effective update is the Broyden–Fletcher–Goldfarb–Shanno (BFGS) formula

> This was proposed separately by all four authors in 1970.

$$B \leftarrow B - \frac{B\mathbf{s}\mathbf{s}^T B}{\mathbf{s}^T B\mathbf{s}} + \frac{\mathbf{y}\mathbf{y}^T}{\mathbf{y}^T\mathbf{s}}$$

where $\mathbf{y} = g(\mathbf{x}+\mathbf{s}) - g(\mathbf{x})$. As the search is along $\alpha\mathbf{p} = -\alpha B^{-1}g(\mathbf{x})$, the BFGS update becomes

$$B \leftarrow B - \frac{g(\mathbf{x})g(\mathbf{x})^T}{g(\mathbf{x})^T\mathbf{p}} + \frac{\mathbf{y}\mathbf{y}^T}{\alpha\mathbf{y}^T\mathbf{p}}. \qquad (A.22)$$

For this method to be viable, we do need B to remain positive-definite. This needs $\mathbf{y}^T\mathbf{s} > 0$, which can be ensured by choosing a sufficiently accurate line search over α.

Quasi-Newton methods generally converge super-linearly (which means $\|\mathbf{x}_{t+1} - \mathbf{x}_t\|/\|\mathbf{x}_t - \mathbf{x}_{t-1}\| \to 0$), but are often more effective than Newton methods away from a local minimum. As with Newton methods, the step length $\alpha = 1$ is preferred, so a line search along $\mathbf{p}$ is only needed in the early stages.

The inverse $C = B^{-1}$ is all that is needed to find the search direction, and (A.22) can be converted to an update for C using the Sherman–Morrison–Woodbury formula to give

> See the glossary.

$$C \leftarrow C + \left[1 + \frac{\mathbf{y}^T C \mathbf{y}}{\mathbf{y}^T \mathbf{s}}\right] \frac{\mathbf{s}\mathbf{s}^T}{\mathbf{y}^T \mathbf{s}} - \left[\frac{\mathbf{s}\mathbf{y}^T C + C \mathbf{y}\mathbf{s}^T}{\mathbf{y}^T \mathbf{s}}\right]. \qquad \text{(A.23)}$$

Some but not all accounts suggest that this is less desirable than updating a factorization of B, for example the Cholesky factorization $B = LDL^T$ for L lower-triangular and D diagonal.

When used on an exactly quadratic function, quasi-Newton methods have conjugate search directions $\mathbf{p}_i$, that is $\mathbf{p}_i^T G \mathbf{p}_j = 0$ for $j \neq i$. There are many other methods based on *conjugate gradients* with this property, and all will find the minimum of a quadratic in at most m steps. We are interested in methods which do not form an $m \times m$ matrix. We can find conjugate search directions by $\mathbf{p}_1 = -g(\mathbf{x}_i)$ and for $2 \leqslant i \leqslant m$,

$$\mathbf{p}_i = -g(\mathbf{x}_i) + \beta_i \mathbf{p}_{i-1}, \qquad \beta_i = \frac{\mathbf{y}_{i-1}^T g(\mathbf{x}_i)}{\|g(\mathbf{x}_{i-1})\|^2} = \frac{\|g(\mathbf{x}_i)\|^2}{\|g(\mathbf{x}_{i-1})\|^2}.$$

The first formula for β_i is the Polak–Ribiere formula, the second the Fletcher–Reeves formula. The two are equal with exact line searches on a quadratic function, but not otherwise. The Polak–Ribiere formula is generally preferred, since if the algorithm is making little progress, $\beta \approx 0$ and steepest descent is used. The theory of conjugate gradient algorithms assumes that they are re-started (by $\beta = 0$) every m steps. However, as they should only be used when m is large, there will be at most a few times m iterations and the theory is not relevant. Fairly accurate line searches are needed for conjugate gradient methods, unlike quasi-Newton ones. Møller (1993) has developed one particular version of conjugate gradients which seems well known in the neural network field; it uses the out-dated Hestenes–Stiefel formula for β_i with a particular line-search algorithm.

An alternative way to find conjugate search directions is to use (A.22) taking $C = I$ and exact line searches. More generally, we can retain only a small number of updates without explicitly forming C. Such methods are known as *limited-memory* quasi-Newton methods. Shanno (1990) refers to other promising methods intermediate between the Polak–Ribiere and quasi-Newton methods; see: Liu & Nocedal (1989); Gilbert & Lemaréchal (1989); Buckley (1994); and Byrd *et al.* (1994).

Non-linear least-squares problems

Many fitting problems amount to minimizing a sum of squares of the form

$$E(\mathbf{w}) = \tfrac{1}{2} \sum_{j=1}^{n} \|y_j - f(\mathbf{x}_j; \mathbf{w})\|^2$$

Code is given by Shanno & Phua (1980).

Møller's comparisons with BFGS are invalid, since he fails to check if $\alpha = 1$ is a sufficiently good value for the step length.

for an m-dimensional vector $\mathbf{w}$ for parameters. Here y and f are univariate, but multivariable problems can be put in this form by making multiple entries for each example.

The gradient and Hessian of E are of a special form, and minimization methods have been developed to exploit this. Let J denote the $n \times m$ matrix whose rows are the vectors $\partial f(\mathbf{x}_j; \mathbf{w})/\partial \mathbf{w}$, and let $r_i = y_i - f(\mathbf{x}_i; \mathbf{w})$ be the residuals. Then

$$\frac{\partial E(\mathbf{w})}{\partial \mathbf{w}} = J(\mathbf{w})^T (r_j)$$

and

$$\frac{\partial^2 E(\mathbf{w})}{\partial \mathbf{w} \partial \mathbf{w}^T} = J(\mathbf{w})^T J(\mathbf{w}) - \sum_j r_j \frac{\partial^2 f(\mathbf{x}_j; \mathbf{w})}{\partial \mathbf{w} \partial \mathbf{w}^T} = J(\mathbf{w})^T J(\mathbf{w}) + Q(\mathbf{w}),$$

say. The specialized methods assume that $Q(\mathbf{w})$ is negligible, so either the residuals or the curvatures of f are small. This is often not the case in statistical problems.

The *Gauss–Newton procedure* is a Newton algorithm with the Hessian replaced by $J(\mathbf{w})^T J(\mathbf{w})$, that is taking $Q(\mathbf{w}) = 0$. This is equivalent to the local linear approximation

$$f(\mathbf{x}_j; \mathbf{w}) \approx f(\mathbf{x}_j; \mathbf{w}_0) + \left[J(\mathbf{w}_0)(\mathbf{w} - \mathbf{w}_0) \right]_j. \qquad (A.24)$$

It is not uncommon to find that $J(\mathbf{w})^T J(\mathbf{w})$ is ill-conditioned, and ridge-regression methods may be used to fit (A.24). In this context they are known as *Levenberg–Marquardt methods*, which replace $J(\mathbf{w})^T J(\mathbf{w})$ by $J(\mathbf{w})^T J(\mathbf{w}) + \lambda I$. If the residuals are not small, it should be better to use general quasi-Newton methods or hybrids between Gauss–Newton and quasi-Newton methods (Gill *et al.*, 1981).

The local linearization (A.24) is also used to find standard errors for the parameters $\mathbf{w}$. The variance of the least-squares solution $\widehat{\mathbf{w}}$ will be approximately

$$\text{Var } \widehat{\mathbf{w}} \approx \sigma^2 \left[J(\mathbf{w})^T J(\mathbf{w}) \right]^{-1} \qquad (A.25)$$

where σ^2 is the variance of the observational errors.

Glossary

AIC ('An Information Criterion') A method developed by Akaike (1973, 1974) to avoid over-fitting, by penalizing the *deviance* by twice the number of free parameters.

back-fitting An iterative method of fitting additive models, by fitting each term to the residuals given the rest. It is a version of the Gauss–Seidel methods of numerical linear algebra.

back-propagation is the method used to calculate the gradient vector of a fitting criterion for a feed-forward neural network with respect to the parameters (weights). Also used for a steepest-descent algorithm with the gradient vector computed in this way.

Bayes formula An elementary formula of probability. If B_i are disjoint events, and $A \subset \bigcup_i B_i$ then

$$\Pr\{B_i \mid A\} = \frac{\Pr\{A \mid B_i\}\Pr\{B_i\}}{\sum_j \Pr\{A \mid B_j\}\Pr\{B_j\}}.$$

Bayes rule is a rule which attains the Bayes *risk*, and so is the 'gold-standard', the best possible for that problem.

bias has two meanings. (a) The bias of an estimator is the difference between its mean and the true value. (b) For a neural network, parameters which are constants (rather than multiplying signals) are often called biases.

BIC has two similar meanings. Akaike (1977, 1978) introduced 'information criterion B'. Schwarz (1978) introduced something which has become known as a 'Bayesian information criterion'. Although most references mean Schwarz's BIC, to avoid confusion this is also known as SBC ('Schwarz Bayes Criterion'). Both penalize the deviance by $\log n$ times the number p of free parameters for n examples, but Akaike's has $O(p)$ terms not depending on n.

Bienaymé–Chebychev inequality For a random variable X with mean μ and variance $\sigma^2 < \infty$ we have

$$\Pr\{|X - \mu| > \epsilon\} \leqslant \frac{\sigma^2}{\epsilon^2}$$

for all $\epsilon > 0$. This follows from Jensen's inequality applied to $(X - \mu)^2$.

bootstrap (Efron, 1979) An idea for statistical inference, using training sets created by re-sampling *with replacement* from the original training set, so examples may occur more than once.

branch-and-bound A technique in combinatorial optimization to rule out solutions without evaluating them.

classification trees Classifiers which partition the examples on one feature at a time. See Chapter 7.

classifier A rule to assign a class (or 'doubt' or 'outlier') to new examples.

codebook vectors Representative examples of a probability distribution. The term comes from *vector quantization*.

compact set A subset $A \subset \mathbb{R}^m$ is compact if it is closed and bounded, that is $A \subset [-K, K]^m$ for some $K > 0$. Compact sets are also called *compacta*.

conjugate gradients A class of methods used in optimization (and solving linear systems). See Section A.5.

consistent An estimator is consistent if in large samples it converges to the true parameter value (when there is one).

concave A function f is concave if $-f$ is convex.

convex A *set* $A \subset \mathbb{R}^m$ is convex if $\alpha a + (1 - \alpha)b \in A$ whenever $a, b \in A$. A function $f : A \to \mathbb{R}$ is convex is $f(\alpha a + (1 - \alpha)b) \geqslant \alpha f(a) + (1 - \alpha)f(b)$.

cross-validation A method of evaluating parameters or classifiers by dividing the training set into several parts, and in turn using one part to test the procedure fitted to the remaining parts. Sometimes used to refer to leave-one-out (or ordinary) cross-validation, where every example is dropped in turn. This term is much abused; it does not mean the use of a *test set* or *validation set*. *Generalized* cross-validation is a measure of the performance of a regularized classifier; see page 141.

deviance A measure of fit of a statistical model. The deviance is twice (log-likelihood of the best model minus log-likelihood of the current model). The best model can be the true model or an exact fit (often called a saturated model).

diagnostic paradigm In the terminology of Dawid (1976), modelling the conditional distribution of the class C given the features X.

Dirichlet distribution A distribution over probability distributions $(\pi_1, \ldots, \pi_K)$ on K classes. Its density (Berger, 1985, p. 561) is, for $\alpha_i > 0$ and $0 \leqslant \pi_i, \sum \pi_i = 1$,

$$p(\pi) \propto \prod_{i=1}^{K} \pi_i^{\alpha_i - 1}$$

which has mean at $(\alpha_i / \sum_l \alpha_l)$, and is increasingly concentrated as (α_i) increases.

Dirichlet tessellation Given a set of points in $\mathbb{R}^p$, associate with each those points of $\mathbb{R}^p$ to which it is nearest. This defines a *tile*, and the tiles partition the space. Also known as Voronoi or Thiessen polygons in $\mathbb{R}^2$. Preparata & Shamos (1985) give algorithmic details.

dissimilarity A measure of the dissimilarity of two examples based on their features. Must be non-negative and symmetric.

early stopping A method of optimization in which the objective used is not the real goal, and optimization is stopped when another measure of fit starts to rise. This may be critically dependent on the starting value chosen.

editing Methods of reducing the training set for use by nearest-neighbour methods.

efficiency A statistical term, measuring the performance of estimators. Unless stated otherwise, efficiency is a measure of $1/(n \times \text{variance in samples of size } n)$ (for large n).

eigendecomposition of a real symmetric matrix A. This is an orthonormal matrix C and a non-negative diagonal matrix D such that $A = CDC^T$.

EM algorithm A device to construct algorithms for maximum likelihood and MAP estimators. See Section A.2.

equivariance An invariant procedure is unchanged under a transformation; an equivariant procedure transforms its answer. For example, if θ is a parameter and $\phi = g(\theta)$, then $\widehat{\phi} = g(\widehat{\theta})$.

estimator A rule to assign a parameter value to a set of observations. The value assigned is called an estimate, and the distinction between estimators and estimates is not always observed.

feature A measurement on an example, so the training set of examples has measured features and a class for each.

feature extraction Creating useful new features by combinations (usually linear) of existing features.

feed-forward network A network in which vertices can be numbered so that all connections go from a vertex to one with a higher number. In practice the vertices are arranged in layers, with connections only to higher layers.

generalization A measure of the ability of a classifier to perform well on future examples, or such a measure applied to a method to design classifiers. The term comes from psychology and refers to the ability to infer the correct structure from examples.

Gibbs sampler A simulation method used in Bayesian inference. See Section A.3.

Hermite polynomials Families of orthogonal polynomials. See Thisted (1988, §5.3.2).

Hessian The second derivative matrix of a function $f(\mathbf{x})$.

hints The idea of 'hints' is to incorporate qualitative information into the classifier.

HME Hierarchical mixtures of experts. A tree-structured way to select a combination of classifiers. See Section 8.5.

information (matrix) The Hessian of the log-likelihood with respect to the parameters (the observed information) or its expected value (the Fisher information).

Jensen's inequality Suppose f is a convex function on a convex domain, and X a random variable on that domain. Then $\mathsf{E} f(X) \geqslant f(\mathsf{E} X)$.

Kullback–Leibler divergence between distributions on the same space with densities p and q is

$$d(p,q) = \int p(x) \log \frac{p(x)}{q(x)} \, \mathrm{d}x.$$

learning Choosing the parameters of a classifier (and perhaps also the family of classifiers) from the training set.

least false parameter value. If the parametric family does not contain the true distribution, this is the best possible parameter value θ in the sense of minimum Kullback–Leibler divergence $d(p, p_\theta)$. See page 32.

likelihood The probability density of the observations, viewed as a function of the parameters, not of the observations.

logistic The logistic distribution has cumulative distribution function $\ell(x) = \exp(x)/[1 + \exp(x)]$, and this function is called the logistic function. Logistic regression and discrimination are based on converting predictions to probabilities through the logistic function.

L_p An L_p space is the space of random variables X such that $\mathsf{E} |X|^p < \infty$ or a space of functions f for which $\int f(x)^p \mathrm{d}x < \infty$. Convergence $X_n \to X$ in L_p means $\mathsf{E} (X_n - X)^p \to 0$.

LVQ Learning Vector Quantization. A method of designing examples for use in nearest-neighbour procedures. See page 201.

Mahalanobis distance Given a positive-definite symmetric matrix Σ (a covariance matrix), the distance between examples $\mathbf{x}$ and $\mathbf{y}$ in feature space is $(\mathbf{x} - \mathbf{y})^T \Sigma^{-1} (\mathbf{x} - \mathbf{y})$.

MAP estimator is the global maximum of the posterior density $p(\theta \mid \mathcal{T})$. See Section A.1 for further discussion. MAP stands for *maximum a posteriori*.

maximum likelihood estimate is a value which maximizes the likelihood function, or, more loosely, is a stationary point of the likelihood function.

missing values are unreported values of features. These could be lost, or they could be deliberate non-measurement and so convey information. (A medical practitioner will not order tests if their value appears low.)

multilayer perceptron is another name for a feed-forward neural network. Despite the name, the 'neurons' are not usually perceptrons.

multivariate analysis is the branch of statistics concerned with multiple observations on each example which are not only of interest to predict or explain just one of the observations.

non-informative prior A prior distribution for a parameter vector θ which is intended to express ignorance about θ. This can be tricky, and often leads to the use of densities which have an infinite integral (known as improper priors), such as a uniform density on $\mathbb{R}$.

normal distribution In our usage, this includes both univariate and multivariate distributions. The density is

$$p(x) = (2\pi)^{-p/2}|\Sigma|^{-1/2}\exp\left[-\tfrac{1}{2}(x-\mu)^T\Sigma^{-1}(x-\mu)\right] \quad \text{for } x \in \mathbb{R}^p.$$

NP-complete It is desirable that algorithms terminate in a time which is polynomial in the size of the problem and the accuracy required. Let P denote all problems for which there is such an algorithm. If we allow *nondeterministic* algorithms (which are allowed to choose the best option whenever there is a choice) the class of problems is called NP. Equivalently, NP is the class of problems for which a solution could be verified in polynomial time. It is widely believed that NP is strictly larger than P, but this remains an open research problem. A problem in NP is called NP-complete if proving it was in P would establish P = NP, which should be regarded as strong evidence that no polynomial algorithm will ever be found. (Cormen *et al.*, 1990; Sedgewick, 1990; Garey & Johnson, 1979.)

NP-hard A NP-hard problem is one that implies a solution to every problem in NP (see NP-complete) but is not known to be in NP. Thus NP-hardness is strong evidence that no polynomial algorithm for the problem will ever be found.

on-line methods of parameter estimation adjust the estimate after each new example is seen.

ordinal feature An ordinal measurement is one of a series of ordered categories, for example income ('poor', 'sufficient', 'well-off', 'rich', ...).

outliers Outliers are examples which did not (or are thought not to have) come from the assumed population of examples. For example, in digit recognition, the segmentation will fail occasionally, so the data will not be from a digit at all.

Parzen windows A name for kernel density estimation once common in the pattern recognition community.

perceptron A simple classifier into two classes which thresholds a linear combination of the features. Much publicized by F. Rosenblatt around 1960.

plug-in classifier A classifier constructed by assuming that estimated parameter values are in fact the true ones.

posterior probability The probability of an event conditional on the observations.

predictive classifier A classifier constructed by averaging over the uncertainty in the estimated parameter values.

principal components are linear combinations of features with high variance. See Section 9.1.

prior probability Probabilities specified before seeing the data, and so based on prior experience or belief. Commonly these are the prior probabilities π_k of the classes.

profile likelihood Suppose we divide the parameters $\theta = (\phi, \psi)$. The profile likelihood for ϕ is the likelihood for θ maximized over ψ.

projection pursuit methods are based on extracting features (linear combinations of the original features). Exploratory projection pursuit (Section 9.1) looks for 'interesting' (non-normal) features, and projection pursuit regression (Section 4.1) uses the extracted features in an additive model.

pruning is the term used for removing parts of trees and networks with the aim of increasing generalization. See Section 7.2.

quasi-Newton methods are methods of optimization which approximate the Hessian using only gradient information. See Section A.5.

radial basis functions are a large class of approximating functions, computed as a linear combination of non-linear functions of the distances to a set of centres:
$$f(\mathbf{x}) = \sum a_i G(\|\mathbf{x} - \mathbf{x}_i\|).$$

rank (of a matrix). The number of linearly independent rows or columns.

regularization A class of methods of avoiding over-fitting to the training set by penalizing the fit by a measure of 'smoothness' of the fitted function.

resistant methods are designed to be little affected by outliers. For example, the median is much more resistant than the mean.

ridge regression See *shrinkage methods*.

risk of a classifier is the expected loss from using it. The *Bayes risk* is the lowest attainable risk (using these features).

robust methods are designed to be resistant, and also to have high efficiency near some target distribution. For example, although the median is resistant, it is inefficient compared to a trimmed mean.

sampling paradigm In the terminology of Dawid (1976), modelling the class-conditional densities $p_k(x)$ and, perhaps, the prior probabilities of the classes π_k.

Sherman–Morrison–Woodbury formula Given a non-singular $n \times n$ matrix A and column vectors b and d we have

$$[A + bd^T]^{-1} = A^{-1} - \frac{A^{-1}bd^T A^{-1}}{1 + d^T A^{-1}b}$$

provided $d^T A^{-1}b \neq -1$. If B and D are $n \times m$ matrices for $m \leqslant n$ then

$$[A + BD^T]^{-1} = A^{-1} - A^{-1}B[I + D^T A^{-1}B]^{-1}D^T A^{-1}$$

provided the $m \times m$ matrix $I + D^T A^{-1}B$ is invertible (Golub & Van Loan, 1989, p. 51).

shrinkage methods of estimation 'shrink' an estimator by moving it towards some fixed value (or an overall mean). Ridge regression shrinks regression coefficients towards zero, apart from the constant. The idea is that the shrunken estimator has more bias but lower variance and hence better generalization. The James–Stein example (Cox & Hinkley, 1974, §11.8) shows that this idea works even for the mean of a normal distribution in $p \geqslant 3$ dimensions.

simulated annealing is a method of combinatorial optimization based on taking a series of random steps in the search space. See Ripley (1987) or Aarts & Korst (1989).

singular value decomposition of a real matrix $X = U \Lambda V^T$, where Λ is a diagonal matrix with decreasing non-negative entries, U is an $n \times p$ matrix with orthonormal columns, and V is a $p \times p$ orthogonal matrix (Golub & Van Loan, 1989).

SOFM, SOM Self-organizing (feature) map of Kohonen. See Section 9.4.

softmax Given outputs $y_1, \dots, y_K$ for each of K classes, assign posterior probabilities as

$$p(k \mid x) = \exp y_k \Big/ \sum_{j=1}^{K} \exp y_j.$$

The term comes from Bridle (1990a, b), but the idea is that of multiple logistic regression.

splines are used in function approximation and smoothing. They are constructed by joining functions defined over a partition of the space: the simplest case is polynomials on adjoining intervals. See Section 4.1.

stacked generalization A method of using cross-validation to choose a combination of classifiers. The term is from Wolpert (1992); the idea goes back at least to M. Stone (1974).

steepest descent A method of minimization which takes steps along the direction to steepest descent, the gradient vector. For maximization the method is known as *steepest ascent* or *hill-climbing*.

stochastic approximation aims to find the value of θ_0 solving $f(\theta) = 0$, but although we can measure $f(\theta)$, the result will measured with error. After taking many measurements for θ with $f(\theta)$ near zero we will be able to find accurate estimators of θ_0. There are also versions which aim to find the maximizer of $f(\theta)$.

supervised learning Choosing a classifier from a training set of correctly classified examples.

t distribution The t distribution in p dimensions with location vector μ and scale matrix Σ is the distribution of $\mu + X/S$ where $X \sim N_p\{0, \Sigma\}$ and $vS^2 \sim \chi_v^2$ (Johnson & Kotz, 1972, §37.3; Mardia *et al.*, 1979, p. 57). For $v > 2$ the mean is μ and the covariance matrix $v\Sigma/(v-2)$. The density is

$$\frac{\Gamma\left(\frac{1}{2}(v+p)\right)}{(v\pi)^{p/2}\,\Gamma(\frac{1}{2}v)}|\Sigma|^{-\frac{1}{2}}\left[1 + \tfrac{1}{v}(x-\mu)^T\Sigma^{-1}(x-\mu)\right]^{-\frac{1}{2}(v+p)}.$$

test set A set of examples used only to assess the performance of a fully-specified classifier.

training set A set of examples used for learning, that is to fit the parameters of the classifier.

uniform convergence A sequence of functions f_n converges uniformly to f if $\max_x |f_n(x) - f(x)| \to 0$ as $n \to \infty$. We have uniform convergence on compacta if this holds whenever the maximum is taken over any compact set K.

unsupervised learning Discovering groupings in the training set when none are pre-specified.

updating Changing the classifier when new examples become available, possibly lacking their true classifications.

validation set A set of examples used to tune the parameters of a classifier, for example to choose the number of hidden units in a neural network.

vector quantization A method of encoding data for signal transmission, in which a vector is replaced by one of a finite number of representatives. See page 201.

weights The parameters in a neural network model. Also weights given to individual examples, for example to indicate multiple copies.

References

The following volumes are abbreviated:

NIPS1 (1989) *Advances in Neural Information Processing Systems. Proceedings of the 1988 Conference*, ed. D. S. Touretzky. San Mateo, CA: Morgan Kaufmann.

NIPS2 (1990) *Advances in Neural Information Processing Systems 2*, ed. D. S. Touretzky. San Mateo, CA: Morgan Kaufmann.

NIPS3 (1991) *Advances in Neural Information Processing Systems 3*, eds R. P. Lippmann, J. E. Moody & D. S. Touretzky. San Mateo, CA: Morgan Kaufmann.

NIPS4 (1992) *Advances in Neural Information Processing Systems 4*, eds J. E. Moody, S. J. Hanson & R. P. Lippmann. San Mateo, CA: Morgan Kaufmann.

NIPS5 (1993) *Advances in Neural Information Processing Systems 5*, eds S. J. Hanson, J. D. Cowan & C. L. Giles. San Mateo, CA: Morgan Kaufmann.

NIPS6 (1994) *Advances in Neural Information Processing Systems 6*, eds J. D. Cowan, G. Tesauro & J. Alspector. San Francisco, CA: Morgan Kaufmann.

Many of the papers are reprinted in one or more of the following volumes of reprints:

Anderson, J. A. & Rosenfeld, E. (eds) (1988) *Neurocomputing: Foundations of Research*. Cambridge, MA: The MIT Press.

Anderson, J. A., Pellionisz, A. & Rosenfeld, E. (eds) (1990) *Neurocomputing 2: Directions for Research*. Cambridge, MA: The MIT Press.

Dasarathy, B. V. (ed.) (1991) *Nearest Neighbor (NN) Norms: NN Pattern Classification Techniques*. Los Alamitos, CA: IEEE Computer Society Press.

Lau, C. (ed.) (1992) *Neural Networks: Theoretical Foundations and Analysis*. New York: IEEE Press.

Shafer, G. & Pearl, J. (eds) (1990) *Readings in Uncertainty Reasoning*. San Mateo, CA: Morgan Kaufmann.

Shavlik, J. W. & Dietterich, T. G. (eds) (1990) *Readings in Machine Learning*. San Mateo, CA: Morgan Kaufmann.

Aarts, E. & Korst, J. (1989) *Simulated Annealing and Boltzmann Machines*. New York: Wiley.

Abramowitz, M. & Stegun, I. A. (1965) *Handbook of Mathematical Functions with Formulas, Graphs and Mathematical Tables*. New York: Dover.

Abu-Mostafa, Y. S. (1989) The Vapnik–Chervonenkis dimension: information versus complexity in learning. *Neural Computation* 1, 312–317.

Abu-Mostafa, Y. S. (1990) Learning from hints in neural networks. *Journal of Complexity* 6, 192–198.

Abu-Mostafa, Y. S. (1993) A method for learning from hints. In *NIPS5*, pp. 73–80.

Abu-Mostafa, Y. S. (1995a) Financial market applications of learning from hints. In *Neural Networks in the Capital Markets*, ed. A.-P. Refenes, pp. 221–232. Chichester: Wiley.

Abu-Mostafa, Y. S. (1995b) Machines that learn from hints. *Scientific American* 272(4), 64–69.

Abu-Mostafa, Y. S. (1995c) Hints. *Neural Computation* 7, 639–671.

Ackley, D. H., Hinton, G. E. & Sejnowski, T. J. (1985) A learning algorithm for Boltzmann machines. *Cognitive Science* 9, 147–169. Reprinted in Anderson & Rosenfeld (1988).

Agosta, J. M. (1990) The structure of Bayes networks for visual recognition. In *Uncertainty in Artificial Intelligence 4*, eds R. D. Shachter, T. S. Levitt, L. N. Kanal & J. F. Lemmer, pp. 397–405. Amsterdam: North Holland.

Agrawala, A. K. (ed.) (1977) *Machine Recognition of Patterns*. New York: IEEE Press.

Aitchison, J. & Aitken, C. G. G. (1976) Multivariate binary discrimination by the kernel method. *Biometrika* **63**, 413–420.

Aitchison, J. & Dunsmore, I. R. (1975) *Statistical Prediction Analysis*. Cambridge: Cambridge University Press.

Aitchison, J., Habbema, J. D. F. & Kay, J. W. (1977) A critical comparison of two methods of statistical discrimination. *Applied Statistics* **26**, 15–25.

Aitken, C. G. G. (1978) Methods of discrimination in multivariate binary data. In *Proceedings of COMPSTAT 1978*, eds L. C. A. Corsten & J. Hermans. pp. 155–161. Vienna: Physica-Verlag.

Aitken, C. G. G. (1983) Kernel methods for the estimation of discrete distributions. *Journal of Statistical Computation and Simulation* **16**, 189–200.

Aizerman, M. A., Braverman, E. M. & Rozonoér, L. I. (1964a) Theoretical foundations of the potential function method in pattern recognition learning. *Automation and Remote Control* **25**, 821–837.

Aizerman, M. A., Braverman, E. M. & Rozonoér, L. I. (1964b) The probability problem of pattern recognition learning and the method of potential functions. *Automation and Remote Control* **25**, 1175–1190.

Aizerman, M. A., Braverman, E. M. & Rozonoér, L. I. (1965) The Robbins–Munro process and the method of potential functions. *Automation and Remote Control* **26**, 1882–1885.

Akaike, H. (1973) Information theory and an extension of the maximum likelihood principle. In *Second International Symposium on Information Theory*, eds B. N. Petrov & F. Cáski, pp. 267–281. Budapest: Akademiai Kaidó. Reprinted in *Breakthroughs in Statistics*, eds Kotz, S. & Johnson, N. L. (1992), volume I, pp. 599–624. New York: Springer.

Akaike, H. (1974) A new look at statistical model identification. *IEEE Transactions on Automatic Control* **19**, 716–723.

Akaike, H. (1977) On entropy maximization principle. In *Applications of Statistics*, ed. P. R. Krishnaiah, pp. 27–42. Amsterdam: North-Holland.

Akaike, H. (1978) A Bayesian analysis of the minimum AIC procedure. *Annals of the Institute of Statistical Mathematics* **30A**, 9–14.

Akaike, H. (1985) Prediction and entropy. In *A Celebration of Statistics. The ISI Centenary Volume*, eds A. C. Atkinson & S. E. Fienberg, pp. 1–24. New York: Springer.

Albert, A. & Anderson, J. A. (1984) On the existence of maximum likelihood estimates in logistic regression models. *Biometrika* **71**, 1–10.

Albert, A. & Lesaffre, E. (1986) Multiple group logistic discrimination. *Computers and Mathematics with Applications* **12A**, 209–224.

Albertini, F., Sontag, E. D. & Maillot, V. (1993) Uniqueness of weights for neural networks. In Mammone (1993), pp. 115–125.

Aleksander, I. & Morton, H. (1990) *An Introduction to Neural Computing*. London: Chapman & Hall.

Alexander, K. S. (1984) Probability inequalities for empirical processes and a law of the iterated logarithm. *Annals of Probability* **12**, 1041–1067.

Almond, R. G. (1995) *Graphical Belief Modeling*. London: Chapman & Hall.

Amari, S.-I. (1967) A theory of adaptive pattern classifiers. *IEEE Transactions on Electronic Computers* **16**, 299–307.

Amari, S.-I. (1993) Mathematical methods of neurocomputing. In *Networks and Chaos—Statistical and Probabilistic Aspects*, eds O. E. Barndorff-Nielsen, J. L. Jensen & W. S. Kendall, pp. 1–39. London: Chapman & Hall.

Amit, D. J. (1989) *Modeling Brain Function. The World of Attractor Neural Networks*. Cambridge: Cambridge University Press.

Anderberg, M. R. (1973) *Cluster Analysis for Applications*. New York: Academic Press.

Andersen, S. K., Olesen, K. G., Jensen, F. V. & Jensen, F. (1989) HUGIN—a shell for building Bayesian belief universes for expert systems. In *Proceedings of the 11th International Joint Conference on Artificial Intelligence*, pp. 1080–1085. San Mateo, CA: Morgan Kaufmann. Reprinted in Shafer & Pearl (1990).

Anderson, J. A. (1972) Separate sample logistic discrimination. *Biometrika* **59**, 19–35.

Anderson, J. A. (1982) Logistic discrimination. In *Handbook of Statistics 2: Classification, Pattern Recognition and Reduction of Dimensionality*, eds P. R. Krishnaiah & L. N. Kanal, Amsterdam: North Holland, pp. 169–191.

Anderson, J. A. & Phillips, P. R. (1981) Regression, discrimination and measurement models for ordered categorical variables. *Applied Statistics* **30**, 22–31.

Anderson, J. A. & Rosenfeld, E. (eds) (1988) *Neurocomputing: Foundations of Research*. Cambridge, MA: The MIT Press.

Anderson, J. A., Pellionisz, A. & Rosenfeld, E. (eds) (1990) *Neurocomputing 2: Directions for Research.* Cambridge, MA: The MIT Press.

Anderson, T. W. (1984) *An Introduction to Multivariate Statistical Analysis.* Second edition. New York: Wiley.

Anderson, T. W. & Bahadur, R. R. (1962) Classification into two multivariate normal distributions with different covariance matrices. *Annals of Mathematical Statistics* **33**, 420–431.

Andreassen, S., Jensen, F. V. & Olesen, K. G. (1991) Medical expert systems based on causal probabilistic networks. *International Journal of Biomedical Computing* **28**, 1–30.

Angluin, D. (1987) Learning regular sets from queries and counterexamples. *Information and Computation* **75**, 87–106.

Angluin, D. (1988) Queries and concept learning. *Machine Learning* **2**, 319–342.

Angluin, D. (1993) Learning with queries. In *Computational Learning and Cognition*, ed. E. B. Baum, pp. 1–28. Philadelphia: SIAM.

Angluin, D. & Valiant, L. G. (1979) Fast probabilistic algorithms for Hamiltonian circuits and matchings. *Journal of Computer and System Sciences* **18**, 155–193.

Anthony, M. & Biggs, N. (1992) *Computational Learning Theory: An Introduction.* Cambridge: Cambridge University Press.

Anthony, M. & Shawe-Taylor, J. (1993) A result of Vapnik with applications. *Discrete Applied Mathematics* **47**, 207–217. Erratum (1994) **52**, 211 (the proof of theorem 2.1 is corrected).

Apolloni, B. & de Falco, D. (1991) Learning by asymmetric parallel Boltzmann Machines. *Neural Computation* **3**, 402–408.

Argentiero, P., Chin, R. & Beaudet, P. (1982) An automated approach to the design of decision tree classifiers. *IEEE Transactions on Pattern Analysis and Machine Intelligence* **4**, 51–57.

Arbib, M. A. (ed.) (1995) *The Handbook of Brain Theory and Neural Networks.* Cambridge, MA: MIT Press.

Arkedev, A. G. & Braverman, E. M. (1966) *Computers and Pattern Recognition.* Washington, DC: Thompson.

Ash, T. (1989) Dynamic mode creation in backpropagation neural networks. *Connection Science: Journal of Neural Computing, Artificial Intelligence and Cognitive Research* **1**, 365–375.

Asimov, D. (1985) The grand tour: a tool for viewing multidimensional data. *SIAM Journal on Scientific and Statistical Computing* **6**, 128–143.

Assouad, P. (1983) Densité et dimension. *Annales de l'Institut Fourier Grenoble* **33**, 233–282.

Averintsev, M. V. (1975) Gibbs description of random fields whose conditional probabilities may vanish. *Problemy Peredaci Informatsii* **11**, 86–96.

Baba, N., Mogami, Y., Kohzaki, M., Shiraishi, Y. & Yoshida, Y. (1994) A hybrid algorithm for finding the global minimum of error function of neural networks and its applications. *Neural Networks* **7**, 1253–1265.

Bahadur, R. R. (1961a) A representation of the joint distribution of responses to *n* dichotomous items. In *Studies in Item Analysis and Prediction*, ed. H. Solomon, pp. 158–167. Palo Alto, CA: Stanford University Press.

Bahadur, R. R. (1961b) On classification based on responses to *n* dichotomous items. In *Studies in Item Analysis and Prediction*, ed. H. Solomon, pp. 169–176. Palo Alto, CA: Stanford University Press.

Bahl, L. R., Brown, P. F., de Souza, P. V. & Mercer, R. L. (1989) A tree-based statistical language model for natural language speech recognition. *IEEE Transactions on Acoustics, Speech and Signal Processing* **37**, 1001–1008.

Bailey, T. & Jain, A. K. (1978) A note on distance-weighted *k*-nearest neighbor rules. *IEEE Transactions on Systems, Man and Cybernetics* **8**, 311–313.

Baird, H. S. (1993) Recognition technology frontiers. *Pattern Recognition Letters* **14**, 327–334.

Baldi, P. & Hornik, K. (1989) Neural networks and principal components analysis: learning from examples without local minima. *Neural Networks* **2**, 53–58. Reprinted in Anderson *et al.* (1990).

Ball, G. B. (1965) Data analysis in the social sciences: what about the details? In *Proceedings of the Fall Joint Computing Conference*, pp. 533–559. Washington, DC: Spartan Books.

Banfield, J. D. & Raftery, A. E. (1993) Model-based Gaussian and non-Gaussian clustering. *Biometrics* **49**, 803–821.

Barlow, R. E., Bartholomew, D., Bremner, J. E. & Brunk, H. M. (1972) *Statistical Inference under Order Restrictions. The Theory and Application of Isotonic Regression.* London: Wiley.

Barron, A. R. (1990) Complexity regularization with application to artificial neural networks. In *Nonparametric Functional Estimation and Related Topics*, ed. G. Roussas, pp. 561–576. Dordrecht: Kluwer Academic Publishers.

Barron, A. R. (1993) Universal approximation bounds for superpositions of a sigmoid function. *IEEE Transactions on Information Theory* **39**, 930–945.

Barron, A. R. (1994) Approximation and estimation bounds for artificial neural networks. *Machine Learning* **14**, 115–133.

Barron, A. R. & Cover, T. M. (1991) Minimum complexity density estimation. *IEEE Transactions on Information Theory* **37**, 1034–1054.

Barry, D. (1986) Nonparametric Bayesian regression. *Annals of Statistics* **14**, 934–953.

Bartlett, P. L. (1993) Vapnik–Chervonenkis dimension bounds for two- and three-layer networks. *Neural Computation* **5**, 371–373.

Bartlett, P. L. & Williamson, R. C. (1996) The VC dimension and pseudodimension of two-layer neural networks with discrete inputs. *Neural Computation* **8**, 625–628.

Basford, K. E. & McLachlan, G. J. (1985) Estimation of allocation rates in a cluster analysis context. *Journal of the American Statistical Association* **80**, 286–293.

Bashkirov, O. A., Braverman, E. M. & Muchnik, I. B. (1964) Potential function algorithms for pattern recognition learning machines. *Automation and Remote Control* **25**, 629–631.

Bates, D. M. & Watts, D. G. (1988) *Nonlinear Regression Analysis and its Applications*. New York: Wiley.

Bather, J. (1996) A conversation with Herman Chernoff. *Statistical Science* **11**, 335–350.

Battiti, R. (1989) Accelerated backpropagation learning: two optimization methods. *Complex Systems* **3**, 331–342.

Battiti, R. (1992) First- and second-order methods for learning: between steepest descent and Newton's method. *Neural Computation* **4**, 141–166.

Battiti, R. & Massuli, F. (1990) BFGS optimization for faster and automated supervised learning. In *Proceedings of the International Neural Network Conference (Paris, 1990)* **2**, 757–760.

Baum, E. B. (1988) On the capabilities of multilayer perceptrons. *Journal of Complexity* **4**, 193–215.

Baum, E. B. & Haussler, D. (1989) What size net gives valid generalization? *Neural Computation* **1**, 151–160. Reprinted in Shavlik & Dietterich (1990).

Baum, L. E., Petrie, T., Soules, G. & Weiss, N. (1970) A maximization technique occurring in the statistical analysis of probabilistic functions of Markov chains. *Annals of Mathematical Statistics* **41**, 164–171.

Baxt, W. G. (1992) Improving the accuracy of an artificial neural network using multiple differently trained networks. *Neural Computation* **4**, 772–780.

Beaulieu, J.-M. & Goldberg, M. (1989) Hierarchy in picture segmentation: a stepwise optimization approach. *IEEE Transactions on Pattern Analysis and Machine Intelligence* **11**, 150–163.

Beeri, C., Fagin, R., Maier, D. & Yannakakis, M. (1983) On the desirability of acyclic database schemes. *Journal of the Association for Computing Machinery* **30**, 479–513.

Begg, C. B. & Gray, R. (1984) Calculation of polychotomous logistic regression parameters using individualized regressions. *Biometrika* **71**, 11–18.

Beigi, H. S. M. & Li, C. J. (1990) Learning algorithms for neural networks based on quasi-Newton with self-scaling. *Intelligent Control Systems* **23**, 23–28.

Beigi, H. S. M. & Li, C. J. (1993) Learning algorithms for neural networks based on quasi-Newton with self-scaling. *Journal of Dynamical Systems, Measurement, and Control – Transactions of the ASME* **115**, 38–43.

Benediktsson, J. A. & Swain, P. H. (1992) Consensus theoretic classification methods. *IEEE Transactions on Systems, Man and Cybernetics* **22**, 688–704.

Berge, C. (1973) *Graphs and Hypergraphs*. Amsterdam: North-Holland.

Berger, J. O. (1985) *Statistical Decision Theory and Bayesian Analysis*. New York: Springer.

Berger, J. O. & Delampady, M. (1987) Testing precise hypotheses (with discussion). *Statistical Science* **2**, 317–352.

Bernardo, J. M. & Smith, A. F. M. (1994) *Bayesian Theory*. Chichester: Wiley.

Besag, J., & Green, P. J. (1993) Spatial statistics and Bayesian computation (with discussion). *Journal of the Royal Statistical Society series B* **55**, 25–37.

Besag, J., Green, P., Higdon, D. & Mengersen, K. (1995) Bayesian computation and stochastic systems (with discussion). *Statistical Science* **10**, 3–66.

Best, D. J. & Rayner, J. C. W. (1988) A test for bivariate normality. *Statistics and Probability Letters* **6**, 407–412.

Bezdek, J. C. (1974) Cluster validity with fuzzy sets. *Journal of Cybernetics* **3**, 58–72.

Bhattacharyya, A. (1943) On a measure of divergence between two statistical populations defined by their probability distributions. *Bulletin of the Calcutta Mathematics Society* **35**, 99–110.

Bichsel, M. & Seitz, P. (1989) Minimum class entropy: a maximum information approach to layered networks. *Neural Networks* **2**, 133–141.

Bienenstock, E., Cooper, L. N. & Munro, W. (1982) Theory for the development of neuron selectivity: orientation specificity and binocular interaction in the visual cortex. *Journal of Neuroscience* **2**, 32–48. Reprinted in Anderson & Rosenfeld (1988).

Binford, T. O., Levitt, T. S. & Mann, W. B. (1989) Bayesian inference in model-based machine vision. In *Uncertainty in Artificial Intelligence 3*, eds L. N. Kanal, T. S. Levitt & J. F. Lemmer. Amsterdam: Elsevier.

Bishop, C. (1991) Improving the generalization properties of radial basis function neural networks. *Neural Computation* **3**, 579–588.

Bishop, C. (1992) Exact calculation of the Hessian matrix for the multilayer perceptron. *Neural Computation* **4**, 494–501.

Bishop, C. M. (1993) Curvature-driven smoothing: a learning algorithm for feedforward networks. *IEEE Transactions on Neural Networks* **4**, 882–884.

Bishop, C. M. (1995a) *Neural Networks for Pattern Recognition*. Oxford: Clarendon Press.

Bishop, C. M. (1995b) Training with noise is equivalent to Tikohonov regularization. *Neural Computation* **7**, 108–116.

Blair, J. R. S. & Peyton, B. (1993) An introduction to chordal graphs and clique trees. In *Graph Theory and Sparse Matrix Computations*, eds A. George, J. R. Gilbert & J. H. U. Liu, pp. 1–29. New York: Springer.

Block, H. D. (1962) The perceptron: a model for brain functioning I. *Reviews of Modern Physics* **34**, 123–135. Reprinted in Anderson & Rosenfeld (1988).

Block, H. D. & Levin, S. A. (1970) On the boundedness of an iterative procedure for solving a system of linear inequalities. *Proceedings of the American Mathematical Society* **26**, 229–235.

Block, H. D., Knight, B. W. Jr & Rosenblatt, F. (1962) Analysis of a four-layer series-coupled perceptron II. *Reviews of Modern Physics* **34**, 135–142.

Blue, J. L., Candela, G. T., Grother, P. J. and Wilson, C. L. (1994) Evaluation of pattern classifiers for fingerprint and OCR applications. *Pattern Recognition* **27**, 485–501.

Blumer, A., Ehrenfeucht, A., Haussler, D. & Warmuth, M. K. (1987) Occam Razor. *Information Processing Letters* **24**, 377–280. Reprinted in Shavlik & Dietterich (1990).

Blumer, A., Ehrenfeucht, A., Haussler, D. & Warmuth, M. K. (1989) Learnability and the Vapnik–Chervonenkis dimension. *Journal of the Association for Computing Machinery* **36**, 926–965.

de Boor, C. (1978) *A Practical Guide to Splines*. New York: Springer.

Bourlard, H. & Kamp, Y. (1988) Auto-association by multilayer perceptrons and singular value decomposition. *Biological Cybernetics* **59**, 291–294.

Bouton, C. & Pagès, G. (1993) Self-organization of the one-dimensional Kohonen algorithm with non-uniformly distributed stimuli. *Stochastic Processes and their Applications* **47**, 249–274.

Bouton, C. & Pagès, G. (1994) Convergence in distribution of the one-dimensional Kohonen algorithms when the stimuli are not uniform. *Advances in Applied Probability* **26**, 80–103.

Box, G. E. P. & Tiao, G. C. (1962) A further look at robustness via Bayes's theorem. *Biometrika* **49**, 419–432.

Box, G. E. P. & Tiao, G. C. (1973) *Bayesian Inference in Statistical Analysis*. New York: Wiley. (Formerly Reading, MA: Addison-Wesley.)

Box, G. E. P., Hunter, W. G. & Hunter, J. S. (1978) *Statistics for Experimenters: An Introduction to Design, Data Analysis and Model Building*. New York: Wiley.

Boyles, R. A. (1983) On the convergence of the EM algorithm. *Journal of the Royal Statistical Society series B* **45**, 47–50.

Bratko, I. & Kononenko, I. (1987) Learning diagnostic rules from incomplete and noisy data. In *Interactions in Artificial Intelligence and Statistical Methods*, ed. B. Phelps, pp. 142–153. Aldershot: Gower Technical Press.

Bratko, I. & Muggleton, S. (1995) Applications of inductive logic programming. *Communications of the Association for Computing Machinery* **38**, 65–70.

Braverman, E. M. (1965) On the method of potential functions. *Automation and Remote Control* **26**, 2130–2138.

Breiman, L. (1991) The Π-method for estimating multivariate functions from noisy data (with discussion). *Technometrics* **33**, 125–160.

Breiman, L. (1992) Stacked regressions. Technical Report 367, Dept of Statistics, University of California, Berkeley.

Breiman, L. (1993) Hinging hyperplanes for regression, classification and function approximation. *IEEE Transactions on Information Theory* **3**, 999–1013.

Breiman, L. & Ihaka, R. (1984) Nonlinear discriminant analysis via ACE and scaling. Technical Report 40, Dept of Statistics, University of California, Berkeley.

Breiman, L., Friedman, J. H., Olshen, R. A. & Stone, C. J. (1984) *Classification and Regression Trees*. Monterey, CA: Wadsworth and Brooks/Cole.

Brent, R. P. (1991) Fast training algorithms for multi-layer neural nets. *IEEE Transactions on Neural Networks* **2**, 346–354.

Bridle, J. S. (1990a) Probabilistic interpretation of feedforward classification network outputs, with relationships to statistical pattern recognition. In *Neuro-computing: Algorithms, Architectures and Applications*, eds F. Fogelman Soulié & J. Hérault, pp. 227–236. Berlin: Springer. (Although the volume was published in 1990, the article gives a 1989 copyright date.)

Bridle, J. S. (1990b) Training stochastic model recognition algorithms as networks can lead to maximum mutual information estimation of parameters. In *NIPS2*, pp. 211–217.

Bridle, J. S. & Cox, S. J. (1991) RecNorm: simultaneous normalisation and classification applied to speech recognition. In *NIPS3*, pp. 234–240.

Brier, G. W. (1950) Verification of forecasts expressed in terms of probabilities. *Monthly Weather Review* **78**, 1–3.

Brockett, R. W. (1991) Dynamical systems that sort lists, diagonalize matrices and solve linear programming problems. *Linear Algebra and its Applications* **146**, 79–91.

Broffit, B., Clarke, W. R. & Lachenbruch, P. A. (1980) The effect of Huberizing and trimming on the quadratic discriminant function. *Communications in Statistics—Theory and Methods* **A9**, 13–25.

Bronowski, J. & Long, W. M. (1951) Statistical methods in anthropology. *Nature* **1168**, 794.

Broomhead, D. S. & Lowe, D. (1988) Multivariable functional interpolation and adaptive networks. *Complex Systems* **2**, 321–355.

Brown, P. J. & Rundell, P. W. K. (1985) Kernel estimates for categorical data. *Technometrics* **28**, 293–299.

Brown, T. A. & Koplowitz, J. (1979) The weighted nearest neighbor rule for class dependent sample sizes. *IEEE Transactions on Information Theory* **25**, 617–619.

Bryan, J. G. (1951) The generalized discriminant function: mathematical foundations and computational routine. *Harvard Educational Review* **21**, 90–95.

Bryant, J. (1989) A fast classifier for image data. *Pattern Recognition* **22**, 45–48.

Bryson, A. E. & Ho, Y.-C. (1969) *Applied Optimal Control*. New York: Blaisdell. (Revised printing New York: Hemisphere, 1975.)

Buckland, S. T. (1992a) Fitting density functions with polynomials. *Applied Statistics* **41**, 63–76.

Buckland, S. T. (1992b) Algorithm AS270. Maximum likelihood fitting of Hermite and simple polynomial densities. *Applied Statistics* **41**, 241–266.

Buckley, A. G. (1994) A Fortran-90 code for unconstrained nonlinear minimization. *ACM Transactions on Mathematical Software* **20**, 354–372.

Buntine, W. L. (1992) Learning classification trees. *Statistics and Computing* **2**, 63–73.

Buntine, W. L. & Weigend, A. S. (1991) Bayesian back-propagation. *Complex Systems* **5**, 603–643.

Buntine, W. L. & Weigend, A. S. (1994) Calculating second derivatives on feed-forward networks: a review. *IEEE Transactions on Neural Networks* **5**, 480–488.

Burrascano, P. (1991) Learning vector quantization for the probabilistic neural network. *IEEE Transactions on Neural Networks* **2**, 458–461.

Byrd, R. H., Nocedal, J. & Schnabel, R. B. (1994) Representations of quasi-Newton matrices and their use in limited memory methods. *Mathematical Programming* **63**, 129–156.

Byth, K. & McLachlan, G. J. (1978) The biases associated with maximum likelihood methods of estimation of the multivariate logistic risk function. *Communications in Statistics—Theory and Methods* **A7**, 877–890.

Cacoullos, T. (1966) Estimation of a multivariate density. *Annals of the Institute of Statistical Mathematics* **18**, 179–189.

Campbell, N. A. (1980a) Shrunken estimators in discriminant and canonical variate analysis. *Applied Statistics* **29**, 5–14.

Campbell, N. A. (1980b) Robust procedures in multivariate analysis I. Robust covariance estimation. *Applied Statistics* **29**, 231–237.

Campbell, N. A. (1982) Robust procedures in multivariate analysis II. Robust canonical variate analysis. *Applied Statistics* **31**, 1–8.

Campbell, N. A. & Mahon, R. J. (1974) A multivariate study of variation in two species of rock crab of genus *Leptograpsus*. *Australian Journal of Zoology* **22**, 417–425.

Candela, G. T. & Chellappa, R. (1993) Comparative performance of classification methods for fingerprints. US National Institute of Standards and Technology report NISTIR 5163.

Cannings, C. & Thompson, E. A. (1981) *Genealogical and Genetic Structure*. Cambridge: Cambridge University Press.

Cannings, C., Thompson, E. A. & Skolnick, M. H. (1978) Probability functions on complex pedigrees. *Advances in Applied Probability* **10**, 26–61.

Carbonell, J. G. (ed.) (1990) *Machine Learning. Paradigms and Methods*. Cambridge, MA: The MIT Press.

Carpenter, G. A. & Grossberg, S. (1987a) A massively parallel architecture for a self-organizing neural pattern recognition machine. *Computer Vision, Graphics, and Image Processing* **37**, 54–115.

Carpenter, G. A. & Grossberg, S. (1987b) ART 2: stable self-organization of stable category recognition codes for analog input patterns. *Applied Optics* **26**, 4919–4930. Reprinted in Anderson *et al.* (1990).

Carpenter, G. A. & Grossberg, S. (1990) ART 3: hierarchical search using chemical transmitters in self-organizing pattern recognition architectures. *Neural Networks* **3**, 129–152.

Carpenter, G. A. & Grossberg, S. (1994) Self-organizing neural networks for supervised and unsupervised learning and prediction. In Cherkassky *et al.* (1994), pp. 319–348.

Carpenter, G. A., Grossberg, S. & Reynolds, J. H. (1991a) ARTMAP: supervised real-time learning and classification of nonstationary data by a self-organizing neural network. *Neural Networks* **4**, 565–588.

Carpenter, G. A., Grossberg, S. & Rosen, D. B. (1991b) Fuzzy ART: fast stable learning and categorization of analog patterns by an adaptive resonance system. *Neural Networks* **4**, 759–771.

Carpenter, G. A., Grossberg, S., Markuzon, N., Reynolds, J. H. & Rosen, D. B. (1992) Fuzzy ARTMAP: a neural network architecture for incremental supervised learning of analog multidimensional maps. *IEEE Transactions on Neural Networks* **3**, 698–713.

Carroll, S. M. & Dickinson, B. W. (1989) Construction of neural nets using the Radon transform. In *Proceedings of the International Joint Conference on Neural Networks* **I**, 607–611. New York: IEEE Press.

Carter, C. & Catlett, J. (1987) Assessing credit card applications using machine learning. *IEEE Expert* **2(3)**, 71–79.

Casey, R. G & Nagy, G. (1984) Decision tree design using a probabilistic model. *IEEE Transactions on Information Theory* **30**, 93–99.

Celeux, G. & Diebolt, J. (1992) A stochastic approximation type EM algorithm for the mixture problem. *Stochastics and Stochastics Reports* **41**, 119–134.

Cestnik, B., Kononenko, I. & Bratko, I. (1987) ASSISTANT 86: a knowledge-elicitation tool for sophisticated users. In *Progress in Machine Learning*, eds I. Bratko & N. Lavrač, pp. 31–45. Wilmslow: Sigma Press.

Chambers, J. M. & Hastie, T. J. (eds) (1992) *Statistical Models in S*. Pacific Grove, CA: Wadsworth and Brooks/Cole.

Chan, C. & Bao, J. (1991) On the design of a tree classifier and its application to speech recognition. *International Journal of Pattern Recognition and Artificial Intelligence* **5**, 677–692.

Chan, K. S. (1993) Asymptotic behaviour of the Gibbs sampler. *Journal of the American Statistical Association* **88**, 320–326.

Chandran, P. S. (1994) Comments on "Comparative analysis of backpropagation and the extended Kalman filter for training multilayer perceptrons". *IEEE Transactions on Pattern Analysis and Machine Intelligence* **16**, 862–863.

Chang, C. L. (1974) Finding prototypes for nearest neighbor classifiers. *IEEE Transactions on Computers* **23**, 1179–1184. Reprinted in Dasarathy (1991).

Charniak, E. (1991) Bayesian networks without tears. *AI Magazine* **12(4)**, 50–63.

Chauvin, Y. (1989) A back-propagation algorithm with optimal use of hidden units. In *NIPS1*, pp. 519–526.

Chavez, R. M. & Cooper, G. F. (1990) A randomized approximation algorithm for probabilistic inference on Bayesian belief networks. *Networks* **20**, 661–685.

Cheeseman, P. (1995) On Bayesian model selection. In Wolpert (1995), pp. 315–330.

Cheeseman, P., Kelly, J., Self, M., Stutz, J., Taylor, W. & Freeman, D. (1988a) AutoClass: a Bayesian classification system. In *Proceedings of the Fifth International Workshop on Machine Learning, Ann Arbor*, pp. 54–64. San Mateo, CA: Morgan Kaufmann.

Cheeseman, P., Self, M., Kelly, J., Taylor, W., Freeman, D. & Stutz, J. (1988b) Bayesian classification. In *Proceedings of the Seventh AAAI National Conference on Artificial Intelligence, St Paul, MN*, pp. 607–611. San Mateo, CA: Morgan Kaufmann.

Chen, D. S. & Jain, R. C. (1994) A robust back propagation learning algorithm for function approximation. *IEEE Transactions on Neural Networks* 5, 467–479.

Chen, S., Cowan, C. F. N. & Grant, P. M. (1991) Orthogonal least squares learning algorithm for radial basis function networks. *IEEE Transactions on Neural Networks* 2, 302–309.

Cheng, Y.-Q., Zhuang, Y.-M. & Yang, J.-Y. (1992) Optimal Fisher discriminant analysis using the rank decomposition. *Pattern Recognition* 25, 101–111.

Cherkassky, V. & Mulier, F. (1994) Self-organizing networks for nonparametric regression. In Cherkassky et al. (1994), pp. 188–212.

Cherkassky, V., Friedman, J. H. & Wechsler, H. (eds) (1994) *From Statistics to Neural Networks. Theory and Pattern Recognition Applications*. Berlin: Springer.

Chernick, M. R., Murthy, V. K. & Nealy, C. D. (1985) Application of bootstrap and other resampling techniques: evaluation of classifier performance. *Pattern Recognition Letters* 3, 167–178.

Chernoff, H. (1952) A measure of asymptotic efficiency for tests of a hypothesis based on the sum of observations. *Annals of Mathematical Statistics* 23, 493–507.

Chernoff, H. (1973) Some measures for discriminating between normal multivariate distributions with unequal covariance matrices. In *Multivariate Analysis III*, ed. P. R. Krishnaiah, pp. 337–344. New York: Academic Press.

Chidananda Gowda, K. & Krishna, G. (1979) The condensed nearest neighbor rule using the concept of mutual nearest neighborhood. *IEEE Transactions on Information Theory* 25, 488–490. Reprinted in Dasarathy (1991).

Chou, P. A. (1989) Recognition of equations using a two-dimensional stochastic context-free grammar. In *Visual Communications and Image Processing IV*, ed. W. A. Pearlman. SPIE Proceedings Series 1199, 852–863.

Chou, P. A. (1991) Optimal partitioning for classification and regression trees. *IEEE Transactions on Pattern Analysis and Machine Intelligence* 13, 340–354.

Chou, W.-S. & Chen, Y.-C. (1992) A new fast algorithm for the effective training of neural classifiers. *Pattern Recognition* 25, 423–429.

Chow, C. K. (1970) On optimum recognition error and reject tradeoff. *IEEE Transactions on Information Theory* 16, 41–46.

Chow, C. K. & Liu, C. N. (1968) Approximating discrete probability distributions with dependence trees. *IEEE Transactions on Information Theory* 14, 462–467.

Ciampi, A., Chang, C.-H., Hogg, S. & McKinney, S. (1987) Recursive partition: a versatile method for exploratory data analysis in biostatistics. In *Biostatistics*, eds I. B. MacNeil & G. J. Umphrey, pp. 23–50. Dordrecht: Reidel.

Clark, L. A. & Pregibon, D. (1992) Tree-based models. Chapter 9 of Chambers & Hastie (1992).

Clark, P. & Niblett, T. (1989) The CN2 induction algorithm. *Machine Learning* 3, 261–283.

Cleveland, W. S., Grosse, E. & Shyu, W. M. (1992) Local regression models. Chapter 8 of Chambers & Hastie (1992).

Clifford, P. (1990) Markov random fields in statistics. In *Disorder in Physical Systems. A Volume in Honour of John M. Hammersley*, eds G. R. Grimmett & D. J. A. Welsh, pp. 19–32. Oxford: Clarendon Press.

Clunies-Ross, C. W. & Riffenburgh, R. H. (1960) Geometry and linear discrimination. *Biometrika* 47, 185–189.

Cohen, E., Hull, J. J. & Srihari, S. N. (1991) Understanding handwritten text in a structured environment: determining ZIP codes from addresses. *International Journal of Pattern Recognition and Artificial Intelligence* 5, 221–264.

Cohn, D. & Tesauro, G. (1992) How tight are the Vapnik–Chervonenkis bounds? *Neural Computation* 4, 249–269.

Cook, D., Buja, A. & Cabrera, J. (1993) Projection pursuit indices based on orthonormal function expansions. *Journal of Computational and Graphical Statistics* 2, 225–250.

Coomans, D. & Broeckaert, I. (1986) *Potential Pattern Recognition in Chemical and Medical Decision Making.* Letchworth: Research Studies Press.

Cooper, G. F. (1984) *NESTOR: a computer-based medical diagnostic aid that integrates causal and probabilistic knowledge,* Ph.D. thesis, Dept of Computer Science, Stanford University.

Cooper, G. F. (1989) Current research directions in the development of expert systems based on belief networks. *Applied Stochastic Models and Data Analysis* **5**, 39–52.

Cooper, G. F. (1990) The computational complexity of probabilistic inference using Bayesian belief networks. *Artificial Intelligence* **42**, 393–405.

Cooper, G. F. & Herskovits, E. (1992) A Bayesian method for the induction of probabilistic networks from data. *Machine Learning* **9**, 309–347.

Cormen, T. H., Leiserson, C. E. & Rivest, R. L. (1990) *Introduction to Algorithms.* Cambridge MA: The MIT Press and New York: McGraw-Hill.

Cortes, C. & Vapnik, V. (1995) Support-vector networks. *Machine Learning* **20**, 273–297.

Cosslett, S. R. (1981) Maximum likelihood estimators for choice-based samples. *Econometrica* **49**, 1289–1316.

Cottrell, G. W. & Metcalfe, J. (1991) EMPATH: face, emotion and gender recognition using holons. In *NIPS3*, pp. 564–571.

Cottrell, M. & Fort, J. C. (1987) Etude d'un algorithme d'auto-organisation. *Annales de l'Institut Henri Poincaré* **23**, 1–20.

Cover, T. M. (1965) Geometrical and statistical properties of systems of linear inequalities with applications in pattern recognition. *IEEE Transactions on Electronic Computers* **14**, 326–334.

Cover, T. M. (1968) Rates of convergence of nearest neighbor procedures. In *Proceedings of the First Annual Hawaii Conference on Systems Theory, Honolulu*, pp. 413–418.

Cover, T. M. (1969) Learning in pattern recognition. In *Methodologies of Pattern Recognition*, ed. S. Watanabe, pp. 111–132. New York: Academic Press.

Cover, T. M. & Hart, P. E. (1967) Nearest neighbor pattern classification. *IEEE Transactions on Information Theory* **13**, 21–27. Reprinted in Anderson *et al.* (1990), Dasarathy (1991) and Lau (1992).

Cowell, R. G. (1992) BAIES—a probabilistic expert reasoning shell with qualitative and quantitative learning. In *Bayesian Statistics 4*, eds J. M. Bernardo, J. O. Berger, A. P. Dawid & A. F. M. Smith, pp. 595–600. Oxford: Clarendon Press.

Cowell, R. G. (1995) A C++ class library for building Bayesian belief networks. In Gammerman (1995), pp. 159–165.

Cowell, R. G. & Dawid, A. P. (1992) Fast retraction of evidence in a probabilistic expert system. *Statistics and Computing* **2**, 37–40.

Cox, D. R. (1958) Two further applications of a model for binary regression. *Biometrika* **45**, 562–565.

Cox, D. R. & Hinkley, D. V. (1974) *Theoretical Statistics.* London: Chapman & Hall.

Cox, D. R. & Snell, E. J. (1989) *Analysis of Binary Data.* Second edition. London: Chapman & Hall.

Cox, T. F. & Cox, M. A. A. (1994) *Multidimensional Scaling.* London: Chapman & Hall.

Craven, P. & Wahba, G. (1979) Smoothing noisy data with spline functions: estimating the correct degree of smoothing by the method of generalized cross-validation. *Numerische Matematik* **31**, 377–403.

Crawford, S. L. (1989) Extensions to the CART algorithm. *International Journal of Man–Machine Studies* **31**, 197–217.

Crevier, D. (1993) *AI. The Tumultuous History of the Search for Artificial Intelligence.* New York: Basic Books.

Cybenko, G. (1988) Continuous valued neural networks with two hidden layers are sufficient. Technical Report, Dept of Computer Science, Tufts University.

Cybenko, G. (1989) Approximation by superpositions of a sigmoidal function. *Mathematics of Control Signals, and Systems* **2**, 303–314.

Darken, C. & Moody, J. (1991) Note on learning rate schedules for stochastic optimization. In *NIPS3*, pp. 832–838.

Dasarathy, B. V. (ed.) (1991) *Nearest Neighbor (NN) Norms: NN Pattern Classification Techniques.* Los Alamitos, CA: IEEE Computer Society Press.

Dattatreya, G. R. & Sarma, V. V. S. (1981) Bayesian and decision tree approaches for pattern recognition including feature measurement costs. *IEEE Transactions on Pattern Analysis and Machine Intelligence* **3**, 293–298.

Dattatreya, G. R. & Sarma, V. V. S. (1985) Decision trees in pattern recognition. In *Progress in Pattern Recognition 2*, eds L. N. Kanal & A. Rosenfeld. Amsterdam: Elsevier.

Dawid, A. P. (1976) Properties of diagnostic data distributions. *Biometrics* **32**, 647–658.

Dawid, A. P. (1979) Conditional independence in statistical theory (with discussion). *Journal of the Royal Statistical Society series B* **41**, 1–31.

Dawid, A. P. (1980) Conditional independence for statistical operations. *Annals of Statistics* **8**, 598–617.

Dawid, A. P. (1982) The well-calibrated Bayesian (with discussion). *Journal of the American Statistical Association* **77**, 605–613.

Dawid, A. P. (1986) Probability forecasting. In *Encyclopedia of Statistical Sciences*, eds S. Kotz, N. L. Johnson & C. B. Read, pp. 210–218. New York: Wiley.

Dawid, A. P. (1992) Applications of a general propagation algorithm for probabilistic expert systems. *Statistics and Computing* **2**, 25–36.

Dawid, A. P. & Lauritzen, S. L. (1993) Hyper Markov laws in the statistical analysis of decomposable graphical models. *Annals of Statistics* **21**, 1272–1317.

Deely, J. J. & Lindley, D. V. (1981) Bayes empirical Bayes. *Journal of the American Statistical Association* **76**, 833–841.

DeMers, D. & Cottrell, G. (1993) Non-linear dimensionality reduction. In *NIPS5*, pp. 580–587.

Dempster, A. P. & Kong, A. (1988) Uncertain evidence and artificial analysis. *Journal of Statistical Planning and Inference* **20**, 355–368. Reprinted in Shafer & Pearl (1990).

Dempster, A. P., Laird, N. M. & Rubin, D. B. (1977) Maximum likelihood from incomplete data via the EM algorithm (with discussion). *Journal of the Royal Statistical Society series B* **39**, 1–38.

Dennis, J. E. & Schnabel, R. B. (1983) *Numerical Methods for Unconstrained Optimization and Nonlinear Equations*. Englewood Cliffs, NJ: Prentice-Hall.

Devijver, P. A. & Kittler, J. V. (1982) *Pattern Recognition. A Statistical Approach*. Englewood Cliffs, NJ: Prentice-Hall.

Devijver, P. A. & Kittler, J. V. (eds) (1987) *Pattern Recognition Theory and Applications*. Berlin: Springer.

Devlin, S. J., Gnanadesikan, R. & Kettenring, J. R. (1981) Robust estimation of dispersion matrices and principal components. *Journal of the American Statistical Association* **76**, 354–362.

DeVore, R. A., Howard, R. & Micchelli, C. A. (1989) Optimal nonlinear approximation. *Manuscripta Mathematica* **63**, 469–478.

Devroye, L. (1981a) On the inequality of Cover and Hart in nearest neighbor discrimination. *IEEE Transactions on Pattern Analysis and Machine Intelligence* **3**, 75–78.

Devroye, L. (1981b) On the almost everywhere convergence of nonparametric regression function estimates. *Annals of Statistics* **9**, 1310–1319.

Devroye, L. (1982) Bounds for the uniform deviation of empirical measures. *Journal of Multivariate Analysis* **12**, 72–79.

Devroye, L. (1988) Automatic pattern recognition: a study of the probability of error. *IEEE Transactions on Pattern Analysis and Machine Intelligence* **10**, 530–543.

Diaconis, P. & Freedman, D. (1984) Asymptotics of graphical projection pursuit. *Annals of Statistics* **12**, 793–815.

Diaconis, P. & Shahshahani, M. (1984) On non-linear functions of linear combinations. *SIAM Journal on Scientific and Statistical Computing* **5**, 175–191.

Diebolt, J. & Robert, C. P. (1994) Estimation of finite mixture distributions through Bayesian sampling. *Journal of the Royal Statistical Society series B* **56**, 363–375.

Dietterich, T. G. (1990) Machine learning. *Annual Review of Computer Science* **4**, 255–306.

Dietterich, T. G. & Bakiri, G. (1991) Error-correcting output codes: a general method for improving multiclass inductive learning programs. In *Proceedings, Ninth AAAI National Conference on Artificial Intelligence*, Menlo Park, CA: AAAI Press, pp. 572–577. (An identical paper appears as pp. 395–407 of Wolpert (1995).)

Dietterich, T. G. & Bakiri, G. (1995) Solving multiclass learning problems via error-correcting output codes. *Journal of Artificial Intelligence Research* **2**, 263–286.

Diggle, P. J. & Hall, P. (1986) The selection of terms in a orthogonal series density estimator. *Journal of the American Statistical Association* **81**, 230–233.

Donoho, D. L. & Johnstone, I. M. (1989) Projection-based approximation and a duality with kernel methods. *Annals of Statistics* **17**, 58–106.

Doyle, P. (1973) The use of automatic interaction detector and similar search procedures. *Operational Research Quarterly* **24**, 465–467.

Doyle, P. & Fenwick, I. (1975) The pitfalls of AID analysis. *Journal of Marketing Research* **12**, 408–413.

Draper, D. (1995) Assessment and propagation of model uncertainty (with discussion). *Journal of the Royal Statistical Society series B* **57**, 45–97.

Duchon, J. (1977) Spline minimizing rotation-invariant semi-norms in Sobolev spaces. In *Constructive Theory of Functions of Several Variables*, eds W. Schempp & K. Zeller. Lecture Notes in Mathematics **571**, 85–100.

Duda, R. O. & Hart, P. E. (1973) *Pattern Classification and Scene Analysis.* New York: Wiley.

Dudani, S. A. (1976) The distance-weighted k-nearest-neighbor rule. *IEEE Transactions on Systems, Man and Cybernetics* **6**, 325–327. Reprinted in Dasarathy (1991).

Dunn, J. C. (1974) A fuzzy relative of the ISODATA process and its use in detecting compact well-separated clusters. *Journal of Cybernetics* **3**, 32–57.

Dyn, N. (1987) Interpolation of scattered data by radial functions. In *Topics in Multivariate Approximation*, eds C. K. Chui, L. L. Schumaker & F. I. Utreras. New York: Academic Press.

Eaton, H. A. C. & Oliver, T. L. (1992) Learning coefficient dependence on training set size. *Neural Networks* **5**, 283–288.

Edwards, D. (1995) *Introduction to Graphical Modelling.* New York: Springer.

Edwards, D. & Havárnek, T. (1985) A fast procedure for model-search in multidimensional contingency tables. *Biometrika* **72**, 339–351.

Efron, B. (1975) The efficiency of logistic regression compared to normal discriminant analysis. *Journal of the American Statistical Association* **70**, 892–898.

Efron, B. (1979) Bootstrap methods: another look at the jackknife. *Annals of Statistics* **7**, 1–26.

Efron, B. (1982) *The Jackknife, the Bootstrap and Other Resampling Plans.* Philadelphia: SIAM.

Efron, B. (1983) Estimating the error rate of a prediction rule. Improvements on cross-validation. *Journal of the American Statistical Association* **78**, 316–331.

Efron, B. (1986) How biased is the apparent error rate of a prediction rule? *Journal of the American Statistical Association* **81**, 461–470.

Efron, B. & Gong, G. (1983) A leisurely look at the bootstrap, the jackknife, and cross-validation. *American Statistician* **37**, 36–48.

Efron, B. & Tibshirani, R. J. (1993) *An Introduction to the Bootstrap.* New York: Chapman & Hall.

Ehrenfeucht, A., Haussler, D., Kearns, M. & Valiant, L. (1989) A general lower bound on the number of examples needed for learning. *Information and Computation* **82**, 247–261.

Eisenberger, I. (1964) Genesis of bimodal distributions. *Technometrics* **6**, 357–363.

Eriksen, P. S. (1987) Proportionality of covariances. *Annals of Statistics* **15**, 732–748.

Erwin, E., Obermayer, K. & Schulten, K. (1992) Self-organizing maps: ordering, convergence properties and energy functions. *Biological Cybernetics* **67**, 47–55.

Eslava-Gómez, G. (1989) *Projection Pursuit and Other Graphical Methods for Multivariate Data.* Unpublished D. Phil thesis, University of Oxford.

Eslava, G. & Marriott, F. H. C. (1994) Some criteria for projection pursuit. *Statistics and Computing* **4**, 13–20.

Evans, M. & Swartz, T. (1995) Methods for approximating integrals in statistics with special emphasis on Bayesian integration problems. *Statistical Science* **10**, 254–272.

Fagin, R. (1977) Multivalued dependencies and a new normal form for relational databases. *ACM Transactions on Database Systems* **2**, 262–278.

Fahlman, S. E. (1989) Faster-learning variations on back-propagation: an empirical study. In *Proceedings of the 1988 Connectionist Models Summer School, Pittsburg*, eds D. Touretzky, G. Hinton & T. Sejnowski, pp. 38–51. San Mateo, CA: Morgan Kaufmann.

Fahlman, S. E. & Lebiere, C. (1990) The cascade-correlation learning architecture. In *NIPS2*, pp. 524–532.

Fauquet, C., Desbois, D., Fargette, D. & Vidal, G. (1988) Classification of furoviruses based on the amino acid composition of their coat proteins. In *Viruses with Fungal Vectors*, eds J. I. Cooper & M. J. C. Asher, pp. 19–38. Edinburgh: Association of Applied Biologists.

Fayyad, U. M. & Irani, K. B. (1992) On the handling of continuous-valued attributes in decision tree generation. *Machine Learning* **8**, 87–102.

Fefferman, C. & Markel, S. (1994) Recovering a feed-forward network from its output. In *NIPS6*, pp. 335–342.

Feldman, J. A. (1985) Connectionist models and their applications: introduction. *Cognitive Science* **9**, 1–2.

Fienberg, S. E. & Holland, P. W. (1973) Simultaneous estimation of multinomial cell probabilities. *Journal of the American Statistical Association* **68**, 683–691.

Finnoff, W., Hergert, F. & Zimmerman, H. G. (1993) Improving model selection by nonconvergent methods. *Neural Networks* **6**, 771–783.

Fisher, R. A. (1936) The use of multiple measurements in taxonomic problems. *Annals of Eugenics* **7**, 179–188.

Fix, E. & Hodges, J. L. (1951) Discriminatory analysis—nonparametric discrimination: consistency properties. Report no. 4, US Air Force School of Aviation Medicine, Random Field, Texas. [Published in Agrawala (1977), Silverman and Jones (1989) and Dasarathy (1991).]

Flanagan, J. K., Morrell, D. R., Frost, R. L., Read, C. J. & Nelson, B. E. (1989) Vector quantization codebook generation using simulated annealing. In *Proceedings of the International Conference on Acoustics, Speech and Signal Processing (Glasgow, May 1989)*, pp. 1759–1762.

Fleiss, J. L. (1981) *Statistical Methods for Rates and Proportions*. Second edition. New York: Wiley.

Fletcher, R. (1987) *Practical Methods of Optimization*. Chichester: Wiley.

Flocchini, P., Gardin, F., Mauri, G., Pensini, M. P., & Stofella, P. (1992) Combining image processing operators and neural networks in a face recognition system. *International Journal of Pattern Recognition and Artificial Intelligence* **6**, 447–467.

Flury, B. (1986) Proportionality of k covariance matrices. *Statistics and Probability Letters* **4**, 29–33.

Flury, B., Schmid, M. J. & Natayanan, A. (1994) Error rates in quadratic discrimination with constraints on the covariance matrices. *Journal of Classification* **11**, 101–120.

Forgy, E. W. (1965) Cluster analysis of multivariate data: efficiency vs interpretability of classifications. *Biometrics* **21**, 768–769.

Fort, M. & Pagès, G. (1993) Sur la convergence *p.s.* de l'algorithme de Kohonen généralisé. *Note aux Compte Rendus de l'Académie des Sciences de Paris* **317**, Série I, 389–394.

Frank, I. E. & Friedman, J. H. (1993) A statistical view of some chemometrics regression tools (with discussion). *Technometrics* **35**, 109–148.

Fraser, D. A. S. (1968) *The Structure of Inference*. New York: Wiley.

Frean, M. (1990) The upstart algorithm: A method for constructing and training feedforward neural networks. *Neural Computation* **2**, 198–209.

Freedman, D. (1971) *Markov Chains*. San Francisco: Holden-Day.

Friedman, H. P. & Rubin, J. (1967) Some invariant criteria for grouping data. *Journal of the American Statistical Association* **62**, 1159–1178.

Friedman, J. H. (1977) A recursive partitioning decision rule for nonparametric classification. *IEEE Transactions on Computers* **26**, 404–408.

Friedman, J. H. (1984) SMART users' guide. Laboratory for Computational Statistics Technical Report No. 1, Dept of Statistics, Stanford University.

Friedman, J. H. (1987) Exploratory projection pursuit. *Journal of the American Statistical Association* **82**, 249–266.

Friedman, J. H. (1989) Regularized discriminant analysis. *Journal of the American Statistical Association* **84**, 165–175.

Friedman, J. H. (1991) Multivariate adaptive regression splines (with discussion). *Annals of Statistics* **19**, 1–141.

Friedman, J. H. & Silverman, B. W. (1989) Flexible parsimonious smoothing and additive modeling (with discussion). *Technometrics* **31**, 3–39.

Friedman, J. H. & Stuetzle, W. (1981) Projection pursuit regression. *Journal of the American Statistical Association* **76**, 817–823.

Friedman, J. H. & Tukey, J. W. (1974) A projection pursuit algorithm for exploratory data analysis. *IEEE Transactions on Computers* **23**, 881–890.

Friedman, J. H., Baskett, F. & Shustek, L. J. (1975) An algorithm for finding nearest neighbors. *IEEE Transactions on Computers* **24**, 1000–1006.

Friedman, J. H., Bentley, J. L. & Finkel, R. A. (1977) An algorithm for finding best matches in logarithmic expected time. *ACM Transactions on Mathematical Software* **3**, 209–226.

Friedman, J. H., Stuetzle, W. & Schroeder, A. (1984) Projection pursuit density estimation. *Journal of the American Statistical Association* **79**, 599–608.

Fu, K.-S. (1982) *Syntactic Pattern Recognition and Applications*. Engelwood Cliffs, NJ: Prentice Hall.

Fukunaga, K. (1990) *Introduction to Statistical Pattern Recognition*. Second edition. Boston: Academic Press. (First edition, 1972).

Fukunaga, K. & Flick, T. E. (1984) An optimal global nearest neighbor metric. *IEEE Transactions on Pattern Analysis and Machine Intelligence* **6**, 314–318. Reprinted in Dasarathy (1991).

Fukunaga, K. & Flick, T. E. (1985) The 2-NN rule for more accurate NN risk estimation. *IEEE Transactions on Pattern Analysis and Machine Intelligence* **7**, 107–112.

Fukunaga, K. & Hummels, D. M. (1987a) Bias of nearest neighbor error estimates. *IEEE Transactions on Pattern Analysis and Machine Intelligence* **9**, 103–112. Reprinted in Dasarathy (1991).

Fukunaga, K. & Hummels, D. M. (1987b) Bayes error estimation using Parzen and *k*-NN procedures. *IEEE Transactions on Pattern Analysis and Machine Intelligence* **9**, 634–643. Reprinted in Dasarathy (1991).

Fukunaga, K. & Kessell, D. L. (1971) Estimation of classification error. *IEEE Transactions on Computers* **20**, 1521–1527.

Fukunaga, K. & Mantock, J. M. (1984) Nonparametric data reduction. *IEEE Transactions on Pattern Analysis and Machine Intelligence* **6**, 115–118.

Fukunaga, K. & Narendra, P. M. (1975) A branch and bound algorithm for computing *k*-nearest neighbors. *IEEE Transactions on Computers* **24**, 750–753.

Funahashi, K. (1989) On the approximate realization of continuous mappings by neural networks. *Neural Networks* **2**, 183–192.

Fung, R. M. & Crawford, S. L. (1990) Constructor: a system for induction of probabilistic models. In *Proceedings, Eighth AAAI National Conference on Artificial Intelligence, Boston*, pp. 762–769. Menlo Park, CA: AAAI Press.

Furman, W. & Lindsay, B. (1994) Testing for the number of components in a mixture of normal distributions using moment estimators. *Computational Statistics and Data Analysis* **17**, 473–492.

Furnival, G. M. & Wilson, R. W. Jr (1974) Regressions by leaps and bounds. *Technometrics* **16**, 499–511.

Gader, P., Forester, B., Ganzberger, M., Gillies, A., Mitchell, B., Whalen, M. & Yocum, T. (1991) Recognition of handwritten digits using template and model matching. *Pattern Recognition* **24**, 421–431.

Gallant, S. I. (1990) Perceptron-based learning algorithms. *IEEE Transactions on Neural Networks* **1**, 179–191.

Gallant, S. I. (1993) *Neural Network Learning and Expert Systems*. Cambridge, MA: The MIT Press.

Gammerman, A. (ed.) (1995) *Probabilistic Reasoning and Bayesian Belief Networks*. Henley-on-Thames: Alfred Waller.

Gammerman, A., Luo, Z., Aitken, C. G. G. & Brewer, M. J. (1995) Exact and approximate algorithms and their implementations in mixed graphical models. In Gammerman (1995), pp. 33–53.

Garey, M. R. & Johnson, D. S. (1979) *Computers and Intractability: A Guide to the Theory of NP-completeness*. New York: Freeman.

Gates, G. W. (1972) The reduced nearest neighbor rule. *IEEE Transactions on Information Theory* **18**, 431–433.

Geiger, D. & Pearl, J. (1990) On the logic of causal models. In *Uncertainty in Artificial Intelligence 4*, eds R. D. Shachter, T. S. Levitt, L. N. Kanal & J. F. Lemmer, pp. 3–14. Amsterdam: North-Holland.

Geiger, D. & Pearl, J. (1993) Logical and algorithmic properties of conditional independence and graphical models. *Annals of Statistics* **21**, 2001–2021.

Geiger, D., Verma, T. & Pearl, J. (1990) Recognizing independence in Bayesian networks. *Networks* **20**, 507–534.

Geisser, S. (1964) Posterior odds for multivariate normal classifications. *Journal of the Royal Statistical Society series B* **26**, 69–76.

Geisser, S. (1966) Predictive discrimination. In *Multivariate Analysis*, ed. P. R. Krishnaiah, pp. 149–163. New York: Academic Press.

Geisser, S. (1975) The predictive sample reuse method with applications. *Journal of the American Statistical Association* **70**, 320–328.

Geisser, S. (1984) On prior distributions for binary trials (with discussion). *American Statistician* **38**, 244–251.

Geisser, S. (1987) Comment on Hodges (1987). *Statistical Science* **2**, 277–279.

Geisser, S. (1993) *Predictive Inference: An Introduction*. New York: Chapman & Hall.

Geisser, S. & Cornfield, J. (1963) Posterior distributions for multivariate normal parameters. *Journal of the Royal Statistical Society series B* **25**, 368–376.

Gelfand, A. E. & Dey, D. K. (1994) Bayes model choice: asymptotics and exact calculations. *Journal of the Royal Statistical Society series B* **56**, 501–514.

Gelfand, A. E. & Smith, A. F. M. (1990) Sampling-based approaches to calculating marginal densities. *Journal of the American Statistical Association* **85**, 398–409.

Gelfand, A. E., Hills, S. E., Racine-Poon, A. & Smith, A. F. M. (1990) Illustration of Bayesian inference in normal data models using Gibbs sampling. *Journal of the American Statistical Association* **85**, 972–985.

Gelfand, S. B. & Delp, E. J. (1991) On tree structured classifiers. In Sethi & Jain (1991), pp. 51–70.

Gelfand, S. B. & Mitter, S. K. (1991) Recursive stochastic algorithms for global optimization in $\mathbb{R}^d$. *SIAM Journal on Control and Optimization* **29**, 999–1018.

Gelfand, S. B., Ravishankar, C. S. & Delp, E. J. (1991) An iterative growing and pruning algorithm for classification tree design. *IEEE Transactions on Pattern Analysis and Machine Intelligence* **13**, 163–174.

Gelman, A., Carlin, J. B., Stern, H. S. & Rubin, D. B. (1995) *Bayesian Data Analysis*. New York: Chapman & Hall.

Geman, D. (1990) Random fields and inverse problems in imaging. In *École d'Été de Probabilités de Saint-Flour XVIII – 1988*, ed. P. L. Hennequin. Lecture Notes in Mathematics **1427**, 113–193.

Geman, S. & Geman, D. (1984) Stochastic relaxation, Gibbs distributions and the Bayesian restoration of images. *IEEE Transactions on Pattern Analysis and Machine Intelligence* **6**, 721–741. Reprinted in Shafer & Pearl (1990).

Geman, S. & Hwang, C.-R. (1982) Nonparametric maximum likelihood estimation by the method of sieves. *Annals of Statistics* **10**, 401–414.

Geman, S., Bienenstock, E. & Doursat, R. (1992) Neural networks and the bias/variance dilemma. *Neural Computation* **4**, 1–58.

George, E. I. & McCulloch, R. E. (1993) Variable selection via Gibbs sampling. *Journal of the American Statistical Association* **88**, 881–889.

Georgiopoulos, M., Heileman, G. L. & Huang, J. (1990) Convergence properties of learning in ART1. *Neural Computation* **2**, 502–509.

Georgiopoulos, M., Heileman, G. L. & Huang, J. (1991) Properties of learning related to pattern diversity in ART1. *Neural Networks* **4**, 751–757.

Gersho, A. & Gray, R. M. (1992) *Vector Quantization and Signal Compression*. Boston: Kluwer Academic Publishers.

Geyer, C. (1992) Practical Markov chain Monte Carlo (with discussion). *Statistical Science* **7**, 473–511.

Ghurye, S. G. & Olkin, I. (1969) Unbiased estimation of some multivariate probability densities and related functions. *Annals of Mathematical Statistics* **40**, 1261–1271.

Gilbert, J. C. & Lemaréchal, C. (1989) Some numerical experiments with variable storage quasi-Newton methods. *Mathematical Programming* **45**, 407–436.

Gill, P. E., Murray, W. & Wright, M. H. (1981) *Practical Optimization*. London: Academic Press.

Girosi, F. & Anzellotti, G. (1993) Rates of convergence for radial basis functions and neural networks. In Mammone (1993), pp. 97–114.

Girosi, F. & Poggio, T. (1990) Networks and the best approximation property. *Biological Cybernetics* **63**, 169–176.

Girosi, F., Jones, M. & Poggio, T. (1995) Regularization theory and neural networks architectures. *Neural Computation* **7**, 219–269.

Glick, N. (1972) Sample-based classification procedures derived from density estimators. *Journal of the American Statistical Association* **67**, 116–122.

Glick, N. (1976) Sample-based classification procedures related to empiric distributions. *IEEE Transactions on Information Theory* **22**, 454–461.

Glymour, C., Scheines, R., Spirtes, P. & Kelly, K. (1987) *Discovering Causal Structure: Artificial Intelligence, Philosophy of Science, and Statistical Modeling*. San Diego: Academic Press.

Goldstein, M. & Dillon, W. R. (1978) *Discrete Discriminant Analysis*. New York: Wiley.

Golomb, B. A., Lawrence, D. T. & Sejnowski, T. J. (1991) SEXNET: A neural network identifies sex from human faces. In *NIPS3*, pp. 572–577.

Golombic, M. C. (1980) *Algorithmic Graph Theory and Perfect Graphs*. New York: Academic Press.

Golub, G. H. & Van Loan, C. F. (1989) *Matrix Computations*. Second edition. Baltimore: Johns Hopkins University Press.

Gonzalez, R. C. & Thomason, M. G. (1978) *Syntactic Pattern Recognition: An Introduction*. Reading, MA: Addison-Wesley.

Good, I. J. (1965) *The Estimation of Probabilities*. Cambridge, MA: The MIT Press.

Good, I. J. (1983) *Good Thinking: The Foundations of Probability and its Applications*. Minneapolis: University of Minnesota Press.

Goodman, R. M. & Smyth, P. (1988) Decision tree design from a communication theory standpoint. *IEEE Transactions on Information Theory* **34**, 979–994.

Gordon, A. D. (1981) *Classification. Methods for Exploratory Analysis of Multivariate Data*. London: Chapman & Hall.

Gori, M. & Tesi, A. (1992) On the problem of local minima in backpropagation. *IEEE Transactions on Pattern Analysis and Machine Intelligence* **14**, 76–86.

Gower, J. C. (1966) Some distance properties of latent root and vector methods used in multivariate analysis. *Biometrika* **53**, 325–328.

Gower, J. C. (1971) A general coefficient of similarity and some of its properties. *Biometrics* **27**, 857–871.

Gower, J. C. & Legendre, P. (1986) Metric and Euclidean properties of dissimilarity coefficients. *Journal of Classification* **3**, 5–48.

Gray, R. M. (1984) Vector quantization. *IEEE ASSP Magazine* **1(2)**, 4–29.

Green, P. J. & Silverman, B. W. (1994) *Nonparametric Regression and Generalized Linear Models. A Roughness Penalty Approach*. London: Chapman & Hall.

Grenander, U. (1981) *Abstract Inference*. New York: Wiley.

Grenander, U., Chow, Y. & Keenan, D. M. (1991) *Hands. A Pattern Theoretic Study of Biological Shapes*. New York: Springer.

Grinold, R. C. (1969) Comment on 'Pattern classification design by linear programming'. *IEEE Transactions on Computers* **18**, 378–379.

Grother, P. J. & Candela, G. T. (1993) Comparison of handprinted digit classifiers. US National Institute of Standards and Technology report NISTIR 5209.

Gu, C. (1990) Adaptive spline smoothing in non-Gaussian regression models. *Journal of the American Statistical Association* **85**, 801–807.

Gu, C. & Wahba, G. (1991) Minimizing GCV/GML scores with multiple smoothing parameters via the Newton method. *SIAM Journal on Scientific and Statistical Computing* **12**, 383–398.

Gu, C., Bates, D. M., Chen, Z. & Wahba, G. (1989) The computation of generalized cross-validation functions through Householder tridiagonalization with applications to the fitting of interaction spline models. *SIAM Journal on Matrix Analysis and Applications* **10**, 459–480.

Guo, H. & Gelfand, S. B. (1992) Classification trees with neural network feature extraction. *IEEE Transactions on Neural Networks* **3**, 923–933.

Guyon, I., Vapnik, V., Boser, B., Bottou, L. & Solla, S. A. (1992) Structural risk minimization for character recognition. In *NIPS4*, pp. 471–479.

Hall, D. J. & Ball, G. B. (1965) ISODATA: a novel method of data analysis and pattern classification. Technical report, Stanford Research Institute, Menlo Park CA.

Hall, D. J. & Khanna, D. (1977) The ISODATA method of computation for relative perception of similarities and differences in complex and real computers. In *Statistical Methods for Digital Computers* 3, eds K. Enslein, A. Ralston & H. S. Wilf, pp. 340–373. New York: Wiley.

Hall, P. (1981) On nonparametric multivariate binary discrimination. *Biometrika* **68**, 287–294.

Hall, P. (1989) On polynomial-based projection indices for exploratory projection pursuit. *Annals of Statistics* **17**, 589–605.

Hall, P. & Wand, M. P. (1988) On nonparametric discrimination using density differences. *Biometrika* **75**, 541–547.

Hampel, F. R., Ronchetti, E. M., Rousseeuw, P. J. & Stahel, W. A. (1986) *Robust Statistics: The Approach Based on Influence Functions*. New York: Wiley.

Hampson, S. E. & Volper, D. J. (1986) Linear function neurons: structure and training. *Biological Cybernetics* **53**, 203–217.

Hand, D. J. (1981) *Discrimination and Classification*. Chichester: Wiley.

Hand, D. J. (1982) *Kernel Discriminant Analysis*. Chichester: Research Studies Press.

Hand, D. J. & Batchelor, B. G. (1978) An edited condensed nearest neighbor rule. *Information Sciences* **14**, 171–180.

Hannan, E. J. & Quinn, B. G. (1979) The determination of the order of an autoregression. *Journal of the Royal Statistical Society series B* **41**, 190–195.

Hansen, L. K. & Salamon, P. (1990) Neural network ensembles. *IEEE Transactions on Pattern Analysis and Machine Intelligence* **12**, 993–1001.

Hanson, S. J. & Pratt, L. Y. (1989) Comparing biases for minimal network construction with backpropagation. In *NIPS1*, pp. 177–185.

Härdle, W. (1990) *Applied Nonparametric Regression*. Cambridge: Cambridge University Press.

Härdle, W. (1991) *Smoothing Techniques with Implementation in S*. New York: Springer.

Hardy, R. L. (1971) Multiquadric equations of topography and other irregular surfaces. *Journal of Geophysical Research* **76**, 1906–1915.

Hardy, R. L. (1990) Theory and applications of the multiquadric-biharmonic method: 20 years of discovery 1968-1988. *Computers and Mathematics with Applications* **19**, 163–208.

Hart, P. E. (1968) The condensed nearest neighbor rule. *IEEE Transactions on Information Theory* **14**, 515–516.

Hartigan, J. A. (1975) *Clustering Algorithms*. New York: Wiley.

Hartigan, J. A. & Wong, M. A. (1979) Algorithm AS136. A *K*-means clustering algorithm. *Applied Statistics* **28**, 100–108.

Hartman, E. J., Keeler, J. D. & Kowalski, J. M. (1990) Layered neural networks with Gaussian hidden units as universal approximations. *Neural Computation* **2**, 210–215.

Hassibi, B. & Stork, D. G. (1993) Second derivatives for network pruning: Optimal Brain Surgeon. In *NIPS5*, pp. 164–171.

Hassibi, B., Stork, D. G., Wolff, G. & Watanabe, T. (1994) Optimal Brain Surgeon: extensions and performance comparisons. In *NIPS6*, pp. 263–270.

Hastie, T. & Mallows, C. (1993) Discussion of Frank & Friedman (1993). *Technometrics* **35**, 140–143.

Hastie, T. & Stuetzle, W. (1989) Principal curves. *Journal of the American Statistical Association* **84**, 502–516.

Hastie, T. J. & Tibshirani, R. J. (1990) *Generalized Additive Models*. London: Chapman & Hall.

Hastie, T. & Tibshirani, R. (1996) Discriminant analysis by Gaussian mixtures. *Journal of the Royal Statistical Society series B* **58,**, 155–176.

Hastie, T., Buja, A. & Tibshirani, R. (1995) Penalized discriminant analysis. *Annals of Statistics* **23**, 73–102.

Hastie, T., Tibshirani, R. & Buja, A. (1994) Flexible discriminant analysis by optimal scoring. *Journal of the American Statistical Association* **89**, 1255–1270.

Hastings, W. K. (1970) Monte Carlo sampling methods using Markov chains and their applications. *Biometrika* **57**, 97–109.

Hathaway, R. J. (1985) A constrained formulation of maximum-likelihood estimation for normal mixture distributions. *Annals of Statistics* **13**, 795–800.

Hauck, W. W. Jr & Donner, A. (1977) Wald's test as applied to hypotheses in logit analysis. *Journal of the American Statistical Association* **72**, 851–853.

Haussler, D. (1992) Decision theoretic generalizations of the PAC model for neural net and other learning applications. *Information and Computation* **100**, 78–150. Reprinted as pp. 37–116 of Wolpert (1995).

Haykin, S. (1994) *Neural Networks. A Comprehensive Foundation*. New York: Macmillan College Publishing.

Hebb, D. O. (1949) *The Organization of Behavior*. New York: Wiley.

Hellman, M. E. (1970) The nearest neighbor classification rule with a reject option. *IEEE Transactions on Systems Science and Cybernetics* **6**, 179–185. Reprinted in Dasarathy (1991).

Henrichon, E. G. Jr & Fu, K.-S. (1968) On mode estimation in pattern recognition. In *Proceedings of the Seventh Symposium on Adaptive Processes, UCLA*, p. 3-a-1.

Henrichon, E. G. Jr & Fu, K.-S. (1969) A nonparametric partitioning procedure for pattern classification. *IEEE Transactions on Computers* **18**, 614–624.

Henrion, M. (1988) Propagating uncertainty in Bayesian networks by probabilistic logic sampling. In *Uncertainty in Artificial Intelligence 2*, eds J. Lemmer & L. N. Kanal, pp. 149–163. Amsterdam: North-Holland.

Henrion, M., Breese, J. S. & Horvitz, E. J. (1991) Decision analysis and expert systems. *AI Magazine* **12(4)**, 64–91.

Hermans, J., Habbema, J. D. F. & Schaefer, J. R. (1982) The ALLOC80 package for discriminant analysis. *Statistical Software Newsletter* **8**, 15–20.

Hertz, J., Krogh, A. & Palmer, R. G. (1991) *Introduction to the Theory of Neural Computation*. Redwood City, CA: Addison-Wesley.

Heskes, T. M. & Kappen, B. (1991) Learning processes in neural networks. *Physical Reviews A* **44**, 2718–2726.

Highleyman, W. H. (1962a) The design and analysis of pattern recognition experiments. *Bell Systems Technical Journal* **41**, 723–744.

Highleyman, W. H. (1962b) Linear decision functions, with application to pattern recognition. *Proceedings of the IRE* **50**, 1501–1514.

Hills, M. (1966) Allocation rules and their error rates (with discussion). *Journal of the Royal Statistical Society series B* **28**, 1–31.

Hinton, G. E. (1986) Learning distributed representations of concepts. In *Proceedings of the Eighth Annual Conference of the Cognitive Science Society (Amherst, 1986)*, pp. 1–12. Hillsdale: Erlbaum.

Hinton, G. E. (1989a) Connectionist learning procedures. *Artificial Intelligence* **40**, 185–234. (Reprinted in Carbonell, 1990.)

Hinton, G. E. (1989b) Deterministic Boltzmann machine learning performs steepest descent in weight space. *Neural Computation* **1**, 143–150.

Hinton, G. E. & Sejnowski, T. J. (1983) Optimal perceptual inference. In *Proceedings of the IEEE Conference on Computer Vision and Pattern Recognition (Washington, 1983)*, pp. 448–453. New York: IEEE Press.

Hjort, N. L. (1986) *Notes on the Theory of Statistical Symbol Recognition*. Norwegian Computing Center Report 778.

Hjort, N. L. & Glad, I. K. (1995) Nonparametric density estimation with a parametric start. *Annals of Statistics* **23**, 882–904.

Hjort, N. L. & Jones, M. C. (1996) Locally parametric nonparametric density estimation. *Annals of Statistics* **24**, 1619–1647.

Ho, Y.-C. & Kashyap, R. L. (1965) An algorithm for linear inequalities and its applications. *IEEE Transactions on Electronic Computers* **14**, 683–688.

Hodges, J. S. (1987) Uncertainty, policy analysis and statistics (with discussion). *Statistical Science* **2**, 259–291.

Hoeffding, W. (1963) Probability inequalities for sums of bounded random variables. *Journal of the American Statistical Association* **58**, 13–30.

Hoerl, A. E. & Kennard, R. W. (1970a) Ridge regression: biased estimation for nonorthogonal problems. *Technometrics* **12**, 55–67.

Hoerl, A. E. & Kennard, R. W. (1970b) Ridge regression: applications to nonorthogonal problems. *Technometrics* **12**, 69–82.

Höffgen, K.-U., Simon, H.-U. & Van Horn, K. S. (1995) Robust trainability of single neurons. *Journal of Computer and System Sciences* **50**, 114–125.

Holt, M. J. J. & Semnani, S. (1990) Convergence of back-propagation in neural networks using a log-likelihood cost function. *Electronics Letters* **26**, 1964–1965.

Hopfield, J. J. (1982) Neural networks and physical systems with emergent collective computational facilities. *Proceedings of the National Academy of Sciences of the USA* **79**, 2554–2558. Reprinted in Anderson & Rosenfeld (1988) and Lau (1992).

Hopfield, J. J. (1987) Learning algorithms and probability distributions in feed-forward and feed-back networks. *Proceedings of the National Academy of Sciences of the USA* **84**, 8429–8433.

Hornik, K., Stinchcombe, M. & White, H. (1989) Multilayer feedforward networks are universal approximators. *Neural Networks* **2**, 359–366. Reprinted in White (1992).

Hornik, K., Stinchcombe, M. & White, H. (1990) Universal approximation of an unknown mapping and its derivatives using feedforward networks. *Neural Networks* **3**, 551–560. Reprinted in White (1992).

Hrycej, T. (1990) Gibbs sampling in Bayesian networks. *Artificial Intelligence* **46**, 351–363.

Hrycej, T. (1992) *Modular Learning in Neural Networks. A Modularized Appproach to Neural Network Classification.* New York: Wiley.

Huang, J., Georgiopoulos, M. & Heileman, G. L. (1995) Fuzzy ART properties. *Neural Networks* **8**, 203–213.

Huber, P. J. (1967) The behavior of maximum likelihood estimates under nonstandard conditions. In *Proceedings of the Fifth Berkeley Symposium on Mathematical Statistics and Probability*, eds L. M. Le Cam & J. Neyman, **1**, pp. 221–233. Berkeley: University of California Press.

Huber, P. J. (1981) *Robust Statistics*. New York: Wiley.

Huber, P. J. (1985) Projection pursuit (with discussion). *Annals of Statistics* **13**, 435–525.

Hunt, E. B., Marin, J. & Stone, P. J. (1966) *Experiments in Induction.* New York: Academic Press.

Hwang, J.-N., Lay, S.-R., Maechler, M., Martin, D. & Schimert, J. (1994a) Regression modeling in back-propagation and projection pursuit learning. *IEEE Transactions on Neural Networks* **5**, 342–353.

Hwang, J.-N., Li, H., Maechler, M., Martin, D. & Schimert, J. (1992a) A comparison of projection pursuit and neural network regression modeling. In *NIPS4*, pp. 1159–1166.

Hwang, J.-N., Li, H., Maechler, M., Martin, D. & Schimert, J. (1992b) Projection pursuit learning networks for regression. *Engineering Applications of Artificial Intelligence* **5**, 193–204.

Hwang, J.-N., Li, H., Martin, D. & Schimert, J. (1991) The learning parsimony of projection pursuit and back-propagation networks. In *25th Asilomar Conference on Signals, Systems and Computers, Pacific Grove, CA*, pp. 491–495. Los Alamitos, CA: IEEE Computer Society Press.

Hwang, J.-N., You, S.-S., Lay, S.-R. & Jou, I.-C. (1994b) What's wrong with a cascaded correlation learning network: a projection pursuit learning perspective. Technical Report, Dept of Electrical Engineering, University of Washington.

Hwang, J.-N., You, S.-S., Lay, S.-R. & Jou, I.-C. (1996) The cascaded correlation learning: A projection pursuit perspective. *IEEE Transactions on Neural Networks* **7**, 278–289.

Hyafil, R. & Rivest, R. L. (1976) Constructing optimal binary trees is NP-complete. *Information Processing Letters* **5**, 15–17.

Impedovo, S., Ottaviano, L. & Occhinegro, S. (1991) Optical character recognition—a survey. *International Journal of Pattern Recognition and Artificial Intelligence* **5**, 1–24.

Ingrassia, S. (1992) A comparison between the simulated annealing and the EM algorithms in normal mixture decompositions. *Statistics and Computing* **2**, 203–211.

Intrator, N. (1990) A neural network for feature extraction. In *NIPS2*, pp. 719–726.

Intrator, N. (1991) Exploratory feature extraction in speech signals. In *NIPS3*, pp. 241–247.

Intrator, N. (1992) Feature extraction using an unsupervised neural network. *Neural Computation* **4**, 98–107.

Intrator, N. & Cooper, L. N. (1992) Objective function formulation of the BCM theory of visual cortical plasticity: statistical connections, stability conditions. *Neural Networks* **5**, 3–17.

Intrator, N. & Gold, J. I. (1993) Three-dimensional object recognition using an unsupervised BCM network: the usefulness of distinguishing features. *Neural Computation* **5**, 61–74.

Isham, V. (1981) An introduction to spatial point processes and Markov random fields. *International Statistical Review* **49**, 21–43.

Jackson, J. E. (1991) *A User's Guide to Principal Components.* New York: Wiley.

Jacobs, R. A. (1988) Increased rates of convergence through learning rate adaptation. *Neural Networks* **1**, 295–307.

Jacobs, R. A., Jordan, M. I., Nowlan, S. J. & Hinton, G. E. (1991) Adaptive mixtures of local experts. *Neural Computation* **3**, 79–87.

Jain, A. K. & Dubes, R. C. (1988) *Algorithms for Clustering Data.* Englewood Cliffs, NJ: Prentice-Hall.

Jain, A. K., Dubes, R. C. & Chen, C.-C. (1987) Bootstrap techniques for error estimation. *IEEE Transactions on Pattern Analysis and Machine Intelligence* **9**, 628–633.

James, M. (1988) *Pattern Recognition.* New York: Wiley.

Jancey, R. C. (1966) Multidimensional group analysis. *Australian Journal of Botany* **14**, 127–130.

Jeffreys, H. (1961) *Theory of Probability.* Third edition. Oxford: Clarendon Press.

Jensen, F. V. (1991) Calculation in HUGIN of probabilities for specific configurations — a trick with many applications. In *Proceedings of the Scandinavian Conference on Artificial Intelligence*, pp. 176–186. Amsterdam: IOS Press.

Jensen, F. V. (1996) *An introduction to Belief Networks.* London: UCL Press (Taylor & Francis Ltd) and New York: Springer.

Jensen, F. V. & Liang, J. (1995) drHugin. A system for hypothesis driven data request. In Gammerman (1995), pp. 109–124.

Jensen, F. V., Lauritzen, S. L. & Olesen, K. G. (1990) Bayesian updating in causal probabilistic networks by local computations. *Computational Statistics Quarterly* **5**, 269–282.

Jiang, Q. & Zhang, W. (1993) An improved method for finding nearest neighbors. *Pattern Recognition Letters* **14**, 531–535.

Johansson, E. M., Dowla, F. U. & Goodman, D. M. (1991) Back-propagation learning for multi-layer feed-forward neural networks using the conjugate gradient method. *International Journal of Neural Systems* **2**, 291–302.

Johnson, N. L. & Kotz, S. (1972) *Distributions in Statistics: Continuous Multivariate Distributions.* New York: Wiley.

Jolliffe, I. T. (1986) *Principal Component Analysis.* New York: Springer.

Jones, L. K. (1987) On a conjecture of Huber concerning the convergence of projection pursuit regression. *Annals of Statistics* **15**, 880–882.

Jones, L. K. (1992) A simple lemma on greedy approximation in Hilbert space and convergence rates for projection pursuit regression and neural network training. *Annals of Statistics* **20**, 608–613.

Jones, M. C. & Sibson, R. (1987) What is projection pursuit (with discussion)? *Journal of the Royal Statistical Society series A* **150**, 1–36.

Jordan, M. I. & Jacobs, R. A. (1992) Hierarchies of adaptive experts. In *NIPS4*, pp. 985–992.

Jordan, M. I. & Jacobs, R. A. (1994) Hierarchical mixtures of experts and the EM algorithm. *Neural Computation* **6**, 181–214.

Kalantari, I. & McDonald, G. (1983) A data structure and an algorithm for the nearest point problem. *IEEE Transactions on Software Engineering* **9**, 631–634.

Kambhatla, N. & Leen, T. K. (1994) Fast non-linear dimension reduction. In *NIPS6*, pp. 152–159.

Kamgar-Parsi, B. & Kanal, L. N. (1985) An improved branch and bound algorithm for computing k-nearest neighbours. *Pattern Recognition Letters* **3**, 7–12.

Kansa, E. J. (1990) Multiquadrics—a scattered data approximation scheme with applications to computational fluid dynamics. 1. *Computers and Mathematics with Applications* **19**, 127–145.

Kappen, H. J. (1995) Deterministic learning rules for Boltzmann machines. *Neural Networks* **8**, 537–548.

Karpinski, M. & Macintyre, A. (1995a) Bounding VC-dimension of neural networks: Progress and prospects. In *Proceedings of the Second European Conference on Computational Learning Theory (Barcelona, Spain)*, ed. P. Vitanyi, number 904 in Lecture Notes in Artificial Intelligence, pp. 337–341. Berlin: Springer.

Karpinski, M. & Macintyre, A. (1995b) Polynomial bounds for VC dimension of sigmoidal neural networks. In *Proceedings of the Twenty-Seventh Annual ACM Symposium on Theory of Computing (Las Vegas)*, pp. 200–208. ACM Press.

Kashyap, R. L. & Blaydon, C. C. (1968) Estimation of probability density and distribution functions. *IEEE Transactions on Information Theory* **14**, 549–556.

Kass, G. V. (1980) An exploratory technique for investigating large quantities of categorical data. *Applied Statistics* **29**, 119–127.

Kass, R. E. & Raftery, A. E. (1995) Bayes factors. *Journal of the American Statistical Association* **90**, 733–795.

Kass, R. E. & Vaidyanathan, S. K. (1992) Approximate Bayes factors and orthogonal parameters, with application to testing equality of two binomial proportions. *Journal of the Royal Statistical Society series B* **54**, 129–144.

Kaufman, L. & Rousseeuw, P. J. (1990) *Finding Groups in Data. An Introduction to Cluster Analysis.* New York: Wiley.

Kendall, M. G. & Stuart, A. (1966) *The Advanced Theory of Statistics, volume III.* London: Griffin.

Kent, J. T., Tyler, D. E. & Vardi, Y. (1994) A curious likelihood identity for the multivariate *t*-distribution. *Communications in Statistics —Simulation and Computation* **23**, 441–453.

Kibler, D. & Aha, D. W. (1987) Learning representative exemplars of concepts: an initial case study. In *Proceedings of the Fourth International Workshop on Machine Learning (Irvine, 1987)*, ed. P. Langley, pp. 24–30. Palo Alto, CA: Morgan Kaufmann. Reprinted in Shavlik & Dietterich (1990).

Kiiveri, H., Speed, T. P. & Carlin, J. B. (1984) Recursive causal models. *Journal of the Australian Mathematical Society (series A)* **36**, 30–52.

Kim, B. S. & Park, S. B. (1986) A fast *k* nearest neighbor finding algorithm based on the ordered partition. *IEEE Transactions on Pattern Analysis and Machine Intelligence* **8**, 761–766. Reprinted in Dasarathy (1991).

Kim, J. H. & Pearl, J. (1983) A computational model for combined causal and diagnostic reasoning in inference systems. In *Proceedings of the Eighth International Joint Conference on Artificial Intelligence (Karlsruhe, 1983)*, pp. 190–193. Menlo Park, CA: AAAI.

King, R. D., Muggleton, S., Lewis, R. A. & Sternberg, M. J. E. (1992) Drug design by machine learning: The use of inductive logic programming to model the structure-activity relationships of trimethoprim analogues binding to dihydrofolate reductase. *Proceedings of the National Academy of Sciences of the USA* **89**, 11322–11326.

Kjærulff, U. (1992) Optimal decomposition of probabilistic networks by simulated annealing. *Statistics and Computing* **2**, 7–17.

Kleijnen, J. P. C. (1987) *Statistical Tools for Simulation Practitioners.* New York: Marcel Dekker.

Kleijnen, J. P. C. & van Groenendaal, W. (1992) *Simulation: A Statistical Perspective.* Chichester: Wiley.

Knerr, S., Personnaz, L. & Dreyfus, G. (1992) Handwritten digit recognition by neural networks with single-layer training. *IEEE Transactions on Neural Networks* **3**, 962–968.

Knuth, D. E. (1968) *The Art of Computer Programming, Volume 1: Fundamental Algorithms.* Reading, MA: Addison-Wesley. (Second edition, 1973.)

Kohonen, T. (1982a) Self-organized formation of topologically correct feature maps. *Biological Cybernetics* **43**, 59–69. Reprinted in Anderson & Rosenfeld (1988).

Kohonen, T. (1982b) Analysis of a simple self-organizing process. *Biological Cybernetics* **43**, 135–140.

Kohonen, T. (1988a) An introduction to neural computing. *Neural Networks* **1**, 3–16.

Kohonen, T. (1988b) Learning vector quantization. *Neural Networks* **1** (suppl. 1), 303.

Kohonen, T. (1989) *Self-Organization and Associative Memory.* Third edition. Berlin: Springer. [First edition, 1984]

Kohonen, T. (1990a) The self-organizing map. *Proceedings of the IEEE* **78**, 1464–1480. Reprinted in Lau (1992).

Kohonen, T. (1990b) Improved versions of learning vector quantization. In *Proceedings of the IEEE International Conference on Neural Networks, San Diego* **I**, 545–550. New York: IEEE Press.

Kohonen, T. (1993) Physiological interpretation of the self-organizing map algorithm. *Neural Networks* **6**, 895–905.

Kohonen, T. (1995) *Self-Organizing Maps*. Berlin: Springer.

Kohonen, T., Barna, G. & Chrisley, R. (1988) Statistical pattern recognition with neural networks: benchmarking studies. In *Proceedings of the IEEE International Conference on Neural Networks, San Diego*, **I**, 61–68. Long Beach, CA: IEEE Press. Reprinted in Anderson *et al.* (1990).

Kohonen, T., Kangas, T., Laaksonen, J. & Torkkola, K. (1992) *LVQ_PAK. The learning vector quantization program package version 2.1*. Laboratory of Computer and Information Science, Helsinki University of Technology. [Version 3.1 became available in 1995.]

Koiran, P. & Sontag, E. D. (1996) Neural networks with quadratic VC dimension. In *Advances in Neural Information Processing Systems 8* eds D. S. Touretzky, M. C. Moser & M. E. Hasselmo, pp. 197–203. Cambridge, MA: MIT Press.

Kong, A. (1991) Efficient methods for computing linkage likelihoods of recessive diseases in inbred pedigrees. *Genetic Epidemiology* **8**, 81–103.

Kononenko, I., Bratko, I. & Roškar, E. (1984) Experiments in the automatic learning of medical diagnosis rules. Technical Report, Josef Stefan Institute, Ljubljana.

Koontz, W. L. G., Narendra, P. M. & Fukunaga, K. (1975) A branch and bound clustering algorithm. *IEEE Transactions on Computers* **24**, 908–915.

Kramer, A. H. & Sangiovanni-Vincentelli, A. (1989) Efficient parallel learning algorithms for neural networks. In *NIPS1*, pp. 40–48.

Kramer, M. A.. (1991) Nonlinear principal component analysis using autoassociative neural networks. *AICHE Journal* **37**, 233–243.

Krishnaiah, P. R. & Kanal, L. N. (eds) (1982) *Handbook of Statistics 2: Classification, Pattern Recognition and Reduction of Dimensionality*. Amsterdam: North Holland.

Kruskal, J. B. (1964a) Multidimensional scaling by optimizing goodness-of-fit to a nonmetric hypothesis. *Psychometrika* **29**, 1–29.

Kruskal, J. B. (1964b) Non-metric multidimensional scaling: a numerical method. *Psychometrika* **29**, 115–129.

Kruskal, J. B. (1969) Toward a practical method which helps uncover the structure of a set of multivariate observations by finding the linear transformation which optimizes a new 'index of condensation'. In *Statistical Computation*, eds R. C. Milton & J. A. Nelder, pp. 427–440. New York: Academic Press.

Kruskal, J. B. (1971) Monotone regression: continuity and differentiability properties. *Psychometrika* **36**, 57–62.

Kruskal, J. B. (1972) Linear transformation of multivariate data to reveal clustering. In *Multidimensional Scaling: Theory and Application in the Behavioural Sciences*, eds R. N. Shephard, A. K. Romney & S. K. Nerlove, pp. 179–191. New York: Seminar Press.

Krzanowski, W. J. (1975) Discrimination and classification using both binary and continuous variables. *Journal of the American Statistical Association* **70**, 782–790.

Kung, S. Y. & Diamantaras, K. I. (1990) A neural network learning algorithm for Adaptive Principal component EXtraction (APEX). In *Proceedings of the IEEE International Conference on Acoustics, Speech and Signal Processing (Albuquerque, NM, 1990)* **2**, pp. 861–864. Long Beach, CA: IEEE Press.

Kůrková, V. (1991) Kolmogorov's theorem is relevant. *Neural Computation* **3**, 617–622.

Kůrková, V. (1992) Kolmogorov's theorem and multilayer neural networks. *Neural Networks* **5**, 501–506.

Kurzynski, M. W. (1983a) Decision rules for a hierarchical classifier. *Pattern Recognition Letters* **1**, 305–310.

Kurzynski, M. W. (1983b) The optimal strategy of a tree classifier. *Pattern Recognition* **16**, 81–87. (Correction page 361).

Kushner, H. (1987) Asymptotic global behavior for stochastic approximation and diffusions with slowly decreasing noise effects: global minimization via Monte Carlo. *SIAM Journal on Applied Mathematics* **47**, 169–185.

Kwok, S. W. & Carter, C. (1990) Multiple decision trees. In *Uncertainty in Artificial Intelligence 4*, eds R. D. Shachter, T. S. Levitt, L. N. Kanal & J. F. Lemmer, pp. 327–335. Amsterdam: North Holland.

Lachenbruch, P. A. (1975) *Discriminant Analysis*. New York: Hafner Press.

Lachenbruch, P. A. & Mickey, M. R. (1968) Estimation of error rates in discriminant analysis. *Technometrics* **10**, 1–11.

Lange, K. L., Little, R. J. A. & Taylor, J. M. G. (1989) Robust statistical modeling using the *t* distribution. *Journal of the American Statistical Association* **84**, 881–896.

Langley, P. (1996) *Elements of Machine Learning*. San Francisco: Morgan Kaufmann.

Langley, P. & Simon, H. A. (1995) Applications of machine learning and rule induction. *Communications of the Association for Computing Machinery* **38**, 54–64.

Lau, C. (ed.) (1992) *Neural Networks: Theoretical Foundations and Analysis*. New York: IEEE Press.

Lauritzen, S. (1989) Mixed graphical association models (with discussion). *Scandinavian Journal of Statistics* **16**, 273–306.

Lauritzen, S. (1992) Propagation of probabilities, means and variances in mixed graphical association models. *Journal of the American Statistical Association* **87**, 1089–1108.

Lauritzen, S. L. (1996) *Graphical Models*. Oxford: Clarendon Press.

Lauritzen, S. & Spiegelhalter, D. J. (1988) Local computations with probabilities on graphical structures and their application to expert systems (with discussion). *Journal of the Royal Statistical Society series B* **50**, 157–224. Reprinted in Shafer & Pearl (1990).

Lauritzen, S. L., Dawid, A. P., Larsen, B. N. & Leimer, H.-G. (1990) Independence properties of directed Markov fields. *Networks* **20**, 491–505.

Lauritzen, S. L., Thiesson, B. & Spiegelhalter, D. J. (1994) Diagnostic systems created by model selection methods—a case study. In *Selecting Models from Data: AI and Statistics IV*, eds P. Cheeseman & R. W. Oldford, pp. 143–152. Lecture Notes in Statistics **89**. New York: Springer.

Lazarsfeld, P. F. (1961) The algebra of dichotomous systems. In *Studies in Item Analysis and Prediction*, ed. H. Solomon, pp. 111–157. Palo Alto, CA: Stanford University Press.

LeBlanc, M. & Tibshirani, R. J. (1993) Combining estimates in regression and classification. Preprint, Depts of Preventive Medicine and Biostatistics and of Statistics, University of Toronto.

Le Cun, Y., Boser, B., Denker, J. S., Henderson, D., Howard, R. E., Hubbard, W. & Jackel, L. D. (1989) Backpropagation applied to handwritten Zip code recognition. *Neural Computation* **1**, 541–551.

Le Cun, Y., Boser, B., Denker, J. S., Henderson, D., Howard, R. E., Hubbard, W. & Jackel, L. D. (1990a) Handwritten digit recognition with a back-propagation network. In *NIPS2*, pp. 396–404.

Le Cun, Y., Denker, J. S. & Solla, S. A. (1990b) Optimal brain damage. In *NIPS2*, pp. 598–605.

Lee, S. & Kil, R. M. (1988) Multi-layer feedforward potential function network. In *Proceedings of the IEEE International Conference on Neural Networks, San Diego*, I, pp. 161–171. Long Beach, CA: IEEE Press.

Lee, T.-C., Peterson, A. M. & Tsai, J. C. (1990) A multi-layer feed-forward neural network with dynamically adjustable structures. In *Proceedings of the IEEE International Conference on Systems, Man and Cybernetics, Los Angeles*, pp. 367–369. Long Beach, CA: IEEE Press.

Lee, Y. (1991) Handwritten digit recognition using K nearest-neighbor, radial-basis function, and back-propagation neural networks. *Neural Computation* **3**, 440–449.

de Leeuw, J. (1984) Differentiability of Kruskal's stress at a local minimum. *Psychometrika* **49**, 111–113.

Lehmann, E. L. (1983) *Theory of Point Estimation*. New York: Wiley.

Lehmann, E. L. (1986) *Testing Statistical Hypotheses*. Second edition. Pacific Grove, CA: Wadsworth & Brooks/Cole. (Formerly New York: Wiley.)

Leonard, J. A., Kramer, M. A. & Ungar, J. H. (1992) Using radial basis functions to approximate a function and its error bounds. *IEEE Transactions on Neural Networks* **3**, 624–627.

Lesaffre, E. & Albert, A. (1989) Partial separation in logistic discrimination. *Journal of the Royal Statistical Society series B* **51**, 109–116.

Levin, A. U., Leen, T. K. & Moody, J. E. (1994) Fast pruning using principal components. In *NIPS6*, pp. 35–42.

Levitt, T. S., Binford, T. O. & Ettinger, G. L. (1990) Utility-based control for computer vision. In *Uncertainty in Artificial Intelligence 4*, eds R. D. Shachter, T. S. Levitt, L. N. Kanal & J. F. Lemmer, pp. 407–422. North Holland, Amsterdam.

Li, X. B. & Dubes, R. C. (1986) Tree classifier design with a permutation statistic. *Pattern Recognition* **19**, 229–235.

Lincoln, W. P. & Skrzypek, J. (1990) Synergy of clustering multiple backpropagation networks. In *NIPS2*, pp. 650–657.

Lindley, D. V. (1980) Approximate Bayesian methods. In *Bayesian Statistics*, eds J. M. Bernardo, M. H. DeGroot, D. V. Lindley & A. F. M. Smith, pp. 223–237. Valencia: Valencia University Press.

Little, R. J. A. & Rubin, D. B. (1987) *Statistical Analysis with Missing Data*. New York: Wiley.

Liu, D. C. & Nocedal, J. (1989) On the limited memory BFGS method for large-scale optimization. *Mathematical Programming* 45, 503–528.

Liu, L., Wilkins, D. C., Ying, X. & Bain, Z. (1991) Minimum error tree decomposition. In *Proceedings of the Conference on Uncertainty in AI (Cambridge, MA)*, pp. 180–185.

Liu, Y. (1993) Neural network model selection using asymptotic jackknife estimator and cross-validation method. In *NIPS5*, pp. 599–606.

Liu, Y. (1994) Robust parameter estimation and model selection for neural network regression. In *NIPS6*, pp. 192–199.

Liu, Y. (1995) Unbiased estimate of generalization error and model selection in neural network. *Neural Networks* 8, 215–219.

Lloyd, S. P. (1957, 1982) Least squares quantization in PCM. Technical Note, Bell Laboratories. Published in 1982 in *IEEE Transactions on Information Theory* 28, 128–137.

Loizou, G. & Maybank, S. J. (1987) The nearest neighbor and the Bayes error rates. *IEEE Transactions on Pattern Analysis and Machine Intelligence* 9, 254–262.

Louis, T. A. (1982) Finding the observed information matrix when using the EM algorithm. *Journal of the Royal Statistical Society series B* 44, 226–233.

Lunts, A. L. & Brailovsky, V. L. (1967) Evaluation of attributes obtained in statistical decision rules. *Engineering Cybernetics* 3, 98–109.

Luttrell, S. P. (1989) Hierarchical vector quantization. *IEE Proceedings I* 136, 405–413.

Maass, W. G. (1994a) Neural networks with superlinear VC dimension. *Neural Computation* 6, 877–884.

Maass, W. G. (1994b) Perspectives of current research about the complexity of learning on neural nets. Chapter 5 of *Theoretical Advances in Neural Computation and Learning*, eds V. Roychowdhury, K.-Y. Siu & A. Orlitsky, pp. 153–172. Boston: Kluwer Academic Publishers.

Maass, W. & Turán, G. (1994) How fast can a threshold gate learn? In *Computational Learning Theory and Natural Learning Systems: Constraints and Prospects*, eds S. J. Hanson, G. A. Drastal & R. L. Rivest, volume I, pp. 381–414. MIT Press.

Macintyre, A. & Sontag, E. D. (1993) Finiteness results for sigmoidal "neural" networks. *Proceedings of the 25th Annual ACM Symposium Theory of Computing, San Diego, 1993*, pp. 325–334. New York: ACM Press.

MacKay, D. J. C. (1992a) Bayesian interpolation. *Neural Computation* 4, 415–447.

MacKay, D. J. C. (1992b) A practical Bayesian framework for backprop networks. *Neural Computation* 4, 448–472.

MacKay, D. J. C. (1992c) Information-based objective functions for active data selection. *Neural Computation* 4, 590–604.

MacKay, D. J. C. (1992d) The evidence framework applied to classification networks. *Neural Computation* 4, 720–736.

MacKay, D. J. C. (1992e) Bayesian model comparison and backprop nets. In *NIPS4*, pp. 839–846.

MacKay, D. M. & McCulloch, W. S. (1952) The limiting information capacity of a neuronal link. *Bulletin of Mathematical Biophysics* 14, 127–135.

MacLeod, J. E. S., Luk, A. & Titterington, D. M. (1987) A re-examination of the distance-weighted k-nearest neighbor classification rule. *IEEE Transactions on Systems, Man and Cybernetics* 17, 689–696. Reprinted in Dasarathy (1991).

Macnaughton-Smith, P., Williams, W. T., Dale, M. B. & Mockett, L. G. (1964) Dissimilarity analysis: a new technique of hierarchical sub-division. *Nature* 202, 1034–1035.

MacQueen, J. (1967) Some methods for classification and analysis of multivariate observations. In *Proceedings of the Fifth Berkeley Symposium on Mathematical Statistics and Probability*, eds L. M. Le Cam & J. Neyman, 1, pp. 281–297. Berkeley, CA: University of California Press.

Madigan, D. & Raftery, A. E. (1994) Model selection and accounting for model uncertainty in graphical models using Occam's window. *Journal of the American Statistical Association* 89, 1535–1546.

Madigan, D. & York, J. (1995) Bayesian graphical models for discrete data. *International Statistical Review* 63, 215–232.

Madych, W. R. & Nelson, S. A. (1990) Multivariate interpolation and conditionally positive definite functions II. *Mathematics of Computation* 54, 211–230.

Mahalanobis, P. C. (1936) On generalized distance in statistics. *Proceedings of the National Inst. Sci. (India)* 12, 49–55.

Maier, D. (1983) *The Theory of Relational Databases.* Rockville, Md: Computer Science Press.

Makram-Ebeid, S., Sirat, J.-A. & Viala, J.-R. (1989) A rationalized back-propagation learning algorithm. In *International Joint Conference on Neural Networks (Washington, 1989)* **II**, 373–380. New York: IEEE Press.

Mammone, R. J. (ed.) (1993) *Artificial Neural Networks for Speech and Vision.* London: Chapman & Hall.

Mangarasian, O. L. (1968) Multisurface methods of pattern separation. *IEEE Transactions on Information Theory* **14**, 801–807.

Mangarasian, O. L., Setiono, R. & Wolberg, W. H. (1990) Pattern-recognition via linear-programming: theory and application to medical diagnosis. In *Large-Scale Numerical Optimization, 1990*, eds T. F. Coleman & Y. Li, pp. 22–31. Philadelphia: SIAM.

Manly, B. F. J. & Rayner, J. C. W. (1987) The comparison of sample covariance matrices using likelihood ratio tests. *Biometrika* **74**, 841–847.

Mansfield, A. J. (1991) Comparison of perceptron training by linear-programming and by the perceptron convergence procedure. *Proceedings of the International Joint Conference on Neural Networks (Seattle 1991)* **II**, 25–30. Long Beach, CA: IEEE Press.

Mardia, K. V., Kent, J. T. & Bibby, J. M. (1979) *Multivariate Analysis.* London: Academic Press.

Maritz, J. S. & Lwin, T. (1989) *Empirical Bayes Methods.* Second edition. London: Chapman & Hall.

Marks, S. & Dunn, O. J. (1974) Discriminant functions when covariance matrices are unequal. *Journal of the American Statistical Association* **69**, 555–559.

Maronna, R. A. (1976) Robust M-estimators of multivariate location and scatter. *Annals of Statistics* **4**, 51–67.

Marriott, F. H. C. (1975) Separating mixtures of normal distributions. *Biometrics* **31**, 767–769.

Martin, G. L. & Pitman, J. A. (1990) Recognizing hand-printed letters and digits. In *NIPS2*, pp. 405–414.

Martin, G. L. & Pitman, J. A. (1991) Recognizing hand-printed letters and digits using backpropagation learning. *Neural Computation* **3**, 258–267.

Massart, D. L., Plastria, F. & Kaufman, L. (1983) Non-hierarchical clustering with MASLOC. *Pattern Recognition* **16**, 507–516.

Mathieson, M. J. (1996) Ordinal models for neural networks. In *Neural Networks in Financial Engineering.* eds A.-P. Refenes, Y. Abu-Mostafa & J. Moody. Singapore: World Scientific, 523–536.

Matúš, F. (1992) On equivalence of Markov properties over undirected graphs. *Journal of Applied Probability* **29**, 745–749.

Max, J. (1960) Quantizing for minimum distortion. *IRE Transactions on Information Theory* **6**, 7–12.

McCullagh, P. & Nelder, J. A. (1989) *Generalized Linear Models.* Second edition. London: Chapman & Hall.

McCulloch, W. S. & Pitts, W. (1943) A logical calculus of ideas immanent in nervous activity. *Bulletin of Mathematical Biophysics* **5**, 115–133. Reprinted in Anderson & Rosenfeld (1988).

McKay, R. J. & Campbell, N. A. (1982a) Variable selection techniques in discriminant analysis. I: Description. *British Journal of Mathematical and Statistical Psychology* **35**, 1–29.

McKay, R. J. & Campbell, N. A. (1982b) Variable selection techniques in discriminant analysis. II: Allocation. *British Journal of Mathematical and Statistical Psychology* **35**, 30–41.

McLachlan, G. J. (1992) *Discriminant Analysis and Statistical Pattern Recognition.* New York: Wiley.

McLachlan, G. J. & Basford, K. E. (1988) *Mixture Models: Inference and Applications to Clustering.* New York: Marcel Dekker.

Meinguet, J. (1979) Multivariate interpolation at arbitrary points made simple. *Journal of Applied Mathematics and Physics (ZAMP)* **30**, 292–304.

Meisel, W. S. (1972) *Computer-Oriented Approaches to Pattern Recognition.* New York: Academic Press.

Metropolis, N., Rosenbluth, A., Rosenbluth, M., Teller, A. & Teller, E. (1953) Equations of state calculations by fast computing machines. *Journal of Chemical Physics* **21**, 1087–1091.

Mhaskar, H. N. & Micchelli, C. A. (1992) Approximation by superposition of sigmoidal function and radial basis functions. *Advances in Applied Mathematics* **13**, 350–373.

Michalski, R. S. (1980) Pattern recognition as rule-guided inductive inference. *IEEE Transactions on Pattern Analysis and Machine Intelligence* **2**, 349–361.

Michie, D. (1989) Problems of computer-aided concept formation. In *Applications of Expert Systems volume 2*, ed. J. R. Quinlan, pp. 310–333. Glasgow: Turing Institute Press/Addison-Wesley.

Michie, D., Spiegelhalter, D. J. & Taylor, C. C. (eds) (1994) *Machine Learning, Neural and Statistical Classification*. New York: Ellis Horwood.

Mingers, J. (1987) Expert systems—rule induction with statistical data. *Journal of the Operational Research Society* **38**, 39–47.

Minnick, R. C. (1961) Linear-input logic. *IRE Transactions on Electronic Computers* **10**, 6–16.

Minsky, M. (1961) Steps towards artificial intelligence. *Proceedings of the IRE* **49**, 8–30.

Minsky, M. L. & Papert, S. A. (1988) *Perceptrons. An Introduction to Computational Geometry*. Expanded edition. Cambridge, MA: The MIT Press.

Møller, M (1993) A scaled conjugate gradient algorithm for fast supervised learning. *Neural Networks* **6**, 525–533.

Moody, J. E. (1989) Fast learning in multi-resolution hierarchies. In *NIPS1*, pp. 29–39.

Moody, J. E. (1991) Note on generalization, regularization and architecture selection in nonlinear learning systems. In *First IEEE-SP Workshop on Neural Networks in Signal Processing*, pp. 1–10. Los Alamitos, CA: IEEE Computer Society Press.

Moody, J. E. (1992) The *effective* number of parameters: an analysis of generalization and regularization in nonlinear learning systems. In *NIPS4*, pp. 847–854.

Moody, J. & Darken, C. J. (1989) Fast learning in networks of locally-tuned processing units. *Neural Computation* **1**, 281–294.

Moody, J. & Utans, J. (1992) Principled architecture selection for neural networks: application to corporate bond rating prediction. In *NIPS4*, pp. 683–690.

Moody, J. & Utans, J. (1995) Architecture selection strategies for neural networks: application to corporate bond rating prediction. In *Neural Networks in the Capital Markets*, ed. A.-P. Refenes, pp. 277–300. Chichester: Wiley.

Moore, B. (1989) ART 1 and pattern clustering. In *Proceedings of the 1988 Connectionist Models Summer School* eds D. Touretzky, G. Hinton & T. Sejnowski, pp. 174–185. San Mateo, CA: Morgan Kaufmann.

Moran, M. A. & Murphy, B. J. (1979) A closer look at two alternative methods of statistical discrimination. *Applied Statistics* **28**, 223–232.

Morgan, J. N. & Messenger, R. C. (1973) *THAID: a Sequential Search Program for the Analysis of Nominal Scale Dependent Variables*. Survey Research Center, Institute for Social Research, University of Michigan.

Morgan, J. N. & Sonquist, J. A. (1963) Problems in the analysis of survey data, and a proposal. *Journal of the American Statistical Association* **58**, 415–434.

Morin, R. L. & Raeside, D. E. (1981) A reappraisal of distance-weighted k-nearest neighbor classification for pattern recognition with missing data. *IEEE Transactions on Systems, Man and Cybernetics* **11**, 241–243.

Mosteller, F. & Wallace, D. L. (1963) Inference in an authorship problem. *Journal of the American Statistical Association* **58**, 275–309.

Moulton, B. R. (1991) A Bayesian-approach to regression selection and estimation with application to a price-index for radio services. *Journal of Econometrics* **49**, 169–193.

Moussouris, J. (1974) Gibbs and Markov random systems with constraints. *Journal of Statistical Physics* **10**, 11–33.

Murata, N., Yoshizawa, S. & Amari, S. (1991) A criterion for determining the number of parameters in an artificial neural network model. In *Artificial Neural Networks. Proceedings of ICANN-91*, eds T. Kohonen, K. Mäkisara, O. Simula & J. Kangas, volume I, pp. 9–14. Amsterdam: North Holland.

Murata, N., Yoshizawa, S. & Amari, S. (1993) Learning curves, model selection and complexity of neural networks. In *NIPS5*, pp. 607–614.

Murata, N., Yoshizawa, S. & Amari, S. (1994) Network information criterion—determining the number of hidden units for artificial neural network models. *IEEE Transactions on Neural Networks* **5**, 865–872.

Muroga, S. (1965) Lower bounds of the number of threshold functions and a maximum weight. *IEEE Transactions on Electronic Computers* **14**, 136–148.

Muroga, S. (1971) *Threshold Logic and its Applications*. New York: Wiley.

Muroga, S., Toda, I. & Takasu, S. (1961) Theory of majority decision elements. *Journal of the Franklin Institute* **271**, 376–418.

Murphy, P. M. & Aha, D. W. (1995) *UCI Repository of Machine Learning Databases* [Machine-readable data repository]. Irvine, CA: University of California, Dept of Information and Computer Science. Available by anonymous ftp from `ics.uci.edu` in directory `pub/machine-learning-databases`.

Murtagh, F. (1985) A survey of algorithms for contiguity-constrained clustering and related problems. *Computer Journal* **28**, 82–88.

Murtagh, F. (1995a) Contiguity-constrained hierarchical clustering. In *Partitioning Data Sets*, eds I. J. Cox, P. Hansen & B. Julesz. DIMACS. pp. 143–152. Providence, RI: American Mathematical Society.

Murtagh, F. (1995b) Interpreting the Kohonen self-organizing feature map using contiguity-constrained clustering. *Pattern Recognition Letters* **16**, 399–408.

Murthy, V. K. (1966) Nonparametric estimation of multivariate densities with applications. In *Multivariate Analysis*, ed. P. R. Krishnaiah, pp. 43–56. New York: Academic Press.

Musavi, M. T., Ahmed, W., Chan, K. H., Faris, K. B. & Hummels, D. M. (1992) On the training of radial basis function classifiers. *Neural Networks* **5**, 595–603.

Myles, J. P. & Hand, D. J. (1990) The multiclass metric problem in nearest neighbour discrimination rules. *Pattern Recognition* **23**, 1291–1297.

Narendra, P. M. & Fukunaga, K. (1977) A branch and bound algorithm for feature subset selection. *IEEE Transactions on Computers* **26**, 917–922.

Nash, J. C. (1990) *Compact Numerical Methods for Computers. Linear Algebra and Function Minimization.* Second edition. Bristol: Adam Hilger.

Neal, R. (1992a) Connectionist learning of belief networks. *Artificial Intelligence* **56**, 71–113.

Neal, R. M. (1992b) Asymmetric parallel Boltzmann machines are belief networks. *Neural Computation* **4**, 832–834.

Neal, R. (1993) Bayesian learning via stochastic dynamics. In *NIPS5*, pp. 475–482.

Neal, R. M. (1996) *Bayesian Learning for Neural Networks.* Lecture Notes in Statistics **118**. New York: Springer.

Neapolitan, E. (1990) *Probabilistic Reasoning in Expert Systems. Theory and Algorithms.* New York: Wiley.

Niblett, T. (1987) Constructing decision trees in noisy domains. In *Progress in Machine Learning*, eds I. Bratko & N. Lavrač, pp. 67–78. Wilmslow: Sigma Press.

Niblett, T. & Bratko, I. (1987) Learning decision rules in noisy domains. In *Research and Development in Expert Systems III. Proceedings of Expert Systems '86, Brighton 1986*, ed. M. A. Bramer, pp. 25–34. Cambridge: Cambridge University Press.

Niemann, H. & Goppert, G. (1988) An efficient branch-and-bound nearest neighbour classifier. *Pattern Recognition Letters* **7**, 67–72. Reprinted in Dasarathy (1991).

Nowlan, S. J. & Hinton, G. E. (1992a) Adaptive soft weight tying using Gaussian mixtures. In *NIPS4*, pp. 993–1000.

Nowlan, S. J. & Hinton, G. E. (1992b) Simplifying neural networks by soft weight-sharing. *Neural Computation* **4**, 473–493. Reprinted with an introduction as pp. 369–394 of Wolpert (1995).

Oja, E. (1982) A simplified neuron model as a principal component analyzer. *Journal of Mathematical Biology* **16**, 267–273.

Oja, E. (1989) Neural networks, principal components and subspaces. *International Journal of Neural Systems* **1**, 61–68.

Oja, E. (1992) Principal components, minor components and linear neural networks. *Neural Networks* **5**, 927–935.

Oja, E. & Karhunen, J. (1985) On stochastic-approximation of the eigenvectors and eigenvalues of the expectation of a random matrix. *Journal of Mathematical Analysis and its Applications* **106**, 69–84.

Olesen, K. G. (1993) Causal probabilistic networks with both discrete and continuous variables. *IEEE Transactions on Pattern Analysis and Machine Intelligence* **15**, 275–279.

Oliver, L. H., Poulsen, R. S., Toussaint, G. T. & Louis, C. (1979) Classification of atypical cells in the automated cytoscreening for cervical cancer. *Pattern Recognition* **11**, 205–212.

Oliver, R. M. & Smith, J. Q. (eds) (1990) *Influence Diagrams, Belief Nets and Decision Analysis.* Chichester: Wiley.

Olkin, I. & Tate, R. F. (1961) Multivariate correlation models with mixed discrete and continuous variates. *Annals of Mathematical Statistics* **32**, 445–465.

van Ooyen, A. & Nienhuis, B. (1992) Improving the convergence of the back-propagation algorithm. *Neural Networks* **5**, 465–471. (See also letter to the editor and response, **6**, 611–612.)

Ott, J. (1989) Computer-simulation methods in human linkage analysis. *Proceedings of the National Academy of Sciences of the USA* **86**, 4175–4178.

Owen, A. (1984) A neighbourhood-based LANDSAT classifier. *Canadian Journal of Statistics* **12**, 191–200.

Owens, A. J. & Filkin, D. L. (1989) Efficient training of the back propagation network by solving a system of stiff ordinary differential equations. In *Proceedings of the International Conference on Neural Networks (Washington, 1989)*, **II**, 381–386. New York: IEEE Press.

Pagallo, G. (1989) Learning DNF by decision trees. In *Proceedings of the Eleventh International Joint Conference on Artificial Intelligence (Detroit, 1989)*, pp. 639–644.

Pagallo, G. & Haussler, D. (1989) Two algorithms that learn DNF by discovering relevant features. In *Proceedings of the Sixth International Workshop on Machine Learning (Ithaca, 1989)*, ed. A. M. Segre, pp. 119–123. San Mateo, CA: Morgan Kaufmann.

Pagallo, G. & Haussler, D. (1990) Boolean feature discovery in empirical learning. *Machine Learning* **5**, 71–99.

Parberry, I. (1994) *Circuit Complexity and Neural Networks*. Cambridge, MA: MIT Press.

Park, J. & Sandberg, I. W. (1991) Universal approximation using radial-basis-function networks. *Neural Computation* **3**, 246–257.

Parrondo, J. M. R. & Van der Broeck, C. (1993) Vapnik-Chervonenkis bounds for generalization. *J. Physics A* **26**, 2211–2223.

Parthasarthy, G. & Chatterji, B. N. (1990) A class of new KNN methods for low sample problems. *IEEE Transactions on Systems, Man and Cybernetics* **20**, 715–718.

Parzen, E. (1962) On the estimation of a probability density function and mode. *Annals of Mathematical Statistics* **33**, 1065–1076.

Patrick, E. A. & Fisher, F. P. II (1969) Nonparametric feature selection. *IEEE Transactions on Information Theory* **15**, 577–584.

Patterson, A. & Niblett, T. (1983) *ACLS User Manual*. Glasgow: Intelligent Terminals Ltd.

Pavlidis, T. (1993) Recognition of printed text under realistic conditions. *Pattern Recognition Letters* **14**, 317–326.

Payne, H. J. & Meisel, W. S. (1977) An algorithm for constructing optimal binary decision trees. *IEEE Transactions on Computers* **26**, 905–916.

Pearl, J. (1979) Capacity and error estimates for Boolean classifiers with limited capacity. *IEEE Transactions on Pattern Analysis and Machine Intelligence* **1**, 350–356.

Pearl, J. (1982) Reverend Bayes on inference engines: a distributed hierarchical approach. In *Proceedings of the AAAI National Conference on Artificial Intelligence (Pittsburgh)*, pp. 133–136. Menlo Park, CA: AAAI.

Pearl, J. (1986) Fusion, propagation, and structuring in belief networks. *Artificial Intelligence* **29**, 241–288. Reprinted in Shafer & Pearl (1990).

Pearl, J. (1987) Evidential reasoning using stochastic simulation of causal models. *Artificial Intelligence* **32**, 245–257.

Pearl, J. (1988) *Probabilistic Inference in Intelligent Systems. Networks of Plausible Inference*. San Mateo, CA: Morgan Kaufmann.

Pearl, J. (1993a) Belief networks revisited. *Artificial Intelligence* **59**, 49–56.

Pearl, J. (1993b) Graphical models, causality and intervention. Contribution to the discussion of Spiegelhalter *et al.* (1993), pp. 266–269.

Pearl, J. (1995) From Bayesian networks to causal networks. In Gammerman (1995), pp. 1–31.

Pearlmutter, B. A. (1994) Fast exact multiplication by the Hessian. *Neural Computation* **6**, 147–160.

Pearlmutter, B. A. & Rosenfeld, R. (1991) Chaitin–Kolmogorov complexity and generalization in neural networks. In *NIPS3*, pp. 925–931.

Peck, R., Fisher, L. & Van Ness J. (1989) Approximate confidence intervals for the number of clusters. *Journal of the American Statistical Association* **84**, 184–191.

Peng, F., Jacobs, R. A. & Tanner, M. A. (1994) Bayesian inference in mixtures-of-experts and hierarchical mixtures-of-experts architectures. Technical report, Dept of Biostatistics, University of Rochester, NY.

Penrod, C. S. & Wagner, T. J. (1977) Another look at the edited nearest neighbor rule. *IEEE Transactions on Systems, Man and Cybernetics* **7**, 92–94.

Peretto, P. (1992) *An Introduction to the Modeling of Neural Networks*. Cambridge: Cambridge University Press.

Perrone, M. P. & Cooper, L. N. (1993) When networks disagree: Ensemble methods for hybrid neural networks. In Mammone (1993), pp. 126–142.

Peskun, P. H. (1973) Optimal Monte-Carlo sampling using Markov chains. *Biometrika* **60**, 607–612.

Peterson, C. & Anderson, J. R. (1987) A mean field learning algorithm for neural networks. *Complex Systems* **1**, 995–1019.

Pitas, I. (ed.) (1993) *Parallel Algorithms for Digital Image Processing, Computer Vision and Neural Networks.* Chichester: Wiley.

Ploughman, L. M. & Boehnke, M. (1989) Estimating the power of a proposed linkage study for a complex genetic trait. *American Journal of Human Genetics* **44**, 543–551.

Poggio, T. & Girosi, F. (1990a) Regularization algorithms for learning that are equivalent to multilayer networks. *Science* **247**, 978–982.

Poggio, T. & Girosi, F. (1990b) Networks for approximation and learning. *Proceedings of the IEEE* **78**, 1481–1497. Reprinted in Lau (1992).

Pollard, D. (1984) *Convergence of Stochastic Processes.* New York: Springer.

Pollard, D. (1986) Rates of uniform almost-sure convergence for empirical processes indexed by unbounded classes of functions. Unpublished paper, Dept of Statistics, Yale University.

Pollard, D. (1990) *Empirical Processes: Theory and Applications.* Hayward, CA: Institute of Mathematical Statistics and American Statistical Association.

Posse, C. (1990) An effective two-dimensional projection pursuit algorithm. *Communications in Statistics—Simulation and Computation* **19**, 1143–1164.

Posse, C. (1995a) Tools for two-dimensional exploratory projection pursuit. *Journal of Computational and Graphical Statistics* **4**, 83–100.

Posse, C. (1995b) Projection pursuit exploratory data analysis. *Computational Statistics and Data Analysis.*

Powell, M. J. D. (1987) Radial basis functions for multivariable interpolation: a review. In *Algorithms for Approximation*, eds J. C. Mason & M. G. Cox, pp. 143–167. Oxford: Clarendon Press.

Powell, M. J. D. (1992) The theory of radial function approximation in 1990. In *Advances in Numerical Analysis* volume II, ed. W. Light, pp. 105–210. Oxford: Clarendon Press.

Prechelt, L. (1994) A study of experimental evaluation of current neural network learning algorithms: current research practice. Technical Report 19/94, Fakultät für Informatik, Universität Kahlsruhe.

Prentice, R. & Pyke, R. (1979) Logistic disease incidence models and case-control studies. *Biometrika* **66**, 403–411.

Preparata, F. P. & Shamos, M. I. (1985) *Computational Geometry. An Introduction.* New York: Springer.

Press, W. H., Flannery, B. P., Teukolsky, S. A. & Vetterling, W. T. (1992) *Numerical Recipes in C.* Second edition. Cambridge: Cambridge University Press.

Preston, C. J. (1974) *Gibbs States on Countable Sets.* London: Cambridge University Press.

Preston, C. J. (1976) *Random Fields.* Lecture Notes in Mathematics **534**. Berlin: Springer.

Przytula, K. W. & Prasanna, V. K. (1993) *Parallel Digital Implementation of Neural Networks.* Englewood Cliffs, NJ: Prentice Hall.

Quenouille, M. H. (1949) Approximate tests of correlation in time series. *Journal of the Royal Statistical Society series B* **11**, 68–84.

Quinlan, J. R. (1979) Discovering rules by induction from large collections of examples. In *Expert Systems in the Microelectronic Age*, ed. D. Michie, pp. 168–201. Edinburgh: Edinburgh University Press.

Quinlan, J. R. (1983) Learning efficient classification procedures and their application to chess endgames. In *Machine Learning*, eds R. S. Michalski, J. G. Carbonell & T. M. Mitchell, pp. 463–482. Palo Alto, CA: Tioga.

Quinlan, J. R. (1986) Induction of decision trees. *Machine Learning* **1**, 81–106. Reprinted in Shavlik & Dietterich (1990).

Quinlan, J. R. (1987a) Simplifying decision trees. *International Journal of Man–Machine Studies* **27**, 221–234.

Quinlan, J. R. (1987b) Generating production rules from decision trees. In *Proceedings of the Tenth International Joint Conference on Artificial Intelligence, Milan*, pp. 304–307.

Quinlan, J. R. (1988) Decision trees and multivalued attributes. In *Machine Intelligence 11*, eds J. E. Hayes, D. Michie & J. Richards, pp. 305–318. Oxford: Clarendon Press.

Quinlan, J. R. (1990) Decision trees and decision making. *IEEE Transactions on Systems, Man and Cybernetics* **20**, 339–346.

Quinlan, J. R. (1993) *C4.5: Programs for Machine Learning.* San Mateo, CA: Morgan Kaufmann.

Raftery, A. E. (1993) Approximate Bayes factors and accounting for model uncertainty in generalized linear models. Technical report 255, Dept of Statistics, University of Washington.

Rao, C. R. (1948) The utilization of multiple measurements in problems of biological classification (with discussion). *Journal of the Royal Statistical Society series B* **10**, 159–203.

Rao, C. R. (1960) Multivariate analysis: an indispensable statistical aid in applied research. *Sankhyā* **22**, 317–338.

Rayens, W. & Greene, T. (1991) Covariance pooling and stabilization for classification. *Computational Statistics and Data Analysis* **11**, 17–42.

Redner, R. A. & Walker, H. F. (1984) Mixture densities, maximum likelihood and the EM algorithm. *SIAM Review* **26**, 195–239.

Reed, R. (1993) Pruning algorithms—a survey. *IEEE Transactions on Neural Networks* **4**, 740–747.

Reilly, D. L., Cooper, L. N. & Elbaum, C. (1982) A neural model for category learning. *Biological Cybernetics* **45**, 35–41. Reprinted in Anderson *et al.* (1990).

Richards, L. E. (1972) Refinement and extension of distribution-free discriminate analysis. *Applied Statistics* **21**, 174–176.

Riffenburgh, R. H. & Clunies-Ross, C. W. (1960) Linear discriminant analysis. *Pacific Science* **14**, 251–256.

Rimey, R. & Brown, C. (1992) Task-oriented vision with multiple Bayes nets. In *Active Vision*, eds A. Blake & A. Yuille, pp. 217–236. Cambridge, MA: The MIT Press.

Ripley, B. D. (1977) Modelling spatial patterns (with discussion). *Journal of the Royal Statistical Society series B* **39**, 172–212.

Ripley, B. D. (1979) Algorithm AS137. Simulating spatial patterns: dependent samples from a multivariate density. *Applied Statistics* **28**, 109–112.

Ripley, B. D. (1987) *Stochastic Simulation*. New York: Wiley.

Ripley, B. D. (1988) *Statistical Inference for Spatial Processes*. Cambridge: Cambridge University Press.

Ripley, B. D. (1993) Statistical aspects of neural networks. In *Networks and Chaos—Statistical and Probabilistic Aspects*, eds O. E. Barndorff-Nielsen, J. L. Jensen & W. S. Kendall, pp. 40–123. London: Chapman & Hall.

Ripley, B. D. (1994a) Neural networks and related methods for classification (with discussion). *Journal of the Royal Statistical Society series B* **56**, 409–456.

Ripley, B. D. (1994b) Neural networks and flexible regression and discrimination. In *Statistics and Images 2*, ed. K. V. Mardia. *Advances in Applied Statistics* **2**, pp. 39–57. Abingdon: Carfax.

Ripley, B. D. (1994c) Flexible non-linear approaches to classification. In Cherkassky *et al.* (1994), pp. 105–126.

Ripley, B. D. (1995) Statistical ideas for selecting network architectures. In *Neural Networks: Artificial Intelligence and Industrial Applications*, eds B. Kappen & S. Gielen. London: Springer.

Ripley, B. D. & Kelly, F. P. (1977) Markov point processes. *Journal of the London Mathematical Society (2)* **15**, 188–192.

Ripley, B. D. & Kirkland, M. D. (1990) Iterative simulation methods. *Journal of Computational and Applied Mathematics* **31**, 165–172.

Rissanen, J. (1983) A universal prior for integers and estimation by minimum description length. *Annals of Statistics* **11**, 416–431.

Rissanen, J. (1987) Stochastic complexity (with discussion). *Journal of the Royal Statistical Society series B* **49**, 223–239.

Rissanen, J. (1989) *Stochastic Complexity in Statistical Inquiry*. Singapore: World Scientific Publishing Co.

Ritter, G. L., Woodruff, H. B., Lowry, S. R. & Isenhour, T. L. (1975) An algorithm for a selective nearest neighbor decision rule. *IEEE Transactions on Information Theory* **21**, 665–669. Reprinted in Dasarathy (1991).

Ritter, H., Martinetz, T. & Schulten, K. (1992) *Neural Computation and Self-Organizing Maps. An Introduction*. Reading, MA: Addison-Wesley.

Roberts, S. & Tarassenko, L. (1995) Automated sleep EEG analysis using an RBF network. In *Neural Network Applications*, ed. A. F. Murray, pp. 305–322. Dordrecht: Kluwer Academic Publishers.

Robinson, R. W. (1977) Counting unlabeled acyclic digraphs. In *Combinatorial Mathematics V*, ed. C. H. C. Little. Lecture Notes in Mathematics **622**, pp. 28–43. Berlin: Springer.

Roeder, K. (1990) Density estimation with confidence sets exemplified by superclusters and voids in galaxies. *Journal of the American Statistical Association* **85**, 617–624.

Roosen, C. B. & Hastie, T. J. (1994) Automatic smoothing spline projection pursuit. *Journal of Computational and Graphical Statistics* **3**, 235–248.

Rose, D. J., Tarjan, R. E. & Lueker, G. S. (1976) Algorithmic aspects of vertex elimination on graphs. *SIAM Journal on Computing* **5**, 266–283.

Rosenblatt, F. (1957) The perceptron—a perceiving and recognizing automaton. Report 85-460-1, Cornell Aeronautical Laboratory.

Rosenblatt, F. (1958) The perceptron: a probabilistic model for information storage and organization in the brain. *Psychological Review* **65**, 386–408. Reprinted in Shavlik & Dietterich (1990).

Rosenblatt, F. (1962) *Principles of Neurodynamics.* Washington, DC: Spartan Books.

Rosenblatt, M. (1956) Remarks on some non-parametric estimates of a density function. *Annals of Mathematical Statistics* **27**, 832–837.

Rounds, E. M. (1980) A combined nonparametric approach to feature selection and binary decision tree design. *Pattern Recognition* **12**, 313–317.

Rousseeuw, P. J. & Leroy, A. M. (1987) *Robust Regression and Outlier Detection.* New York: Wiley.

Rousseeuw, P. J. & van Zomeren, B. C. (1990) Unmasking multivariate outliers and leverage points (with discussion). *Journal of the American Statistical Association* **85**, 633–651.

Ruck, D. W., Rogers, S. K., Kabrisky, M., Maybeck, P. S. & Oxley, M. E. (1992) Comparative analysis of backpropagation and the extended Kalman filter for training multilayer perceptrons. *IEEE Transactions on Pattern Analysis and Machine Intelligence* **14**, 686–691.

Ruiz, E. V. (1986) An algorithm for finding nearest neighbours in (approximately) constant average time. *Pattern Recognition Letters* **4**, 145–158.

Rumelhart, D. E. & McClelland, J. L. (eds) (1986) *Parallel Distributed Processing: Explorations in the Microstructure of Cognition. Volume 1. Foundations.* Cambridge, MA: The MIT Press.

Rumelhart, D. E., Hinton, G. E. & Williams, R. J. (1986) Learning representations by back-propagating errors. *Nature* **323**, 533–536. Reprinted in Anderson & Rosenfeld (1988).

Russell, S. J. & Norvig, P. (1995) *Artificial Intelligence. A Modern Approach.* Englewood Cliffs, NJ: Prentice-Hall.

Růžička, P. (1993) On the convergence of learning algorithm for topological maps. *Neural Network World* **4**, 413–424.

de Sa, V. R. & Ballard, D. H. (1993) A note on learning vector quantization. In *NIPS5*, pp. 220–227.

Saarinen, S., Bramley, R. & Cybenko, G. (1993) Ill-conditioning in neural network training problems. *SIAM Journal on Scientific Computing* **14**, 693–714.

Safavian, S. R and Landgrebe, D. (1991) A survey of decision tree classifier methodology. *IEEE Transactions on Systems, Man and Cybernetics* **21**, 660–674.

Sakurai, A. (1993) Tighter bounds of the VC-dimension of three-layer networks. In *Proceedings of the 1993 World Congress on Neural Networks*, volume 3, pp. 540–543. Hillsdale, NJ: Erlbaum.

Salomon, R. (1991) Improved convergence rate of back-propagation with dynamic adaption of the learning rate. In *Parallel Problem Solving From Nature (Dortmund, 1990)*. Lecture Notes in Computer Science **496**, 269–273.

Samal, A. & Iyengar, P. A. (1992) Automatic recognition and analysis of human faces and facial expressions: a survey. *Pattern Recognition* **25**, 65–77.

Sammon, J. W. Jr (1969) A non-linear mapping for data structure analysis. *IEEE Transactions on Computers* **18**, 401–409.

Sanger, T. D. (1989) Optimal unsupervised learning in a single-layer linear feedforward network. *Neural Networks* **2**, 459–473.

Sankar, A. & Mammone, R. J. (1993) Growing and pruning neural tree networks. *IEEE Transactions on Computers* **42**, 291–299.

Santer, T. J. & Duffy, D. E. (1986) A note on A. Albert and J. A. Anderson's conditions for the existence of maximum likelihood estimates in logistic regression models. *Biometrika* **73**, 755–758.

Schalkoff, R. J. (1992) *Pattern Recognition: Statistical, Structural and Neural Approaches.* New York: Wiley.

Schlimmer, J. C. & Fisher, D. H. (1986) A case study of incremental concept induction. In *Proceedings of the Fifth National Conference on Artificial Intelligence, Philadelphia*, pp. 496–501. San Mateo, CA: Morgan Kaufmann.

Schlimmer, J. C. & Granger, R. H. Jr (1986) Incremental learning from noisy data. *Machine Learning* **1**, 317–354.

Schmidhuber, J. (1989) Accelerated learning in back-propagation nets. In *Connectionism in Perspective*, pp. 439–445. Amsterdam: Elsevier.

Schoenberg, I. J. (1935) Remarks to Maurice Fréchet's article "Sur la définition axiomatique d'une classe d'espaces distanciés vectoriellement applicable sur l'espace de Hilbert". *Annals of Mathematics* **36**, 724–732.

Schuermann, J. & Doster, D. (1984) A decision-theoretic approach in hierarchical classifier design. *Pattern Recognition* **17**, 359–369.

Schwarz, G. (1978) Estimating the dimension of a model. *Annals of Statistics* **6**, 461–464.

Schwemer, G. T. & Dunn, O. J. (1980) Posterior probability estimators in classification simulations. *Communications in Statistics–Simulation and Computation* **B9**, 133–140.

Scott, A. J. & Symons, M. J. (1971) Clustering methods based on likelihood ratio criteria. *Biometrics* **27**, 387–397.

Scott, A. J. & Wild, C. J. (1986) Fitting logistic models under case-control or choice based sampling. *Journal of the Royal Statistical Society series B* **48**, 170–182.

Scott, D. W. (1992) *Multivariate Density Estimation. Theory, Practice and Visualization.* New York: Wiley.

Seber, G. A. F. & Wild, C. J. (1989) *Nonlinear Regression.* New York: Wiley.

Sebestyen, G. S. (1962) Pattern recognition by an adaptive process of sample set construction. *IEEE Transactions on Information Theory* **8**, S 82–S 91.

Sedgewick, R. (1990) *Algorithms in C.* Reading, MA: Addison-Wesley.

Sen, A. & Srivastava, M. (1990) *Regression Analysis. Theory, Methods and Applications.* New York: Springer.

Sethi, I. K. (1990) Entropy nets: from decision trees to neural networks. *Proceedings of the IEEE* **78**, 1605–1613. Reprinted in Lau (1992).

Sethi, I. K. (1991) Decision tree performance enhancement using an artificial neural network implementation. In Sethi & Jain (1991), pp. 71–88.

Sethi, I. K. & Jain, A. K. (eds) (1991) *Artificial Neural Networks and Statistical Pattern Recognition. Old and New Connections.* Amsterdam: North Holland.

Sethi, I. K. & Sarvarayudu, G. P. R. (1982) Hierarchical classifier design using mutual information. *IEEE Transactions on Pattern Analysis and Machine Intelligence* **4**, 441–445.

Shachter, R. D. & Peot, M. A. (1990) Simulation approaches to general probabilistic inference on belief networks. In *Uncertainty in Artificial Intelligence 5*, eds M. Henrion, R. D. Shachter, L. N. Kanal & J. F. Lemmer, pp. 221–231. Amsterdam: North-Holland.

Shafer, G. (1996) *Probabilistic Expert Systems.* Number 67 in CBMS-NSF Regional Conference Series in Applied Mathematics. Philadelphia, PA: SIAM.

Shafer, G. & Pearl, J. (eds) (1990) *Readings in Uncertainty Reasoning.* San Mateo, CA: Morgan Kaufmann.

Shafer, G. & Shenoy, P. P. (1986) Propagating belief functions with local computations. *IEEE Expert* **1**(3), 43–52.

Shanno, D. F. (1990) Recent advances in numerical techniques for large-scale optimization. In *Neural Networks for Control*, eds W. T. Miller III, R. S. Sutton & P. J. Werbos, pp. 171–178. Cambridge, MA: The MIT Press.

Shanno, D. F. & Phua, K. H. (1980) Remark on algorithm 500: Minimization of unconstrained multivariable functions. *ACM Transactions on Mathematical Software* **6**, 618–622.

Shavlik, J. W. & Dietterich, T. G. (eds) (1990) *Readings in Machine Learning.* San Mateo, CA: Morgan Kaufmann.

Shawe-Taylor, J. & Anthony, M. (1991) Sample sizes for multiple-output threshold networks. *Network* **2**, 107–117.

Sheehan, N. & Thomas, A. (1993) On the irreducibility of a Markov chain defined on a space of genotype configurations by a sampling scheme. *Biometrics* **49**, 163–175.

Shenoy, P. P. (1989) A valuation-based language for expert systems. *International Journal of Approximate Reasoning* **3**, 383–411.

Shenoy, P. P. & Shafer, G. (1990) Axioms of probability and belief-function propagation. In *Uncertainty in Artificial Intelligence 4*, eds R. D. Shachter, T. S. Levitt, L. N. Kanal & J. F. Lemmer, pp. 169–198. Amsterdam: North-Holland. Reprinted in Shafer & Pearl (1990).

Shenoy, P. P., Shafer, G. & Mellouli, K. (1988) Propagation of belief functions: a distributed approach. In *Uncertainty in Artificial Intelligence 2*, eds J. F. Lemmer & L. N. Kanal, pp. 325–335. Amsterdam: North-Holland.

Shepanski, J. F. (1987) Fast learning in artificial neural systems: multilayer perceptron training using optimal estimation. In *Proceedings of IEEE First International Conference on Neural Networks, San Diego, 1987*, eds M. Caudill & C. Butler **I**, 465–472. Long Beach, CA: IEEE Press.

Shepard, R. N. (1962a) The analysis of proximities: multidimensional scaling with an unknown distance function I. *Psychometrika* **27**, 125–139.

Shepard, R. N. (1962b) The analysis of proximities: multidimensional scaling with an unknown distance function II. *Psychometrika* **27**, 219–246.

Shibata, R. (1976) Selection of the order of an autoregressive model by Akaike's Information Criterion. *Biometrika* **63**, 117–126.

Shibata, R. (1980) Asymptotically efficient selection of the order of the model for estimating parameters of a linear process. *Annals of Statistics* **8**, 147–164.

Shibata, R. (1981) An optimal selection of regression variables. *Biometrika* **68**, 45–54.

Short, R. D. & Fukunaga, K. (1980) A new nearest neighbor distance measure. In *Proceedings of the Fifth IEEE International Conference on Pattern Recognition (Miami Beach, 1980)*, pp. 81–86. Los Alamitos, CA: IEEE Computer Society Press.

Short, R. D. & Fukunaga, K. (1981) The optimal distance measure for nearest neighbor classification. *IEEE Transactions on Information Theory* **27**, 622–627. Reprinted in Dasarathy (1991).

Sietsma, J. & Dow, R. J. F. (1991) Creating artificial neural networks that generalize. *Neural Networks* **4**, 67–79.

Silva, F. M. & Almeida, L. B. (1990) Speeding up back-propagation. In *Advanced Neural Computers*, ed. R. Eckmiller, pp. 151–158. Amsterdam: Elsevier.

Silvapulle, M. J. & Burridge, J. (1986) Existence of maximum likelihood estimates in regression models for grouped and ungrouped data. *Journal of the Royal Statistical Society series B* **48**, 100–106.

Silverman, B. W. (1985) Some aspects of the spline smoothing approach to non-parametric regression curve fitting (with discussion). *Journal of the Royal Statistical Society series B* **47**, 1–52.

Silverman, B. W. (1986) *Density Estimation for Statistics and Data Analysis*. London: Chapman & Hall.

Silverman, B. W. & Jones, M. C. (1989) E. Fix and J. L. Hodges (1951): An important contribution to nonparametric discriminant analysis and density estimation. *International Statistical Review* **57**, 233–247.

Simard, P., Le Cun, Y. & Denker, J. (1993) Efficient pattern recognition using a new transformation distance. In *NIPS5*, pp. 50–58.

Simmons, G. F. (1963) *Introduction to Topology and Modern Analysis*. New York: McGraw-Hill.

Singer, Y. & Tishby, N. (1994) Decoding cursive scripts. In *NIPS6*, pp. 833–840.

Singhal, S. & Wu, L. (1989) Training multilayer perceptrons with the extended Kalman filter. In *NIPS1*, pp. 133–140.

Smith, A. F. M. (1991) Discussion of 'Posterior Bayes factors'. *Journal of the Royal Statistical Society series B* **53**, 132–133.

Smith, A. F. M. & Roberts, G. O. (1993) Bayesian computation via the Gibbs sampler and related Markov chain Monte Carlo methods (with discussion). *Journal of the Royal Statistical Society series B* **55**, 3–23.

Smith, A. F. M. & Spiegelhalter, D. J. (1980) Bayes factors and choice criteria for linear models. *Journal of the Royal Statistical Society series B* **42**, 213–220.

Smith, C. A. B. (1947) Some examples of discrimination. *Annals of Eugenics* **13**, 272–282.

Smith, E. E. & Medin, D. L. (1981) *Categories and Concepts*. Cambridge, MA: Harvard University Press.

Smith, F. W. (1968) Pattern classifier design by linear programming. *IEEE Transactions on Computers* **17**, 367–372.

Smith, F. W. (1969) Design of multicategory pattern classifiers with two-category classifier design procedures. *IEEE Transactions on Computers* **18**, 548–551.

Smith, J. Q. (1989) Influence diagrams for statistical modelling. *Annals of Statistics* **17**, 654–672.

Smith, J. W., Everhart, J. E., Dickson, W. C., Knowler, W. C. & Johannes, R. S. (1988) Using the ADAP learning algorithm to forecast the onset of diabetes mellitus. In *Proceedings of the Symposium on Computer Applications in Medical Care (Washington, 1988)*, ed. R. A. Greenes, pp. 261–265. Los Alamitos, CA: IEEE Computer Society Press.

Solla, S. A., Levin, E. & Fleisher, M. (1988) Accelerated learning in layered neural networks. *Complex Systems* **2**, 625–639.

Sontag, E. D. (1992) Feedback stabilization using two-hidden-layer nets. *IEEE Transactions on Neural Networks* **3**, 981–990.

Spackman, K. A. (1992) Maximum likelihood training of connectionist models: comparison with least-squares back propagation and logistic regression. In *Proceedings of the Fifteenth Annual Symposium on Computer Applications in Medical Care, Washington 1991*, ed. P. D. Clayton, pp. 285–289. New York: McGraw-Hill.

Späth, H. (1985) *Cluster Dissection and Analysis. Theory, FORTRAN programs, examples*. Chichester: Ellis Horwood.

Specht, D. F. (1967a) Vectorcardiographic diagnosis using the polynomial discriminant method of pattern recognition. *IEEE Transactions on Bio-medical Engineering* **14**, 90–95.

Specht, D. F. (1967b) Generation of polynomial discriminant functions for pattern recognition. *IEEE Transactions on Electronic Computers* **16**, 308–319.

Specht, D. F. (1990a) Probabilistic neural networks. *Neural Networks* **3**, 109–118.

Specht, D. F. (1990b) Probabilistic neural networks and the polynomial Adaline as complementary techniques for classification. *IEEE Transactions on Neural Networks* **1**, 111–121.

Specht, D. F. (1991) A general regression neural network. *IEEE Transactions on Neural Networks* **2**, 568–576.

Speed, T. (1990) Complexity, calibration and causality in influence diagrams. In Oliver & Smith (1990), pp. 49–63.

Spiegelhalter, D. J. (1990) Fast algorithms for probabilistic reasoning in influence diagrams, with applications in genetics and expert systems. In Oliver & Smith (1990), pp. 361–384.

Spiegelhalter, D. J. & Lauritzen, S. L. (1990) Sequential updating of conditional probabilities on directed graphical structures. *Networks* **20**, 579–605.

Spiegelhalter, D. J. & Smith, A. F. M. (1982) Bayes factors for linear and log-linear models with vague prior information. *Journal of the Royal Statistical Society series B* **44**, 377–387.

Spiegelhalter, D. J., Dawid, A. P., Lauritzen, S. L. & Cowell, R. G. (1993) Bayesian analysis in expert systems (with discussion). *Statistical Science* **8**, 219–283.

Spirtes, P., Glymour, C. & Scheines, R. (1993) *Causality, Prediction, and Search.* Lecture Notes in Statistics **81**. New York: Springer.

Srihari, S. N. (1992) High-performance reading machines. *Proceedings of the IEEE* **80**, 1120–1132.

Srinvas, S. & Breese, J. (1990) IDEAL: a software package for the analysis of influence diagrams. In *Uncertainty in Artificial Intelligence 6*, eds L. N. Kanal, J. Lemmer & T. S. Levitt, pp. 212–219. Amsterdam: North-Holland.

Stace, C. (1991) *New Flora of the British Isles.* Cambridge: Cambridge University Press.

Stanfill, C. & Waltz, D. (1986) Toward memory-based reasoning. *Communications of the Association for Computing Machinery* **29**, 1213–1228.

Stewart, L. (1987) Hierarchical Bayesian analysis using Monte Carlo integration: computing posterior distributions when there are many possible models. *The Statistician* **36**, 211–219.

Stinchcombe, M. & White, H. (1989) Universal approximation using feedforward networks with non-sigmoid hidden layer activation functions. In *Proceedings of the International Joint Conference on Neural Networks* **I**, 613–617. Long Beach, CA: IEEE Press.

Stinchcombe, M. & White, H. (1990) Approximating and learning unknown mappings using multilayer feedforward networks with bounded weights. In *Proceedings of the International Joint Conference on Neural Networks, San Diego*, **III**, 7–16. Long Beach, CA: IEEE Press.

Stone, C. J. (1977) Consistent nonparametric regression (with discussion). *Annals of Statistics* **5**, 595–645.

Stone, C. J. (1985) Additive regression and other nonparametric models. *Annals of Statistics* **13**, 689–705.

Stone, C. J. (1986) The dimensionality reduction principle for generalized additive models. *Annals of Statistics* **14**, 590–606.

Stone, M. (1974) Cross-validatory choice and assessment of statistical predictions (with discussion). *Journal of the Royal Statistical Society series B* **36**, 111–147.

Stone, M. (1977a) Asymptotics for and against cross-validation. *Biometrika* **64**, 29–35.

Stone, M. (1977b) An asymptotic equivalence of choice of model by cross-validation and Akaike's criterion. *Journal of the Royal Statistical Society series B* **39**, 44–47.

Stone, M. (1979) Comments on model selection criteria of Akaike and Schwarz. *Journal of the Royal Statistical Society series B* **41**, 276–278.

Streit, R. L. & Luginbuhl, T. E. (1994) Maximum likelihood training of probabilistic neural networks. *IEEE Transactions on Neural Networks* **5**, 764–783.

Strömberg, J. E., Zrida, J. & Isaksson, A. (1991) Neural trees—using neural nets in a tree classifier structure. In *IEEE International Conference on Acoustics, Speech and Signal Processing (Toronto, 1991)*, pp. 137–140. Long Beach, CA: IEEE Press.

Styblinski, M. A. & Tang, T.-S. (1990) Experiments in nonconvex optimization: stochastic approximation and simulated annealing. *Neural Networks* **3**, 467–483.

Suen, C. Y., Legault, R., Nadal, C., Cheriet, M. & Lam, L. (1993) Building a new generation of handwriting recognition systems. *Pattern Recognition Letters* **14**, 303–315.

Suen, C. Y., Nadal, C., Legault, R., Mai, T. A. & Lam, L. (1992) Computer recognition of unconstrained handwritten numerals. *Proceedings of the IEEE* **80**, 1162–1180.

Sussmann, H. J. (1992) Uniqueness of the weights for minimal feedforward nets with a given input–output map. *Neural Networks* **5**, 589–593.

Swain, P. H. & Hauska, H. (1977) The decision tree classifier: design and potential. *IEEE Transactions on Geoscience Electronics* **15**, 142–147.

Swayne, D. F., Cook, D. & Buja, A. (1991) XGobi: interactive dynamic graphics in the X window system with a link to S. In *Proceedings of the ASA Section on Statistical Graphics*, pp. 1–8. Alexandria, VA: American Statistical Association.

Swonger, C. W. (1972) Sample set condensation for a condensed nearest neighbor decision rule for pattern recognition. In *Frontiers of Pattern Recognition*, ed. S. Watanabe, pp. 511–519. Orlando: Academic Press. Reprinted in Dasarathy (1991).

Tarassenko, L., Hayton, P., Cerneaz, N. & Brady, M. (1995) Novelty detection for the identification of masses in mammograms. In *Proceedings of the Fourth International IEE Conference on Artificial Neural Networks (Cambridge, 1995)*. IEE Conference Publication **409**, 442–447. IEE Press.

Tarjan, R. E. & Yannakakis, M. (1984) Simple linear-time algorithms to test chordality of graphs, test acyclicity of hypergraphs, and selectively reduce acyclic hypergraphs. *SIAM Journal of Computing* **13**, 566–579.

Tarter, M. E. & Lock, M. D. (1993) *Model-Free Curve Estimation*. New York: Chapman & Hall.

Therrien, C. W. (1989) *Decision, Estimation, and Classification: An Introduction to Pattern Recognition and Related Topics*. New York: Wiley.

Thisted, R. A. (1988) *Elements of Statistical Computing. Numerical Computation*. New York: Chapman & Hall.

Thompson, E. A. (1985) *Pedigree Analysis in Human Genetics*. Baltimore, MD: Johns Hopkins University Press.

Thornton, C. J. (1992) *Techniques in Computational Learning. An Introduction*. London: Chapman & Hall.

Tibshirani, R. (1992) Principal curves revisited. *Statistics and Computing* **2**, 183–190.

Tierney, L. (1994) Markov chains for exploring posterior distributions (with discussion). *Annals of Statistics* **22**, 1701–1762.

Tierney, L. & Kadane, J. B. (1986) Accurate approximations for posterior moments and marginal densities. *Journal of the American Statistical Association* **81**, 82–86.

Titterington, D. M. (1976) Updating a diagnostic system using unconfirmed cases. *Applied Statistics* **25**, 238–247.

Titterington, D. M. (1980) A comparative study of kernel-based density estimates for categorical data. *Technometrics* **22**, 259–268.

Titterington, D. M. (1984) Recursive parameter estimation using incomplete data. *Journal of the Royal Statistical Society series B* **46**, 257–267.

Titterington, D. M., Murray, G. D., Murray, L. S., Spiegelhalter, D. J., Skene, A. M., Habbema, J. D. F. & Gelpka, G. J. (1981) Comparison of discrimination techniques applied to a complex data set of head injured patients (with discussion). *Journal of the Royal Statistical Society series A* **144**, 145–174.

Titterington, D. M., Smith, A. F. M. & Makov, U. E. (1985) *Statistical Analysis of Finite Mixture Distributions*. Chichester: Wiley.

Todd, B. S. (1995) Weighted inference rules and Bayesian belief networks. In Gammerman (1995), pp. 205–225.

Tollenaere, T. (1990) SuperSAB: fast adaptive back propagation with good scaling properties. *Neural Networks* **3**, 561–573.

Tomek, I. (1976a) A generalization of the k-NN rule. *IEEE Transactions on Systems, Man and Cybernetics* **6**, 121–126. Reprinted in Dasarathy (1991).

Tomek, I. (1976b) An experiment with the edited nearest-neighbor rule. *IEEE Transactions on Systems, Man and Cybernetics* **6**, 448–452. Reprinted in Dasarathy (1991).

Tomek, I. (1976c) Two modifications of CNN. *IEEE Transactions on Systems, Man and Cybernetics* **6**, 769–772.

Torgerson, W. S. (1952) Multidimensional scaling I. Theory and method. *Psychometrika* **17**, 401–419.

Torgerson, W. S. (1958) *Theory and Methods of Scaling*. New York: Wiley.

Tråvén, H. G. C. (1991) A neural network approach to statistical pattern classification by "semiparametric" estimation of probability density functions. *IEEE Transactions on Neural Networks* **2**, 366–377.

Tsypkin, Ya. Z. (1966) Use of the stochastic approximation method in estimating unknown distribution densities from observations. *Automation and Remote Control* **27**, 432–434.

Tutz, G. (1986) An alternative choice of smoothing for kernel-based density estimates in discrete discriminant analysis. *Biometrika* **73**, 405–411.

Tutz, G. (1988) Smoothing for discrete kernels in discrimination. *Biometrical Journal* **6**, 729–739.

Tutz, G. (1989) On cross-validation for discrete kernel estimates in discrimination. *Communications in Statistics—Theory and Methods* **18**, 4145–4162.

Ullmann, J. R. (1974) Automatic selection of reference data for use in a nearest-neighbor method of pattern classification. *IEEE Transactions on Information Theory* **20**, 541–543.

Ultsch, A. (1993a) Knowledge extraction from self-organizing neural networks. In *Information and Classification*, eds O. Opitz, B. Lausen & R. Klar, pp. 301–306. Berlin: Springer.

Ultsch, A. (1993b) Self-organizing neural networks for visualization and classification. In *Information and Classification*, eds O. Opitz, B. Lausen & R. Klar, pp. 307–313. Berlin: Springer.

Upton, G. J. G. (1991) The exploratory analysis of survey data using log-linear models. *The Statistician* **40**, 169–182.

Usui, S., Nakauchi, S. & Nakano, M. (1991) Internal color representation acquired by a five-layer neural network. In *Artificial Neural Networks. Proceedings of ICANN-91*, eds T. Kohonen, K. Mäkisara, O. Simula & J. Kangas, volume I, pp. 867–872. Amsterdam: North Holland.

Utgoff, P. E. (1988a) ID5: an incremental ID3. In *Proceedings of the Fifth International Conference on Machine Learning*, ed. J. Laird pp. 107–120. San Mateo, CA: Morgan Kaufmann.

Utgoff, P. E. (1988b) Perceptron trees: a case study in hybrid concept representations. In *Proceedings of the Seventh AAAI National Conference on Artificial Intelligence, St Paul*, pp. 601–606. San Mateo, CA: Morgan Kaufmann.

Utgoff, P. E. (1989) Improved training via incremental learning. In *Proceedings of the Sixth International Workshop on Machine Learning (Ithaca, 1989)*, ed. A. M. Segre, pp. 362–365. San Mateo, CA: Morgan Kaufmann.

Utgoff, P. E. (1990) Incremental induction of decision trees. *Machine Learning* **4**, 161–186.

Utgoff, P. E. & Brodley, C. E. (1990) An incremental method for multivariate splits in decision trees. In *Proceedings of the Seventh International Workshop on Machine Learning*, eds B. W. Porter & R. J. Mooney, pp. 58–65. San Mateo, CA: Morgan Kaufmann.

Valiant, L. G. (1984) A theory of the learnable. *Communications of the Association for Computing Machinery* **27**, 1134–1142. Reprinted in Shavlik & Dietterich (1990).

Van de Welde, W. (1989) IDL, or taming the multiplexer. In *Proceedings of the Fourth European Working Session on Learning*, ed. K. Morik, pp. 211–226. London: Pitman.

Van de Welde, W. (1990) Incremental induction of topologically minimal trees. In *Proceedings of the Seventh International Workshop on Machine Learning*, eds B. W. Porter & R. J. Mooney, pp. 66–74. San Mateo, CA: Morgan Kaufmann.

Van Ryzin, J. (1966) Bayes risk consistency of classification procedures using density estimation. *Sankhyā* **A28**, 261–270.

Vapnik, V. N. (1982) *Estimation of Dependencies based on Empirical Data*. New York: Springer.

Vapnik, V. (1992) Principles of risk minimization for learning theory. In *NIPS4*, pp. 831–838.

Vapnik, V. N. (1995) *The Nature of Statistical Learning Theory*. New York: Springer.

Vapnik, V. N. (1998) *Statistical Learning Theory*. New York: Wiley.

Vapnik, V. N. & Chervonenkis, A. Ya. (1971) On the uniform convergence of relative frequencies of events to their probabilities. *Theory of Probability and its Applications* **16**, 264–280.

Venables, W. N. & Ripley, B. D. (1994) *Modern Applied Statistics with S-Plus*. New York: Springer.

Verma, T. & Pearl, J. (1990) Causal networks: semantics and expressiveness. In *Uncertainty in Artificial Intelligence 4*, eds R. D. Shachter, T. S. Levitt, L. N. Kanal & J. F. Lemmer, pp. 69–76. Amsterdam: North-Holland.

Verma, T. S. & Pearl, J. (1991) Equivalence and synthesis of causal models. In *Uncertainty in Artificial Intelligence 6*, eds P. P. Bonissone, M. Henrion, L. N. Kanal & J. F. Lemmer, pp. 255–268. Amsterdam: North Holland.

Villegas, C. (1969) On the a priori distribution of the covariance matrix. *Annals of Mathematical Statistics* **40**, 1098–1099.

Vinod, H. (1969) Integer programming and the theory of grouping. *Journal of the American Statistical Association* **64**, 506–517.

Vlachonikolos, I. (1990) Predictive discrimination and classification with mixed binary and continuous variables. *Biometrika* **77**, 657–662.

Wagner, T. J. (1973) Convergence of the edited nearest neighbor. *IEEE Transactions on Information Theory* **19**, 696–697.

Wahba, G. (1990) *Spline Models for Observational Data*. Philadelphia: SIAM.

Wahba, G. (1995) Generalization and regularization in nonlinear learning systems. In *The Handbook of Brain Theory and Neural Networks*, ed. M. Arbib, pp. 426–430. Cambridge, MA: The MIT Press.

Wahba, G. & Wold, S. (1975) A completely automatic French curve. *Communications in Statistics* **4**, 1–17.

Wahba, G., Gu, C., Wang, Y. & Chappell, R. (1995) Soft classification a.k.a. risk estimation via penalized log likelihood and smoothing spline analysis of variance. In Wolpert (1995), pp. 331–359.

Wakahara, T. (1993) Towards robust handwritten character recognition. *Pattern Recognition Letters* **14**, 345–354.

Wallace, C. S. & Freeman, P. R. (1987) Estimation and inference by compact encoding (with discussion). *Journal of the Royal Statistical Society series B* **49**, 240–265.

Wand, M. P. & Jones, M. C. (1995) *Kernel Smoothing*. London: Chapman & Hall.

Wang, C., Venkatesh, S. S. & Judd, J. S. (1994) Optimal stopping and effective machine complexity in learning. In *NIPS6*, pp. 303–310.

Wang, Q. R. & Suen, C. Y. (1984) Analysis and design of a decision tree based on entropy reduction and its application to large character set recognition. *IEEE Transactions on Pattern Analysis and Machine Intelligence* **6**, 406–417.

Wang, Q. R. & Suen, C. Y. (1987) Large tree classifier with heuristic search and global training. *IEEE Transactions on Pattern Analysis and Machine Intelligence* **9**, 91–102.

Ward, J. H. Jr (1963) Hierarchical grouping to optimize an objective function. *Journal of the American Statistical Association* **58**, 236–244.

Warner, H. R., Toronto, A. F., Veasey, L. R. & Stephenson, R. (1961) A mathematical model for medical diagnosis—application to congenital heart disease. *Journal of the American Medical Association* **177**, 177–184.

Wasserman, P. D. (1993) *Advanced Methods in Neural Computing*. New York: Van Nostrand Reinhold.

Watanabe, S. (1969) *Knowing and Guessing*. New York: Wiley.

Waterhouse, S. R. & Robinson, A. J. (1994) Classification using hierarchical mixtures of experts. In *Proceedings of the 1994 IEEE Workshop on Neural Networks for Signal Processing IV*, pp. 177–186. Long Beach, CA: IEEE Press.

Watrous, R. L. (1987) Learning algorithms for connectionist networks: applied gradient methods of nonlinear optimization. In *Proceedings of the IEEE First International Conference on Neural Networks (San Diego, 1987)*, eds M. Caudill & C. Butler **II**, 619–627. New York: IEEE Press.

Webb, A. R. (1994) Functional approximation by feed-forward networks: a least-squares approach. *IEEE Transactions on Neural Networks* **5**, 363–371.

Weigend, A. S. & Gershenfeld, N. A. (eds) (1993) *Time Series Prediction: Forecasting the Future and Understanding the Past*. Reading, MA: Addison-Wesley.

Weigend, A. S., Huberman, B. A. & Rumelhart, D. E. (1990) Predicting the future: a connectionist approach. *International Journal of Neural Systems* **1**, 193–209.

Weigend, A. S., Huberman, B. A. & Rumelhart, D. E. (1992) Predicting sunspots and exchange rates with connectionist networks. In *Nonlinear Modeling and Forecasting*, eds M. Casdagli & S. Eubank, pp. 395–432. Redwood City, CA: Addison-Wesley.

Weigend, A. S., Rumelhart, D. E. & Huberman, B. A. (1991) Generalization by weight-elimination with application to forecasting. In *NIPS3*, pp. 875–882.

Weiss, S. M. (1991) Small sample error rate estimation for *k*-NN classifiers. *IEEE Transactions on Pattern Analysis and Machine Intelligence* **3**, 285–289.

Weiss, S. M. & Kulikowski, C. A. (1991) *Computer Systems that Learn: Classification and Prediction Methods from Statistics, Neural Nets, Machine Learning and Expert Systems*. San Mateo, CA: Morgan Kaufmann.

Wen, W. X. (1990) Optimal decomposition of belief functions. In *Proceedings of the Sixth Workshop on Uncertainty in Artificial Intelligence (Cambridge, MA)*, pp. 245–256.

Werbos, P. J. (1974) *Beyond Regression: New Tools for Prediction and Analysis in the Behavioural Sciences*. Ph.D. thesis, Harvard University. Reprinted in Werbos (1994).

Werbos, P. J. (1988) Backpropagation: past and future. In *Proceedings of the IEEE International Conference on Neural Networks, San Diego, 1988* **I**, 343–353. Long Beach, CA: IEEE Press.

Werbos, P. J. (1994) *The Roots of Backpropagation. From Ordered Derivatives to Neural Networks and Political Forecasting*. New York: Wiley.

West, M. & Harrison, P. J. (1989) *Bayesian Forecasting and Dynamic Models*. New York: Springer.

Wetterschereck, D. & Dietterich, T. (1992) Improving the performance of radial basis function networks by learning center locations. In *NIPS4*, pp. 1133–1140.

White, H. (1982) Maximum-likelihood estimation of mis-specified models. *Econometrica* **50**, 1–25.

White, H. (1989a) Learning in artificial neural networks: a statistical perspective. *Neural Computation* **1**, 425–464. Reprinted in White (1992).

White, H. (1989b) Some asymptotic results for learning in single hidden-layer feedforward networks. *Journal of the American Statistical Association* **84**, 1003–1013. Reprinted in White (1992). Correction: **87**, 1252.

White, H. (1990) Connectionist nonparametric regression: multilayer feedforward networks can learn arbitrary mappings. *Neural Networks* **3**, 535–549. Reprinted in White (1992).

White, H. (1992) *Artificial Neural Networks: Approximation and Learning Theory*. Oxford: Blackwell.

White, H. & Woolridge, J. (1991) Some results on sieve estimation with dependent observations. In *Nonparametric and Semi-Parametric Methods in Econometrics and Statistics*, eds W. Barnett, J. Powell & G. Tauchen. New York: Cambridge University Press. Reprinted in White (1992).

Widrow, B. & Hoff, M. E. Jr. (1960) Adaptive switching circuits. *IRE WESCON Convention Record* **4**, 96–104. Reprinted in Anderson & Rosenfeld (1988).

Williams, W. T. & Lambert, J. M. (1959) Multivariate methods in plant ecology. I. Association-analysis in plant communities. *Journal of Ecology* **47**, 83–101.

Wilson, D. L. (1972) Asymptotic properties of nearest neighbor rules using edited data. *IEEE Transactions on Systems, Man and Cybernetics* **2**, 408–421. Reprinted in Dasarathy (1991).

Winston, P. H. (1992) *Artificial Intelligence*. Third edition. Reading, MA: Addison-Wesley.

Wolberg, W. H. & Mangarasian, O. L. (1990) Multisurface method of pattern separation for medical diagnosis applied to breast cytology. *Proceedings of the National Academy of Sciences of the USA* **87**, 9193–9196.

Wolfe, J. H. (1970) Pattern clustering via multivariate mixture analysis. *Multivariate Behavioural Research* **5**, 329–350.

Wolpert, D. H. (1992) Stacked generalization. *Neural Networks* **5**, 241–259.

Wolpert, D. H. (1993) On the use of evidence in neural networks. In *NIPS5*, pp. 539–546.

Wolpert, D. H. (1994a) Bayesian backpropagation over I-O functions rather than weights. In *NIPS6*, pp. 200–207.

Wolpert, D. H. (1994b) Discussion of Ripley (1994a). *Journal of the Royal Statistical Society series B* **56**, 450–451.

Wolpert, D. H. (ed.) (1995) *The Mathematics of Generalization*. Reading, MA: Addison-Wesley.

Wu, C. F. J. (1983) On the convergence properties of the EM algorithm. *Annals of Statistics* **11**, 95–103.

Xu, L., Kryzak, A. & Suen, C. Y. (1992) Methods of combining multiple classifiers and their applications to handwriting recognition. *IEEE Transactions on Systems, Man and Cybernetics* **22**, 418–435.

Xu, L., Kryzak, A. & Yuille, A. (1994) On radial basis function nets and kernel regression: statistical consistency, convergence rates, and receptive field sizes. *Neural Networks* **7**, 609–628.

Yair, E. & Gersho, A. (1990a) The Boltzmann perceptron network: a soft classifier. *Neural Networks* **3**, 203–221.

Yair, E. & Gersho, A. (1990b) Maximum *a posteriori* decision and evaluation of class probabilities by Boltzmann perceptron classifiers. *Proceedings of the IEEE* **78**, 1620–1678. Reprinted in Lau (1992).

Yannakakis, M. (1981) Computing the minimal fill-in is NP-complete. *SIAM Journal of Algebraic and Discrete Methods* **2**, 77–79.

York, J. (1992) Use of the Gibbs sampler in expert systems. *Artificial Intelligence* **56**, 115–130, 397–398.

Young, G. & Householder, A. S. (1938) Discussion of a set of points in terms of their mutual distances. *Psychometrika* **3**, 19–22.

Young, T. Y. & Calvert, T. W. (1974) *Classification, Estimation and Pattern Recognition*. New York: American Elsevier.

Zador, P. L. (1982) Asymptotic quantization error of continuous signals and the quantization dimension. *IEEE Transactions on Information Theory* **28**, 139–149.

Zeger, K., Vaisey, J. & Gersho, A. (1992) Globally optimal vector quantization design by stochastic relaxation. *IEEE Transactions on Signal Processing* **40**, 310–322.

Zhao, Y. & Atkeson, C. G. (1992) Some approximation properties of projection pursuit learning networks. In *NIPS4*, pp. 936–943.

Author Index

Subject Index

Page numbers in **bold** refer to entries in the glossary.

Printed in the United States
By Bookmasters

Printed in the United States
By Bookmasters